I0051653

Management of Wastewater and Sludge

Management of micropollutants and disinfection of byproducts in municipal waste-water and extraction of energy from the sludge produced in wastewater treatment plants is under constant focus. This book presents a detailed know-how regarding sustainable management of waste produced in municipal and industrial activities through novel state-of-the-art techniques used for the treatment of toxic industrial wastes and municipal wastewater. It deals with the management of municipal sludge and solid waste including leachates produced from landfill sites. It also provides detailed information for achieving the stringent standards set by regulatory bodies for municipal and industrial effluents.

Features:

- Covers development of new novel reactor configurations for wastewater treatment.
- Describes handling and removal of emerging contaminants like pharmaceutical compounds, endocrine disruptors, and disinfection byproducts.
- Deliberates combination of wastewater and micropollution.
- Contains an in-depth discussion on treatment and disposal of fecal sludge.
- Highlights new economically feasible techniques to enhance biogas recovery from treatment plant sludges.

This book is aimed at researchers and graduate students in environmental engineering, wastewater treatment, mechanical engineering, chemical engineering, and energy engineering.

Science and Engineering of Air Pollution and Waste Management

Series Editor:

Ashok Kumar

Department of Civil and Environmental Engineering,
The University of Toledo, Unites States

Environmental issues continue to be a global concern because of their impact on human health. Therefore, the governments around the world are involved in managing air quality, water quality, and waste in their countries for the welfare of their citizens. The management of air, water, and waste is complex, and the procedures to control air quality and water quality and for the disposal of waste have been developed over time. The management of air pollution involves understanding air pollution sources, monitoring contaminants, modeling air quality, performing laboratory experiments, the use of satellite images for quantifying air quality levels and indoor air pollution, and elimination of contaminants through control. Research activities are being performed on every aspect of air and waste throughout the world in order to respond to public concerns. Aimed at professionals, graduate students, and undergraduate students, the proposed series will focus on air pollution, water pollution, waste and land pollution, climate change, environmental law and policy, impact assessment and education, and so forth.

Management of Wastewater and Sludge

New Approaches

Edited by Izharul Haq Farooqi and Saif Ullah Khan

For more information about this series, please visit www.routledge.com/Science-and-Engineering-of-Air-Pollution-and-Waste-Management/book-series/SEAPWM.

Management of Wastewater and Sludge
New Approaches

Edited by
Izharul Haq Farooqi
and
Saif Ullah Khan

CRC Press
Taylor & Francis Group
Boca Raton London New York

CRC Press is an imprint of the
Taylor & Francis Group, an **informa** business

Designed cover image: ©shutterstock

First edition published 2023
by CRC Press
6000 Broken Sound Parkway NW, Suite 300, Boca Raton, FL 33487-2742

and by CRC Press
4 Park Square, Milton Park, Abingdon, Oxon, OX14 4RN

CRC Press is an imprint of Taylor & Francis Group, LLC

© 2023 selection and editorial matter, Izharul Haq Farooqi and Saif Ullah Khan; individual chapters, the contributors

Reasonable efforts have been made to publish reliable data and information, but the author and publisher cannot assume responsibility for the validity of all materials or the consequences of their use. The authors and publishers have attempted to trace the copyright holders of all material reproduced in this publication and apologize to copyright holders if permission to publish in this form has not been obtained. If any copyright material has not been acknowledged please write and let us know so we may rectify in any future reprint.

Except as permitted under U.S. Copyright Law, no part of this book may be reprinted, reproduced, transmitted, or utilized in any form by any electronic, mechanical, or other means, now known or hereafter invented, including photocopying, microfilming, and recording, or in any information storage or retrieval system, without written permission from the publishers.

For permission to photocopy or use material electronically from this work, access www.copyright. com or contact the Copyright Clearance Center, Inc. (CCC), 222 Rosewood Drive, Danvers, MA 01923, 978-750-8400. For works that are not available on CCC please contact mpkbookspermissions@tandf. co.uk

Trademark notice: Product or corporate names may be trademarks or registered trademarks and are used only for identification and explanation without intent to infringe.

ISBN: 9781032064635 (hbk)
ISBN: 9781032064680 (pbk)
ISBN: 9781003202431 (ebk)

DOI: 10.1201/9781003202431

Typeset in Times
by codeMantra

Contents

Preface...vii

Editors..ix

Contributors xi

Chapter 1   An Overview of Developments in Wastewater Treatment
Technologies... 1

Ahmar Khateeb, Hummaid Zafar, and Saif Ullah Khan

Chapter 2   Introduction to Aerobic Granulation Technology:
A Breakthrough in Wastewater Treatment System 37

Ghazal Srivastava and Absar Ahmad Kazmi

Chapter 3   Modified Sequencing Batch Reactors for Wastewater Treatment...... 53

Sakina Bombaywala and Vinay Pratap

Chapter 4   Zero Liquid Discharge in Industries 75

Aiman Malik, Mohd Salim Mahtab, and Izharul Haq Farooqi

Chapter 5   Advanced Oxidation Processes and Their Applications.................... 85

*Mohd Salim Mahtab, Izharul Haq Farooqi, Anwar Khursheed,
and Mohd Azfar Shaida*

Chapter 6   An Overview Exploring Electrochemical Technologies for
Wastewater Treatment 103

Saif Ullah Khan, Rahat Alam, and Amir Aslam

Chapter 7   Constructed Wetlands for Wastewater Treatment........................... 119

Rahat Alam and Saif Ullah Khan

Chapter 8   Introduction to Micropollutants and Their Sources........................ 139

*Mohd Azfar Shaida, Soumita Talukdar, Mohd Salim Mahtab,
and Izharul Haq Farooqi*

Chapter 9   Effects of Micropollutants on Human Health 151

Areeba Aziz and Asad Aziz

Chapter 10 Biodegradability of Micropollutants in Wastewater and Natural
Systems 167

*Monika Dubey, Bhanu Prakash Vellanki, and
Absar Ahmad Kazmi*

Chapter 11 Biodegradation Technology for the Removal of Micropollutants:
A Critical Review 181

*Kaushar Hussain, Nadeem Ahmad Khan, Izharul Haq Farooqi,
and Sirajuddin Ahmed*

Chapter 12 The Wholistic Approach for Sewage Sludge Management 203

*Ashish Dehal, K. Rathika, Bholu Ram Yadav, and
A. Ramesh Kumar*

Chapter 13 Enhanced Biogas Production from Treatment Plant Sludges 233

*Mohd Imran Siddiqui, Izharul Haq Farooqi, Hasan Rameez,
and Farrukh Basheer*

Chapter 14 Overview of Thermal Based Pre-Treatment Methods for
Enhancing Methane Production of Sewage Sludge 257

*Gowtham Balasundaram, Pallavi Gahlot, Absar Ahmad Kazmi,
and Vinay Kumar Tyagi*

Chapter 15 Management and Disposal of Solid Waste: Practices and
Legislations in Different Countries 271

Arshad Husain

Chapter 16 Sources, Characteristics, Treatment Technologies and Disposal
Methods for Faecal Sludge 297

*Gulafshan Tasnim, Mohd Dawood Khan, Izharul Haq Farooqi,
and Farrukh Basheer*

Index 339

Preface

Deterioration in the water quality of surface water sources has been a matter of great concern, and billions of dollars are being spent on their remediation. Strict legislation on waste disposal has led to the development of new technologies worldwide. Over the past decade, the effluent discharge standards for biological oxygen demand (BOD), suspended solids, and nutrients have been reduced, which has prompted researchers/consultants worldwide to explore new treatment technologies that may enhance the performance of existing wastewater treatment plants. Moreover, in order to safeguard the existing water resources, the regulatory bodies have enforced zero effluent discharge for many polluting industries which requires membrane separation followed by thermal evaporation.

Management of micropollutants, namely, pharmaceutical compounds, daily care products, and disinfection byproducts in municipal wastewater has been a matter of great concern in the past couple of decades. Scientists worldwide are working on the development of feasible technologies for the removal of these compounds. In developing countries, such technologies are at their infancy stage and need to be addressed.

Another subject of concern these days is the extraction of energy from the sludge produced in wastewater treatment plants. Anaerobic digestion is generally practiced for the treatment of sludge. However, due to rate-limiting hydrolysis, only a partial fraction of energy is captured. Scientists worldwide are working on the development of feasible low-cost techniques for overcoming rate-limiting hydrolysis, leading to enhanced biogas recovery which may partially fulfill the energy requirement of the operation of wastewater treatment plants. Another major area for discussion is the management of fecal sludge generated by onsite/decentralized sewage treatment facilities such as septic tanks. Fecal sludge is a toxic waste as it contains a large number of pathogens apart from organics and suspended solids. Implementation of appropriate treatment technologies for fecal sludge also needs to be addressed mainly in developed countries.

This book aims at presenting detailed knowledge described in different chapters regarding the sustainable management of waste of all kinds produced in different municipal and industrial activities. It presents novel state-of-the-art recent techniques along with conventional procedures being used for the treatment of toxic industrial waste and municipal wastewater. The book also deals with the management of municipal and fecal sludge.

Out of a total of 16, the first seven chapters mainly focus on different treatment technologies used for industrial and municipal wastewater. However, in the next four chapters, the fate and occurrence of micropollutants in water bodies along with their health concerns are discussed. These chapters also highlight the approaches adopted for the removal of micropollutants from municipal wastewater treatment plants. Lastly, the concluding five chapters review the technologies used for the management of treatment plant sludges. These include detailed discussions on the different pre-treatment methods used for enhanced energy recovery from the sludges. Also, the approaches and guidelines used for the handling of fecal sludge produced from onsite/decentralized sewage treatment systems are focused.

I hope that this book will provide detailed information for academicians, students, consultants, and practicing engineers for achieving the stringent standards set by regulatory bodies for municipal and industrial effluents.

Editors

Izharul Haq Farooqi has been working in the environmental engineering field for the past 32 years. His areas of specialization are water and wastewater treatment including biological treatment, biodegradation of toxic waste, and corrosion control. Moreover, he has mobilized consultancy for the department, has established a state-of-the-art advanced environmental engineering laboratory through research projects, and has been an expert member of various national bodies. Dr. Farooqi has supervised six Ph.D. scholars, and another seven are under supervision. Dr. Farooqi has supervised 73 MTech dissertations. Besides, Dr. Farooqi has received six research projects from different organizations. He has published more than a hundred research papers in international/national journals and conferences having citations of more than 2,400 (h-index 26 and i10-index 50) so far uptill now. Apart from this, he has presented research papers in international conferences in the U.S.A., Switzerland, Norway, The Netherlands, Malaysia, Singapore, Italy, Saudi Arabia, Denmark, London, and different places in India. He also received a Young Scientist Award from the UP Council of Science and Technology in 2002.

Saif Ullah Khan has been a member of the faculty of Department of Civil Engineering at Aligarh Muslim University since 2016. He received a B.Tech. in Civil Engineering and M.Tech. in Environmental Engineering from Aligarh Muslim University. He received PhD degree from Aligarh Muslim University in 2019. He was awarded the national-level CSIR Senior Research Fellowship for pursuing his doctoral studies. His field of interest is the treatment of water and wastewater using advanced treatment techniques including nanotechnology, electrochemistry, advanced oxidation, etc. He has been an active researcher and has published more than 35 research papers in national/international journals of repute having citations of more than 1,200 (h-index 16 and i10-index 21) uptill now. He is also a leading reviewer of more than 15 top international journals. Besides, he has been invited for several talks at national as well as international levels. His subjects of interest include environmental engineering, industrial pollution control, hydrology, etc.

Contributors

Sirajuddin Ahmed
Department of Civil Engineering
Jamia Millia Islamia
New Delhi, India

Rahat Alam
Department of Civil Engineering
Zakir Husain College of Engineering &
    Technology
Aligarh Muslim University
Aligarh, India

Amir Aslam
Department of Civil Engineering
Zakir Husain College of Engineering &
    Technology
Aligarh Muslim University
Aligarh, India

Areeba Aziz
Department of Biochemistry
Aligarh Muslim University
Aligarh, India

Asad Aziz
Department of Civil & Environmental
    Engineering
State University of New York
Buffalo, New York

Gowtham Balasundaram
Environmental Engineering Group
Department of Civil Engineering
Indian Institute of Technology Roorkee
Roorkee, India

Farrukh Basheer
Environmental Engineering Section
Department of Civil Engineering
Zakir Husain College of Engineering &
    Technology
Aligarh Muslim University
Aligarh, India

Sakina Bombaywala
CSIR-National Environmental
    Engineering Research Institute
    (CSIR-NEERI)
Nagpur, India
and
Academy of Scientific and Innovative
    Research (AcSIR)
Ghaziabad, India

Ashish Dehal
Chemical and Hazardous Waste
    Management Division (CHWMD)
CSIR-National Environmental
    Engineering Research Institute
    (CSIR-NEERI)
Nagpur, India
and
Academy of Scientific and Innovative
    Research (AcSIR)
Ghaziabad, India

Monika Dubey
Environmental Engineering Group
Department of Civil Engineering
Indian Institute of Technology Roorkee
Roorkee, India

Pallavi Gahlot
Environmental Engineering Group
Department of Civil Engineering
Indian Institute of Technology Roorkee
Roorkee, India

Arshad Husain
Civil Engineering Section, University
    Polytechnic
Zakir Husain College of Engineering
    and Technology
Aligarh Muslim University
Aligarh, India

Kaushar Hussain
Department of Civil Engineering
Mewat Engineering College
Nuh, India

Absar Ahmad Kazmi
Department of Civil Engineering
Indian Institute of Technology Roorkee
Roorkee, India

Mohd Dawood Khan
Department of Civil Engineering
Zakir Husain College of Engineering &
    Technology
Aligarh Muslim University
Aligarh, India

Nadeem Ahmad Khan
Department of Civil Engineering
Mewat Engineering College
Nuh, India

Ahmar Khateeb
Department of Civil Engineering
Zakir Husain College of Engineering &
    Technology
Aligarh Muslim University
Aligarh, India

Anwar Khursheed
Department of Civil Engineering
College of Engineering
King Saud University
Riyadh, Kingdom of Saudi Arabia

A. Ramesh Kumar
Chemical and Hazardous Waste
    Management Division (CHWMD)
CSIR-National Environmental
    Engineering Research Institute
    (CSIR-NEERI)
Nagpur, India
and
Academy of Scientific and Innovative
    Research (AcSIR)
Ghaziabad, India

Aiman Malik
Department of Electronics Engineering
Zakir Husain College of Engineering &
    Technology
Aligarh Muslim University
Aligarh, India

Mohd Salim Mahtab
Environmental Engineering Section
Department of Civil Engineering
Zakir Husain College of Engineering &
    Technology
Aligarh Muslim University
Aligarh, India

Vinay Pratap
CSIR-National Environmental
    Engineering Research Institute
    (CSIR-NEERI)
Nagpur, India
and
Academy of Scientific and Innovative
    Research (AcSIR)
Ghaziabad, India

Hasan Rameez
Department of Civil Engineering
Zakir Husain College of Engineering &
    Technology
Aligarh Muslim University
Aligarh, India

K. Rathika
Waste Reprocessing Division
CSIR-National Environmental
    Engineering Research Institute
    (CSIR-NEERI)
Nagpur, India
and
Academy of Scientific and Innovative
    Research (AcSIR)
Ghaziabad, India

Mohd Imran Siddiqui
Department of Civil Engineering
Zakir Husain College of Engineering &
    Technology
Aligarh Muslim University
Aligarh, India

Ghazal Srivastava
Department of Civil Engineering
Indian Institute of Technology Roorkee
Roorkee, India

Mohd Azfar Shaida
Department of Civil Engineering
Zakir Husain College of Engineering &
    Technology
Aligarh Muslim University
Aligarh, India

Soumita Talukdar
School of Engineering and Sciences
G. D. Goenka University
Haryana, India

Gulafshan Tasnim
Department of Civil Engineering
Zakir Husain College of Engineering &
    Technology
Aligarh Muslim University
Aligarh, India

Vinay Kumar Tyagi
Department of Civil Engineering
Indian Institute of Technology Roorkee
Roorkee, India

Bhanu Prakash Vellanki
Environmental Engineering Group
Department of Civil Engineering
Indian Institute of Technology Roorkee
Roorkee, India

Bholu Ram Yadav
Waste Reprocessing Division
CSIR-National Environmental
    Engineering Research Institute
    (CSIR-NEERI)
Nagpur, India
and
Academy of Scientific and Innovative
    Research (AcSIR)
Ghaziabad, India

Hummaid Zafar
Department of Civil Engineering
Zakir Husain College of Engineering &
    Technology
Aligarh Muslim University
Aligarh, India

1 An Overview of Developments in Wastewater Treatment Technologies

Ahmar Khateeb, Hummaid Zafar, and Saif Ullah Khan
Aligarh Muslim University

CONTENTS

1.1 Introduction 1
1.2 Adopting and Implementing Suitable Treatment Technology 2
1.3 Advancements of Treatment Technologies 2
    1.3.1 Biological Treatment 2
        1.3.1.1 Aerobic Treatment 3
        1.3.1.2 Anaerobic Treatment.. 6
    1.3.2 Physiochemical Treatment Methods.. 7
        1.3.2.1 Membrane-Based Techniques... 7
        1.3.2.2 Ion Exchange... 12
        1.3.2.3 Chemical Precipitation... 14
        1.3.2.4 Electrochemical Method of Treating Water and Wastewater 17
        1.3.2.5 Adsorption-Based Treatment 22
1.4 Conclusion 26
References...27

1.1 INTRODUCTION

With increasing industrialization and urbanization, various pollutants and their derivatives are discharged and disposed of into the environment, causing pollution of aquatic bodies (Khan et al., 2015). Most of these pollutants are either sources of organics, inorganic contaminants, nutrients and other emerging contaminants (ECs) with low to moderate concentrations. Generally, these include dyes, heavy metals, antibiotics, pesticides, endocrine-disrupting compounds, pharmaceutical drugs, and their derivatives disinfection byproducts, which prove toxic to human beings as well as the aquatic environment (Mousazadeh et al., 2021).

DOI: 10.1201/9781003202431-1

1

Even though pollution control boards state that all industries should treat the wastewater they produce, the rule is seldom followed due to the high cost of wastewater treatment. Conventional wastewater treatment plant requires higher investment, larger water footprint, intense operation and maintenance with inefficient recycling process water (Usmani et al., 2021). This potentially stops industries, especially small- and medium-scale ones from cleaning their wastewater at the source point itself (Khan et al., 2020). Small- and medium-scale industries are often unable to bear the costs of treatment, and therefore, they dispose of their wastewater without treatment. Practically we are not moving toward sustainability. In terms of average worldwide freshwater usage, consumptive use of water has grown almost three times the global population leading to widespread public health issues hampering economic and agricultural development worldwide. For instance, take example of India that occupies almost 3.29 km^2 of the geographical area which is around 2.4% of the total world land area on which almost 15% of the world population lives. The annual utilizable water resources of India are only 4% of the global water resources. With increasing water demand, water resources in India are stressed out with future projections, indicating the availability much lower than estimated by year 2050 (Kumar et al., 2005).

1.2 ADOPTING AND IMPLEMENTING SUITABLE TREATMENT TECHNOLOGY

With emerging advanced technologies and considering the source and nature of wastewater to be treated, there is a need to select and adopt that particular treatment, which is operationally and economically feasible as well as efficient enough to meet the effluent standards laid down by environmental bodies and pollution control boards worldwide. Moreover, this technology must have low energy input and mechanization so as to reduce the risk of malfunctioning. Furthermore, it would be even more fruitful if it has the ability to be incrementally upgraded with time as per user's requirement or water quality standards and guidelines that may be stricter considering health issues and safety guidelines. Considering all these issues and ongoing technological development worldwide, some of the conventional treatment technologies are discussed in detail here.

1.3 ADVANCEMENTS OF TREATMENT TECHNOLOGIES

1.3.1 BIOLOGICAL TREATMENT

With the increase in the industrialization and production of domestic and industrial wastewater across the globe in the last 40 years, significant efforts have been made for the modifications of the conventional biological treatment process and to find new treatment methods. The traditional treatment methods are not able to keep up with ECs, thus having adverse effects on the environment (Taheran et al., 2018). This opens up opportunities for researchers in the field of the biological treatment. Mainly, there are two methods for the biological treatment of wastewater, *aerobic* treatment, which requires oxygen for the degradation of organic matter, and *anaerobic* treatment process, which does not require any oxygen for the degradation of organic matter.

1.3.1.1   Aerobic Treatment

Since the early 19th century, numerous researchers have tried different methods to develop an efficient economic method without affecting the environment (Alam et al., 2021a). One of such researchers (Tripathi and Shukla, 1991) uses aquatic plants for the treatment of wastewater from the sewage of Varanasi city. The researchers used certain algae and microphytes such as *Microcystis aeruginosa* and *Chlorella vulgaris*, and tested them for their role in the treatment of wastewater. Sewage wastewater with effluent from more than 100 different small industries was used to perform the experiment. The experiment which lasts three stages and 24 days results in the reduction of biochemical oxygen demand (BOD) (about 96.9%), suspended solids (78.1%) and total alkalinity (74.6%) with an increase of the dissolved oxygen (70%). Most of the research work related to the biological treatment of wastewater has been done by aerobic methods.

1.3.1.1.1   Activated Sludge Process

The activated sludge process has been used for the biological treatment of organic waste since the early 19th century in England. Since its invention in 1913, because of continuous research and advancement, it was widely accepted as a conventional method for the treatment of wastewater in other parts of Europe (McKinney, 1965). In the early stage of the development of the activated sludge process, many researchers tried modifications to increase the efficiency and reduce the economy by controlling the aeration in the process. One such design was made by *Jones and Attwood* in 1916 in which the two researchers provide a mechanical valve in a manner that controls cyclic periods of the aeration process; the process results in a significant reduction of net aeration demand (Alleman and Prakasam, 1983). Another effort was made by Gould (1942) to overcome the problems in the conventional sludge process and to conserve the capacity of the aeration tank. He proposed a step aeration technique which uses three or four multiple-pass tanks in which the sewage is added stepwise. Step aeration tries to regulate the oxygen demand in aeration tanks to a uniform level. This technique was widely accepted and had been used till the mid-19th century because of its ability to reduce the volume of the tank and the production of well-settling sludge (Sawyer, 1965).

In the late 19th century, the problem associated with the production of extra sludge and its harmful effects on the environment drew researcher's attention. Numerous researchers tried to create a module to reduce the excess production of sludge while increasing the efficiency. One such research was carried out by Yasui and Shibata (1994), in which the researchers tried to enhance the biological degradation of organic waste by conducting both treatment of water and extra sludge digestion simultaneously in the same aeration tank. A significant reduction was observed in the amount of excess sludge as the excess sludge reduced to nearly zero when biomass in a three-aeration tank of volume 1.2 kg/m was recirculated for a day from the biological stage to ozonation stage at a BOD loading of 1 kg/m^3/day. A difference in the quality of effluent produced was observed between the new and conventional methods. The step feed-activated sludge process was also introduced to reduce the solid loading, where the sludge loading was shifted by the step feed from the clarifiers to the first section of the reactor tank. Tallmans Island treatment plant in New York City utilized this method, and a reduction of 44% and 46% in the loading of sludge storage to the secondary clarifier was observed in the two experimental runs (Randall, 1998). In the last two decades, chemical reduction of excess sludge production draws attention; chemical-using sludge reduction processes such as alkaline-thermal sludge treatment, activated sludge-ozonation process, chlorine-combined activated sludge process were used, and it was noted that excess sludge production can be reduced up to 100% without affecting the stability and efficiency of the process (Liu, 2003). One of the recent advancements in the field of activated sludge is the process called *Nereda*, which is a modified version of activated sludge process (ASP) invented by *Mark van Loosdrecht*. The process produces a quick-settling sludge of granular size, which is transferred into the aeration tank instead of a separation unit; this process has a high reduction rate of excess sludge production with significantly low energy and chemical consumption (van der Roest et al., 2011). One of the other leading-edge techniques at present is the *integrated fixed film activated sludge process (IFAS)*. This process has an advantage over other conventional processes by enhanced nutrient removal, and complete nitrification, offering more than 90% removal of chemical oxygen demand and enhanced sludge-settling properties (Waqas et al., 2020).

1.3.1.1.2   Moving Bed Bioreactors (MBBRs)

Moving bed bioreactor was first introduced in Norway in the late 19th century, and since then, it has been widely used all over the globe for the treatment of both domestic and industrial wastewater. MBBRs consist of small moving biofilm made up of polyethylene, held inside an aeration tank by media retention vessels. These biofilms are available in the market in different sizes and shapes. The moving biofilms in MBBR allow it to treat water with BOD in a small area; it can be used as a standalone unit for treatment, but in most of the cases, it is used in combination with other processes in treatment plants with the least cost possible (Qiqi et al., 2012). In the last 10–12 years, numerous research work and modifications have been done in MBBRs to enhance their capacity to treat organic waste and achieve better efficiency. A combination of a moving bed biofilm reactor with the membrane biomass separation reactor was developed; the results were quite astonishing as the treatment efficiency was high with the production of high-quality effluent, irrespective of the operating mode and rate of loading (Leiknes and Ødegaard, 2007). The recent research on MBBRs is more focused on its efficiency in the removal of BOD and nitrogen along with nutrients.

In a research carried out by Barwal and Chaudhary (2014), MBBRs were found to be an economic and effective replacement of the existing conventional method as MBBRs were found to be applicable to a wide range of flow around 10,000–150,000 m³/day while effectively removing 90% of chemical oxygen demand and 95% biochemical oxygen demand along with nutrients in the effluent. To perform the stable operation of partial nitrification anammox (PNA) in the process of removal of nitrogen from the wastewater with extremely low-strength ammonia, a moving bed reactor with function Carries was invented, and a synthetic effluent was treated with 50 mg/L of ammonia (Chen et al., 2019). Significant removal of nitrogen with efficiency (75 + 10%) and stable operation were achieved in a short period with a retention time of two hours; optimal co-immobilization of anammox bacteria and ammonium-oxidizing bacteria was found to be affected by dissolved oxygen control and formation of biofilm.

1.3.1.1.3 Fixed Bed Bioreactors (FBBRs)

Fixed bed bioreactor (FBBR) was developed in the late 1980s as an industrial treatment system. It consists of multiple chambers filled with porous foam and plastic media. In all the biological wastewater treatment methods, FBBRs are considered ideal for the treatment of wastewater with medium and medium-high BOD because they can hold most microbes in the smallest space, making them space saving and energy efficient. With the help of a unique bacterial population, advanced biological processes such as nitrification, desalination and anammox can also be performed (Wiesmann, 1994). The process has been widely used in textile and food-processing industries (Paździor et al., 2019). Many researchers tried from time to time to develop an efficient model of FBBRs for wastewater of different types and from different industries. One such research was carried out for the treatment of wastewater containing synthetic dye from textile industries in Taiwan. The study was carried out using gel-entrapped cells of *Pseudomonas luteola* for the treatment of wastewater containing azo dye using FBBRs (Chen et al., 2005). It was found that at the fixed feeding rate of 50 mg/L of concentration of dye, the best decolonization rate was not achieved, and because of mass transfer diffusion limitations, it was found to be strongly hydraulic retention time dependent. The optimum decolonization rate with cellulose acetate (CA)-immobilized cells was found to be at the dye loading rate of 2.25 mg/L and a hydraulic retention time of 1.12 hours, when as for PAA cells transfer time was much longer around eight hours proving that mass transfer with PAA cells is less effective.

Studies have been done to understand and counter the effects of saline water on wastewater treatment plants. In one such study, the effect of salinity on enzymatic activities such as acid phosphates, alkaline phosphates and esterase was studied with four different experiments using NaCl concentrations of 0, 3.7, 24.1 and 44.1 gm/L (Cortés-Lorenzo et al., 2012). The result indicates a decrease in the enzymatic activities on increasing the salinity of water due to which the biodegradation of organic matter also decreased. In another study, nitrogen removal capacity and bacterial community structure were determined using four different FBBRs with a salt concentration of 0, 3, 25 and 45, respectively, of NaCl with hydraulic retention time (HRT) of six and 12 hours (García-Ruiz et al., 2018). The results indicate that nitrite-oxidizing bacteria and ammonia decreases drastically at concentrations of 25 and 45 g/L NaCl, whereas adaptation of the anammox microorganisms was observed at concentration of 3g/L NaCl. The nitrogen removal efficiency of the reactor was recorded as 87.68%, 64.25%, 38.79% and 19.74% for salt loading of 0, 9, 25 and 45 gm/L of NaCl for HRT of six hours, respectively.

1.3.1.2 Anaerobic Treatment

Anaerobic process for digestion of organic waste has been used since ages and is gaining popularity at present. The anaerobic digestion process for the treatment of wastewater was first used by a Frenchman named Mouras by inventing a version of Septic tank in 1881 (Abbasi and Abbasi, 2012). In the anaerobic treatment of wastewater, anaerobic microorganisms are used to degrade organic matter in the absence of oxygen. Anaerobic methods of wastewater treatment are being adopted in food and beverage, paper and textile industry. The anaerobic treatment of wastewater consists of two processes: first, the phase of acidification and second, the phase of methane formation. In developing countries like India and China where energy supply is expensive and short, the treatment of domestic wastewater has been focused on anaerobic digestion.

1.3.1.2.1 Anaerobic Digesters

Anaerobic digesters use anaerobic bacteria for the treatment of wastewater by breaking down the organic waste and producing bio gases. There are different versions of anaerobic digesters available in the market, which perform the same process with different functions. Anaerobic treatment methods have been used for 100+ years for the treatment of domestic wastewater using different types of filters and processes (McCarty et al., 2001). Anaerobic methods are considered as one of the best methods for high organic effluent treatment. Different types of methods are applicable for different types of wastewater having preference over one another over different factors such as pre-treatment, methods of operating, etc. In a study carried out by Rajeshwari et al. (2000) on wastewater from slaughterhouse using an anaerobic digester and upflow anaerobic sludge blanket (UASB), the results indicate that an anaerobic digester can be used for slaughterhouse wastewater without the need for pre-treatment, but in the case of UASB, a pre-treatment is required for the removal of fat and suspended solids. Anaerobic digesters have been widely used for treating wastewater containing palm oil. Because of its high Chemicals Oxygen Demand (COD) and BOD concentrations, palm oil produces wastewater, which is highly polluted and harmful to the environment (Poh and Chong, 2009). The research was carried out by Yacob et al. (2006) to evaluate the efficiency of anaerobic digesters in the treatment of wastewater containing palm oil. A closed digester with volume 500 m^3 was developed to study the efficiency of palm oil mill effluent (POME). The results show a high efficiency of COD removal (97%) and alkalinity between 0.1 and 0.3 with the lowest HRT of 17 days were achieved in three months.

1.3.1.2.2 Anaerobic Sludge Blanket Reactors

Anaerobic sludge blankets have been used for ages for the anaerobic treatment of domestic wastewater in villages. Anaerobic sludge blankets treat wastewater by passing it through a free-floating sludge blanket. In the last two decades, anaerobic digestion processes have gone under significant modifications (Bal and Dhagat, 2001). The anaerobic microorganisms present in the sludge blanket multiply with organic waste by digestion and settle at the bottom of the reactor tank. In recent time, UASB reactors have been widely accepted a suitable option for the anaerobic treatment of wastewater because of their less space requirements, low maintenance and operation cost, and green energy generation. Daud et al. (2018). They are also efficient in the treatment of wastewater from textile industries and food-processing industries containing fatty oil and dyes. From time to time, researchers have tried to find an efficient process for the anaerobic treatment of different types of wastewater by doing comparison between different processes. A research was carried out by Peña et al. (2000) on the treatment of domestic wastewater using anaerobic ponds and UASBs under the same environmental conditions in Columbia. The results show the COD, BOD and TSS removal efficiency of USSBs was 66%, 78% and 69%, respectively, and the efficiency of anaerobic pond for the same parameters were found to be 68%, 59% and 73%, respectively. From the above results, it was concluded that removal efficiencies of AP and USSBs are not the key factors for the selection of a particular method; it is based on social and economic criteria and management capability. In another study carried out on the treatment of winery effluent with three UASBs, the first was seeded with *Enterobacter sakazakii enriched sludge*, second with brewery granules and third with just sludge (Keyser et al., 2003). The first sludge blanket shows a COD removal efficiency of 90% in 17 days with an HRT of 24 hours; the second experiment shows a COD removal efficiency of 85% within 50 days. The third test was needed to be continuously reseeded as it showed problem associated with conventional sludge seeding. In another study, the treatment of synthetic textile wastewater containing indigo dye was evaluated, followed by poetry clay absorption (Conceição et al., 2013). The results indicate a color and organic matter removal with the effectiveness of 69% and 81.2%, respectively, while using poetry clay as an alternative absorbent, the color removal was 97% for the concentration of 200 g/L, indicating that poetry clay as an alternative absorbent has effective results and can be used as post-treatment unit in the treatment of wastewater containing dyes. In another study using an anaerobic-aerobic-anoxic sequencing batch reactor (AOB-SBR) system removal of carbon, nitrogen and phosphorus were studied by changing cycle time (Liu et al., 2020). The results demonstrated the high removal efficiency for COD, total nitrogen (TN) and total phosphorus (TP) of 96.8%, 96.32% and 94.33%, respectively, for a cycle time of six hours. The anaerobic release rate and aerobic and anoxic removal rate of TP was recorded at peak with 104.31% and 81.81% mg/g MLVSS/day, respectively. Microbiological analysis showed that *Bacteroidetes, Candidatus saccharibacteria* and *proteobacteria* at the phylum level and *sphingobacteria* at the class level benefitted the AOB-SBR process. Some other recent studies based on biological techniques for wastewater treatment are enlisted in Table 1.1.

1.3.2 PHYSIOCHEMICAL TREATMENT METHODS

1.3.2.1 Membrane-Based Techniques

Membrane-based treatment technologies have gained significant industrial development in the last 50 years, although the history of membrane technique using synthetic membrane traces back to the late 17th century (Wenten, 2003). In a world standing on the verge of water crisis, membrane technologies have been proven as an efficient tool in wastewater treatment processes because of their satisfactory results from both

TABLE 1.1

Few Recent Research Studies Based on Biological Treatment Techniques

S. No.	Technology Used	Types of Impurities Removed	Removal (%)	Remark	References
1.	Sand-filled biofilm reactors	Calcium, fluorides and nitrates	56.3%, 96.3%, and 96.95%	16SrRNA sequencing was used	Ali et al. (2021)
2.	Activated sludge process	Micro plastics (MP)	64.4%	MP decline from 79.9 to 28.4 n/L	Liu et al. (2019)
3.	Activated sludge process	Absorbed dye	95% with (D/B) ratio 0.72		Haddad et al. (2018)
4.	Total reflux sludge reactor and MBBR	Pharmaceutical waste, COD, nitrogen and ammonia		MBBRs were more efficient	Wang et al. (2020)
5.	Moving bed biofilm reactors	Nitrogen	94.43%		Jia et al. (2020)
6.	MBBR	Congo red dye	95.7%	Flow rates 25–100 mL/ hours over a period of 564 hours	Sonwani et al. (2020)
7.	Up-flow and fixed bed bioreactors	Phenol-containing raw olive oil wastewater	75% in the first and 45% in the second stage		Dalis et al. (1996)
8.	UASB reactors	Domestic wastewater containing COD	85%		Behling et al. (1997)
9.	UASB system	SARS-COV-2 RNA		Reduction more than 1.3log 10 in SARS-COV-2 RN	Kumar et al. (2021)

economic and technical points of view (Peters, 2010). After being used in the laboratory and then industries for the treatment of wastewater, in the last two decades, membrane technologies have been widely accepted as an efficient tool for the treatment of municipal wastewater as well as for the production of clean drinking water (Nicolaisen, 2003). Membrane filtration technologies can be classified on the basis of the type of force applied, type of membrane used, separation process, etc. (Fane et al., 2011). Recently, pressure-driven membrane processes such as reverse osmosis, ultrafiltration, microfiltration and nanofiltration are widely used as a single unit or in combination for the treatment of domestic as well as industrial wastewater.

Membrane treatment for the treatment of domestic wastewater and saline water using reverse osmosis has been used for a long time. Osmosis is a natural phenomenon where liquid flows through a semipermeable membrane from low solute concentration to high solute concentration under osmatic pressure. In the reverse osmosis process, liquid flows through a semipermeable membrane from high-concentration solute to low-concentration solute. The first RO membrane for the separation of saline water was demonstrated in 1962 when Loeb and Sourirajan developed the first CA membrane capable of separating salt from water (Fane et al., 2011). In recent years, reverse osmosis has been accepted as a suitable method for the treatment of saline water for drinking purposes as well as industrial purposes; therefore, numerous researches were carried out to find the best and most efficient method of treatment of saline wastewater using RO or in combination with other membrane methods. A combined system of RO and ultrafiltration was developed to treat spent metalworking lubricant aluminum-converting works; the results indicate that fouling can be overcome by adding 50–100 mg/L of surfactant with three-step cleaning and back flushing at a capacity of 100,000 gal/day of oil waste (Sonksen et al., 1978). In another research carried out by Sadr Ghayeni et al. (1998), a combination of hollow fiber microfiltration and reverse osmosis was used to evaluate the production of water from secondary effluent. In this research for low-pressure operating membranes, polyvinylidene fluoride (PVD), cellulose triacetate (CTA), thin film composite (TFCL) and nanofiltration (NF45) were used at an operating pressure of 230, 750, 550 and 360 kpa at a target flux of 20 L/m²/hour. The results indicate the total removal of orthophosphates from secondary effluent and the removal of ionic species with efficiency of 99.2% and 41% for TFCL and NF45, respectively, but nitrites and nitrates were partially removed. In different research studies on the treatment of wastewater containing heavy metals carried out by Qdais and Moussa (2004), they used both NF and RO to treat wastewater with copper and cadmium ions and to investigate the environmental factors affected. The results indicate removal efficiency of 98% and 99% by RO for copper and cadmium ions, respectively, whereas NF shows an efficiency of 90%. The researchers also checked membranes to wastewater containing more than one heavy metals; the results indicate a removal efficiency of RO 99.4% from the initial concentration of 500 ppm to 3 ppm, and the average removal efficiency of NF was 97%. The major problem associated with the use of RO is membrane fouling; it can decrease the efficiency of the membrane and can affect the cost. Numerous researchers have been trying to develop different types of membrane to resist fouling. In a research carried out by Kim et al. (2018), by integration of graphene oxide into a highly cross-linked network, an anti-swelling membrane was developed for excellent desalination efficiency. In the results, a NaCl rejection of 98.5% at 10 bar was recorded under the harsh environment at a water flux of 33.5 L/m² hour.

Ultrafiltration is another pressure-driven membrane process which can separate pathogen, viruses and suspend particles from the wastewater except dissolved solids. Since as early as 1920, ultrafiltration has been studied at small-scale laboratory and industry level. Introduced by millipore and amicon in the 1960s, the initial goal of ultrafiltration membrane was to work as a pre-treatment to develop a high-flux reverse osmosis membrane. Efforts have been made to develop different membranes to achieve high efficiency and a variety of applications. In a study carried out by Kunst and Sourirajan (1974), efforts were made to develop a cellulose acetate ultrafiltration membrane for different types of applications. It was found that the concept of the structure evaporation rate of CA development of RO is also valid for the development of CA ultrafiltration membrane. The size of pores on the surface of the membrane tends to increase by the factors such as a decrease in S/P, increase in N/S, increase in the solvent evaporation period and increase in the N/P casting (S: solvent constant, P: Polymer constant; and N: nonsolvent constant). The effective number of pores tends to increase with the increase in the S/P in the casting solution. The results evaluated give an idea about developing cellulose acetate membranes for different applications. Recently researchers have been trying to decrease the fouling of membrane; in one study carried out by Khajouei et al. (2019) with the help of grapheme nanosheets, graphene oxide was derived into the membrane surface, which then inferred by scanning electron microscopy (SEM) and fourier transform infrared spectroscopy (FTIR). This increases the antifouling properties of the membrane as well as the flux by decreasing membrane surface roughness and checking the hydrophilicity both checked by contact angle and atomic force microscope (AFM), respectively. Two different analyses show that the use of graphene oxide (GO) solution in the production process helps the membrane against membrane fouling, which gives a new range of membrane for future use.

The term nanofiltration was coined around the 1980s for a membrane that permits the ionic solute present in feed water to pass through the membrane. Nanofiltration membrane exists between ultrafilteration (UF) and reverse osmosis (RO) and can operate at low pressure and high flux saving energy consumption (Fane et al., 2011). Nanofiltration membranes have high removal efficiency for monovalent salts; however, they are completely efficient in removing multivalent salts and organic matter. In a study carried out by Watson and Hornburg (1989), they discussed the use of low-energy nanofiltration membrane in the treatment of saline and blackish wastewater and its future aspects. In the early 2000s, plants in France and other countries were using nanofiltration membrane for the treatment of wastewater; however, there was a problem of membrane fouling affecting the efficiency of the membrane (Her et al., 2007). With an increase in the use of nanofiltration membrane, efforts have been made to manufacture an antifouling nanofiltration membrane. An antifouling cellulose NF membrane was developed (Li et al., 2011) using 1-ally-3 methylimidozoliumchloride as a solvent through the phase inversion method with 8% of cellulose solution. When analyzed by X-ray diffraction and SEM, the membrane shows a relatively dense structure with remarkable removal of crystallinity. At a pressure of around 0.4 Mpa, the pure water flux reached 128.5 L/m² hour with molecular weight <700 da. The membrane shows good stability and ability against antifouling.

Microfiltration membrane technique is used for separating impurities such as bacteria, viruses and particles of size 0.1–10 μm with the help of the sieving method/sieving effect under a quite low pressure. The first commercial microfiltration was used in the pharmaceutical industry in the 1960s for the removal of microorganisms; since then, microfiltration membrane process has been used in the wine industry, semiconductor industries and most importantly in the late 20th century for the treatment of wastewater. With the increasing demand for water, efforts are being made to develop effective and efficient membrane technologies for the reclamation of water. Numerous researchers have tried to develop an effective membrane for the treatment of wastewater. In one such research using laser interference and silicon micro machine technology, an inorganic microfiltration membrane with pore size 0.1 μm was developed by Kuiper et al. (1998); the membrane shows high porosity and small flow thickness, relatively high resistance to fouling and smooth surface. The results show flux that is about 40 times higher than conventional earth filtration and very low fouling tendency. In the present time, membrane filtration is being used in an integrated system for pre-treatment of wastewater because of low economy, easy availability, process reliability and feasibility; different types of microfiltration membranes are being designed, one of which is ceramic microfiltration membranes (Hakami et al., 2020). Ceramic microfiltration membranes can easily operate in extreme operating conditions and have good cleaning protocols; ceramic microfiltration membranes are given advantage over organic or polymer membranes. In another research carried out by Gu et al. (2020), to enhance the antifouling potential and permeability of the membrane, graphene oxide quantum dots (GOQDs) were grafted into the alumina membrane surface by covalent bonding. The effects of GOQDs on membrane were then studied in terms of properties such as roughness, permeability, pore size and hydrophilicity along with the antifouling abilities. The study shows that the pore size was decreased from 200.8 to 191.3 nm with an increase in pore uniformity. The modified ceramic membrane shows an increase in water flux of 30% and 15% reduction in surface resistance, with an improvement in antifouling properties. Besides this, a few other studies based on membrane treatment are enlisted in Table 1.2.

TABLE 1.2
Few Recent Studies Carried Out Using Membrane-Based Techniques

S. No.	Membrane Used	Impurity Retained	Retention (%)	Remark	References
1.	Reverse osmosis	Ciprofloxin	Removal rate > 90%	Spiral wound RO membrane was employed	Alonso et al. (2018)
2.	NF270 nanofiltration membrane	Heavy metal ions	100%, 99% and 89% and 74% for copper, cadmium, manganese and lead	Rejection rate of NF270 decreases with increased concentration	Al-Rashdi et al. (2013)
3.	Micellar-enhanced ultrafiltration (MEUF)	Acid dye from aqueous solution	90%		Purkait et al. (2004)
4.	Cross-flow nanofiltration	Cyanide	94% cyanide rejection		Kumar et al. (2011)
5.	Low-pressure reverse osmosis, CRE2421	Ampicillin and amoxicillin	73.5%–99.9% and 75.1%–98.5% for amoxicillin and ampicillin	Permeate flux for both antibiotics was 12–18 L/m^2	Gholami et al. (2012)
6.	Pellet precipitator in reverse osmosis	Ca^{2+}, Mg^{2+} and Si ions	90%–95%	Pellet reactor enhanced the performance of RO by 92%–94%	Sahinkaya et al. (2018)

1.3.2.2 Ion Exchange

Ion exchange is a wastewater treatment method which is based upon exchanging undesirable ionic contaminants with other more acceptable or non-objectionable ionic substances. The ions of contaminants and exchanged substances must contain a similar electric charge (either positive or negative) in order for this process to occur. It is one of the most suitable technologies for removing dissolved inorganic ions effectively, also efficient for the removal of toxic metals from water, though it's not effective in removing bacteria. The first observation of ion-exchange phenomenon dates back to 1850 when two agriculturists, Thomas and Way, performed scientific research in which they passed ammonium sulfate through soil and found that the filtrate contained calcium sulfate instead of ammonium sulfate. The discovery was not of much significance until later that decade it was found that this reaction was reversible.

E.B. Showell in his study for water treatment mentioned the first commercial use of ion exchange (Showell, 1951). Glauconite-rich green sand found in the soil of coastal plains from Long Island to Texas, commercially known as "zeolite", was first used for water softening in which calcium and magnesium, which are weak bases, were replaced by sodium which is a strong base. After this, several modifications were made, zeolite was used with other artificially made siliceous-sodium aluminum silicates from 1924, in 1936 with coal, shale was used for first time, in 1940 phenol-formaldehyde resins and in 1946, non-phenol styrene base resin was used for the first time.

In his study, G.E. Boyd discussed the production of synthetic organic ion-exchange polymers, which in turn gave rise to large-scale applications of water softening and de-ionization. The very first of these polymers was synthesized by Adam and Holmes (England). They made class C Bakelite-type resins by condensing either poly-phenol or aromatic polyamines with formaldehyde, which had the ability to absorb bases or acids, respectively, though it only worked in strong alkaline solutions and was not much useful for water. After that, Holmes and others modified this by combining it with strongly acid nuclear or methylene sulfonic groups and synthesized sulfite phenol formaldehyde and polyamine formaldehyde. Such cation-exchanging materials were commercially available in that country; these cation exchangers showed a practical exchange capacity equal to or more than that of the older mineral zeolites (Boyd, 1951).

When water containing calcium, magnesium and iron is passed through cation-exchanger resin, these metals are replaced by sodium. After passing water through several times, the ability of the bed to produce soft water reduces and eventually gets exhausted; in order to use it again, it has to be regenerated using common salt, which removes calcium, magnesium and iron in the form of chlorides and at the same time restores the sodium in the bed to be usable again. It has been found that the rate of ion-exchange treatment is 6–8 gpm/ft^2 as compared to coagulation and filtration, which have 1–2 gpm/ft^2 and 2–3 gpm/ft^2, respectively (Alsentzer, 1963).

A study on the removal of metals and anions from water was carried out in which seven different ion-exchanging resins were tested. Out of those, two were inorganic and the remaining five were organic, namely, sodium titanate, synthetic zeolite, amino-phosphonate, strong acid cation, weak acid cation, strong base anion and weak base anion. It was concluded that amino-phosphonate resin and sodium titanate were found to be several times more effective than conventional resins in removing toxic metals (Zn, Cu, Mn, Ni). The weak acid cation exchanger (WAC) was also efficient for these metals, though it was affected by the flow rates. The breakthrough level that is the percentage of impurity in the treated water with respect to the feed water was found to be <1% in the case of Mn when sodium titanate was used even when the feed was as high as 1 mg/L. The processing capacity of strong acid cation resin which was used to remove Mn from water was found to be lower than other exchangers. Sodium titanate was also found to be quite efficient in removing iron, and a breakthrough level was found to be as low as 10% and for strong acid cation exchanger (SAC) and WAC breakthrough values were a bit higher than 38% and 16%–41%, respectively (Vaaramaa and Lehto, 2003).

Synthesis of a novel lignin-based ion-exchange resin (LBR) was done by condensation polymerization of sodium lignosulfonate with glucose under acidic conditions for the removal of heavy metals (Liang et al., 2013). The maximum adsorption capacity for Pb(ll) was found to be 194.533 mg/g, for Cu(ll) was 59.998 mg/g, for Cd(ll) was 48.80 mg/g, for Ni(ll) was 42.450 mg/g and for Cr(lll) was 41.847 mg/g. These absorption capacities were affected by pH as the pH increases; the adsorbent surface becomes negatively charged, resulting in increased removal efficiency. Also, these resins can be easily recharged using simple acid treatment.

Removal of reactive dyes from dye house effluents was tried using a number of adsorbents, namely, clay-type adsorbents, ion-exchange resins, biomass-based adsorbents and magnetic ion-exchange adsorbents (Hassan and Carr, 2018). Clay-type adsorbents possess good hydrodynamic properties and are also cheap. But their dye-binding capacity is not comparable to ion-exchange resins and, hence, cannot be used alone for the complete removal of dyes. Ion-exchange resins have good dye-binding capacity; they can be easily recycled and are also easy to handle. But they are expensive; therefore, cheap alternatives such as modified and unmodified cellulosic biomasses are being studied as an alternative to ion-exchange resins. Biomass-based adsorbents have biodegradability and therefore are easily disposable; they are also cheap, but they have poor dye-binding capacities because they are weakly anionic in nature. Therefore, they do not come near to commercial ion-exchange resins. On the other hand, chitosan is cationic, and it can bind dyes using ionic bonds; it was concluded that chitosan derivatives are potential adsorbents that can be used as a replacement for commercial ion-exchange resins for the removal of reactive dyes. The removal of adsorbent from treated water is complicated, and for that, research has been done to make it easy. In result, they produced magnetic nanoparticles, and using a strong magnet, they can be easily separated from treated water. The advantage of using magnetic ion-exchange adsorbents is the easy removal from treated water. However, they showed relatively less dye-binding capacity compared to other adsorbents.

Removal of ammonia from wastewater can be done by biological processes of nitrification and denitrification, which require a lot of energy in pumping and aeration. The use of ion exchange, however, has the advantage that it requires a short contact time, does not require much energy and is simple to operate. In a study, two water samples were taken and several zeolites and resins were used on both including Dowex 50Wx8, purolite MN500, purolite C104Plus, purolite C100H, zeolite NM-Ca, zeolite NV-Na, zeolite AZLB-Ca, zeolite NV-Na *TM, zeolite AZLV-Ca, DIAION PK228, DIAION SK18, DIAION PK216, TUSLION 5–42, TULSION T-52, LEWATIT and KMI zeolite (Al-Sheikh et al., 2021). These resins showed different adsorption capabilities, but two of the acidic resins, i.e. Dowex and purolite C100H were found to be highly effective, showing 95% and 90% efficiency, respectively.

In another study (Jiang et al., 2015), the removal of ibuprofen (IBU), diclofenac (DC) and sulfadiazine (SDZ) was investigated by two magnetic resins, namely, MEIR1 and MEIR2. Parameters influencing the adsorption of compounds onto resins were coexisting ions and the pH of the solution. The presence of cations is found to have a significant effect on the adsorption of pharmaceutical compounds on resins, but the presence of chlorides and sulfates reduces the capacity of both MIER1 and MEIR2. For both resins, the adsorption capacities were reported higher in the neutral pH range (6.5–7.5) than in alkaline conditions (pH = 10–10.3). Thus, it may be concluded that the larger surface area of MIER1 resulted in higher adsorption values of IBU, and in case of SDZ and DC, MIER2 showed more adsorption capacity because of a large number of surface functional groups on the resin. There are other applications of ion exchange in wastewater treatment with further ongoing development in the technology that may further improve the efficiency of the processes involved. Few more studies that have been done on the applications of ion exchange for the removal of impurities from wastewater are enlisted below:

1.3.2.3   Chemical Precipitation

Chemical precipitation in water treatment is the process of converting the dissolved impurities into an insoluble form or precipitate to facilitate the removal of the same from water or wastewater.

It is mainly used for the removal of metallic cations, also for anions such as phosphate, cyanide and fluoride, and organic impurities that are phenols.

Acid mine drainage contains a wide range of heavy metals including Fe, Mg, Mn, Pb, Al, Cu, etc. A study was done to produce a ligand to specifically bind heavy metals; the ligand synthesized was 1,3-benzenediamidoethanethiol ($BDETH_2$) which can be used as a di-alkali salt, $Na_2(BDET)$ (Matlock et al., 2002). It has been found that BDET can reduce the concentration of a broad range of divalent metals from water to below environmental protection agency (EPA) limits; the metal-BDET precipitates are insoluble in an aqueous solution. It was found that BDET ligand was able to reduce metal concentration effectively, initial concentration of Fe, Mg, Mn, Al and Cu was found to be 194, 57.4, 4.65, 0.483 and 0.012 ppm, respectively, and after an hour of addition of BDET ligand, the concentration of Fe, Mg, Mn, Al and Cu was found to be <0.009, 0.873, <0.001, <0.041 and <0.009 ppm, respectively. The advantage of using ligands is the reduction of metal waste and secondary waste generated.

In another study, the synthesis of magnesium hydroxy carbonate from low-grade magnesite was done, and precipitation of heavy metals was studied (Zhang and Duan, 2020). It was found that the removal efficiencies of heavy metals increased as the dose increased, though the applied dose for this study was 6,000 mg/L. The reaction proceeded within 20 minutes, and the remaining concentrations of VO_2^+, Cr^{3+}

S. No.	Technology/Resin/Adsorbent Used	Impurities Removed	Remark	References
1	Ni-Al bimetallic adsorbent (Raghav et al., 2019)	fluoride		Raghav et al. (2019)
2	Cellulose-alpha zirconium phosphate ion-exchange membrane	Heavy metals	Up to 97%, 98% and 99% removal of Cu(II), Zn(II) and Ni(II) were achieved	Ibrahim et al. (2019)
3	Ion-exchange membranes	Heavy metals		Pan and An (2019)
4	Ion-exchange nanofiber membranes			Swanckaert et al. (2022)
5	Amberlite IRC 747, Lewatit TP 208 and Lewatit TP260 (chelating resins)	Ca^{2+} and Mg^{2+}	Adsorption of Ca^{2+} and Mg^{2+} in desulfurization of wastewater	Zhang et al. (2019)
6	Modified biochar (Mg-coated biochar)	Cd and Pb	The absorption capacity of Pb and Cd on Mg-coated biochar was increased by 20–30 times as compared to unmodified biochar	Wu et al. (2021)
7	FF-IP, IRA 910 and N-IP resins	Phosphate, nitrates and colors		Gregory and Dhond (1972)
8	Bentonite clay	Heavy metals (lead and copper)	Optimum pH = 5 Maximum efficiency of 87.7% and 89% achieved for removal of copper and lead at 10ppm initial concentration	Hussain and Khaleefa Ali (2020)
9	MN 500 and PPC 104 purolite cation-exchange resins	Cadmium	Maximum efficiency achieved in the pH range of 2–6 and with an initial Cd concentration of 100 mg/L	Modrogan et al. (2015)
10	Charcoal-based ion exchanger made using 4% amberlite in benzene	Fluoride	pH range = 7–10 reduction in fluoride from 10 to 1 mg/L was achieved	Singh et al. (1999)

and Fe^{3+} were found to be 0.01, 0.05 and 1.12 mg/L, respectively, which were under the limits set by China (0.5–2.0 mg/L, GB 8978-1996), and the final pH was 7.1. The precipitate so formed was composed of Fe_2O_3, V_2O_5 and Cr_2O_3; these large amounts of metals found in the precipitate can be reused in the metallurgical industry.

Another study was done for the removal of heavy metals Cu(ll) and Zn(ll) from wastewater using reagents, namely, lime $Ca(OH)_2$, caustic soda (NaOH) and soda ash (NA_2CO_3) (Benalia et al., 2021). The removal efficiencies were found to be increasing as the dosage of the reagent was increased (10–400 mg/L) and an efficiency of over 90% was achieved. Removal efficiencies were also related to the pH and were higher at high pH values of around 8–10. The remaining concentration of metals so found was in line with the industrial discharge standards. The sludge products of Zn and Cu were precipitated as amorphous hydroxides including $Zn(OH)_2$ and $Cu(OH)_2$. Among all reagents of wastewater treatment, treatment with soda ash was found to produce a lower volume and a large product size of sludge, as a result of this, drying steps could be less expensive. Thus, soda ash may be considered as a cost-effective precipitating agent for the removal of Cu(ll) and Zn(ll) from industrial wastewater.

Phosphate removal from wastewater was carried out using metallic salts, namely, calcium oxide, aluminum sulfate, iron sulfates and a mixture of iron sulfates and calcium oxide (Nassef, 2012). Various parameters were studied such as pH and dosage of each reagent to investigate the relation of these with the removal efficiencies. The results were as follows: the removal efficiency of CaO was found to be 90% with the dosage of 40 mg/L and at a pH value of 8.5–10.

In the case of aluminum sulfate, the removal efficiency was found to be 85% at the same dosage of 40 mg/L and at a pH value of 4. And for iron sulfate, the removal efficiency came out to be 80% at a dosage of 20 mg/L and at a pH range of 8.5–10. Lastly, the results of the combined use of calcium oxide and aluminum sulfate showed no increasing effect on the removal of phosphates. All these tests were performed at room temperature. As evident lime was found to be the best option if phosphate content is high as removing phosphate with lime is a function of pH and is highly independent of phosphate concentration.

Another study was carried out on fluoride removal using calcium-containing precipitant and precipitant fluorite (Wang et al., 2019). The study was done on two water samples; first one was simulated water prepared by sodium fluoride and deionized water at 300 mg/L fluoride concentration with a pH of 6.6, and the second one was smelting wastewater containing fluoride of 300 mg/L. Fluorite and calcite used were 98.89% and 97.54%, respectively. The influencing parameters were the dosage of precipitant, oscillation rate and dosage of acid. The percentage removal of fluoride for simulated water with a reaction time of 30 minutes, calcite dosage of 2 g/L, hydrochloric acid dosage of 21.76 g/L, oscillation rate of 160 r/minutes and 2 g fluorite was noted to be 96.20%, with precipitant settling quickly. And the results of the removal of fluoride for smelting wastewater containing fluoride with calcite dosage of 12 g/L, pH of 2.08, reaction time of 30 minutes, oscillation rate of 160 r/minutes and 2 g fluorite reached 95%. It was noted that calcite can efficiently remove fluoride from wastewater, and the aided precipitant increased the sedimentation or settling of precipitates.

Various coagulants were tested for the removal of arsenic from contaminated water including ferric chloride, hydrated lime, sodium sulfide and alum (Harper and Kingham, 1992). Contaminated water contained 48 mg/L of arsenic, and discharge concentration required for arsenic was <1 mg/L. The use of lime followed by ferric chloride addition was found to be most effective method; the results were very satisfactory as 96% of arsenic was removed during the testing. Increasing the dose of ferric chloride beyond 200 mg/L did not show a significant difference, though in experiments in which two dosages of ferric chloride of concentration 200 mg/L were used, the removal rates were increased to 98%, and after filtration of treated water, additional arsenic was removed. The treatment system which included the addition of lime and ferric chloride, clarification, filtration and carbon adsorption showed arsenic removal rates of 97%–98%. The coagulation mechanism was also used for the removal of impurities or contaminants from wastewater and was a part of chemical precipitation. Some of the coagulants used and contaminants removed from wastewater are as follows: $MgCl_2$ as a coagulant was used for the removal of dyes from wastewater (Gao et al., 2007). Plant-based coagulants (banana peel powder, banana stem juice, papaya seed powder and neem leaf powder) were examined for removal of turbidity, COD and total suspended solids from municipal wastewater (Maurya and Daverey, 2018).

Several other studies that have been done on applications of chemical precipitation for the removal of impurities from wastewater are enlisted in Table 1.3.

1.3.2.4 Electrochemical Method of Treating Water and Wastewater

Electrochemical treatment of wastewater can be done using several methods including electrocoagulation, electroflotation, electrooxidation and electrodialysis.

1.3.2.4.1 Electrocoagulation

This process was first proposed in 1889 in London in which a sewage treatment plant was built and electrochemical treatment was done by mixing the saline water with domestic wastewater. Electrocoagulation was first patented in 1906 by A. E. Dietrich and was used for treating bilge water from ships (Naje and Abbas, 2013). Several studies have been carried out using electrocoagulation for the target pollutants including heavy metals, pharmaceuticals, dyes, COD, etc. For instance, in a study carried out by Patel (2000), the removal of color and COD was done from wastewater using electrocoagulation. The sample contained Direct Black and Acid Red 97 dyes. The parameters that affected the removal of color as well as COD were found to be initial pH, time of electrolysis, current density and electrode materials. The electrodes used were iron and aluminum with direct current. For aluminum and iron electrodes, the removal efficiency was found to be 90.88% and 89.55%, respectively, at an initial pH of 8, which was the highest as compared to all the other pH values. As electrolysis time increased, the color and COD removal efficiency were found to be increasing for aluminum, but for iron, it was maximum at 5 minutes. An increase in current density showed better results for both iron and aluminum electrodes. It was found that Fe-Fe electrodes performed better than the same size of Al-Al electrodes.

Heavy metal removal from wastewater has been reported to be quite successful in several research studies through electrocoagulation (Khan et al., 2019).

TABLE 1.3

Few Recent Studies Based on Applications of Chemical Precipitation

S. No.	Technique/Precipitant/Coagulant Used	Impurities Removed	Remark	Removal Efficiency	References
1	Sodium sulfide, polyferric sulfate, ferric chloride etc	Heavy metals	pH = 5		Wu (2019)
2	Combined use of precipitation, adsorption and reverse osmosis	Lead, zinc, cadmium and copper		91%–98% removal from sulfide precipitation	Gashi et al. (1988)
3	Electrocoagulation with aluminum, iron and hybrid Al/Fe anode	Arsenide and arsenate	pH = 2–10	99%	Deniel et al. (2008)
4	Chemical precipitation with electrocoagulation	Chromium		97% for chemical precipitation and 97.76% for electrocoagulation with aluminum anode	Mella, Glanert, and Gutterres (2016)
5	Fe and Al coagulants combined with an anionic polymer	Phosphate	Optimum dosages of Fe(III), Al and polymer were 18.4, 2.60 and 1.64 mg/L, respectively	Remaining concentration of up to 0.2 mg/L was achieved	Lee et al. (2016)
6	BaCl$_2$	Sulfate	Optimum dosage was 1.6 mg/L	97.22%	Alnakeeb and Rasheed (2021)
7	Calcium salts with polyaluminum chloride and polyacrylic acid coagulants	Fluoride	pH = 6.5–8.5		Chang and Liu (2007)
8	Magnesium salts	Total ammonia nitrogen (TAN), phosphate and fluoride	Optimal pH = 9.5	97% for phosphate, 58% for TAN and 91% for fluoride	Huang et al. (2017)
9	Ammonia nitrogen	Magnesium ammonium phosphate (MAP)	Optimal pH = 10, Temperature = 35°C	46%	(Zhu et al. 2012)
10	BaCl$_2$	Sodium sulfate		Nearly 100%	Navamani Kartic, Aditya Narayana, and Arivazhagan (2018)

TABLE 1.4

Removal Efficiencies of Different Heavy Metals Carried Out by Electrocoagulation

S. No.	Metals	Concentration (mg/L)	Anode-Cathode	Removal Efficiency (%)	References
1	Cr^{3+}, Cr^{6+}	887.2, 1495.2	Fe-Fe	100	Verma et al. (2013)
2	Cu^{2+}, Cr, Ni^{2+}	45, 44.5, 394	Al-Fe	100	Akbal and Camcidotless (2011)
3	Cd^{2+}	20	Al-Al	97.5	Vasudevan et al. (2011)
4	Pb^{2+}, Zn^{2+}, Cd^{2+}	170, 50, 1.5	Al-SS	95, 68, 66	Pociecha and Lestan (2010)
5	As	150	Al-Al, Fe-Fe	93.5, 94	Kobya et al. (2011b)

The parameters which affect the removal efficiency include initial pH, distance between electrodes, current density and electrolysis time (Alam et al., 2021b). Recovery of metals and reuse of treated water are the advantages of the electrocoagulation process, the quantity of sludge produced is lesser and the flocs generated are larger and heavier and settle down better as compared to chemical precipitation (Bazrafshan et al., 2015). Some of the studies carried out by electrochemical methods are enlisted in Table 1.4.

Removal of fluoride and Fe ions simultaneously was carried out in a study (Das and Nandi, 2020), and the electrodes used were aluminum. Along with removal efficiency, several parameters such as pH, current density, inter-electrode distance and NaCl dose were also closely monitored, and it was found that pH of 7.0, current density of 4.31 mA/cm^2, inter-electrode distance of 1 cm and NaCl dose of 0.33 g/L gave the best removal efficiency of 96% for fluoride and 98.88% for Fe(ll) ions after 60 minutes of electrocoagulation. Apart from fluoride and Fe ions, salinity, total dissolve solids (TDS) conductivity and turbidity were also decreased in the sample after performing the electrocoagulation. It was also noted that the presence of fluoride ions increased the removal of Fe ions and Fe ions increased the removal efficiency of fluoride ions.

In conclusion, electrocoagulation was found to be quite efficient in the simultaneous removal of fluoride and Fe ions, and the overall energy consumption was found to be 1.716 kWh/m^3 (Das and Nandi, 2020).

1.3.2.4.2 Electroflotation

In this method of wastewater treatment, tiny bubbles are generated as a result of the electrochemical reaction so as to lift up the colloidal or finely dispersed particles and are thus separated out easily. Initially, this technology was mainly used in countries like Germany, Japan, UK and USSR (Raju and Khangaonkar, 1984). It is also noticed that the electroflotation method is applied along with the electrocoagulation method for the removal and separation of target pollutants.

A study was done in on textile wastewater treatment using electroflotation (Belkacem et al., 2008). It was done in two parts: first affecting parameters were studied, the electrodes used were of aluminum, and it was concluded that for feed tension of 20 V, inter-electrode distance of 1 cm, and period time of treatment of 20 minutes, the BOD_5 removal was found to be 93.5%, COD removal was 90.3%, turbidity removal was 78.7%, suspended solid removal was 93.3% and color removal was more than 93%. In the second part, the removal of heavy metals such as Ni, Cu, Zn, Pb and Cd was investigated. A polymetallic solution was prepared containing Pb, Zn, Ni, Cu and Cd ions with a concentration of 100 mg/L in the presence of sulfates, with an initial pH of 8 and a conductivity of 2.7 mS/cm. Results showed the removal of efficiency of all the metals at about 99% for the treatment period of 15 minutes.

In another study, emulsified oil removal from oily wastewater was studied (Genc and Goc, 2018). The study was done on two water samples: the first was the synthetic sample prepared from tap water, and the second sample was bilge water. The parameters that affected the removal efficiency were electrode bed lengths and pH; the oil removal efficiency of 85% for the synthetic water sample was achieved by applying a 15-V potential difference to stainless steel sponge bed electrodes. The ideal bed length was found to be 18 cm for higher removal efficiency and low power consumption. Ninety percent turbidity removal was also achieved at 18 cm bed length. Though the results were not the same in the case of the real water sample that is bilge water, only 65% oil removal efficiency was achieved; the reason for this poor efficiency is the adherence of the oils in bilge water to the electrodes, which did not occur in case of bor oil, hence the high efficiency of removal.

Removal of organic contaminants from slaughterhouse wastewater obtained from Turkey's slaughterhouse was done using electrocoagulation and electroflotation as a combined process. Chemical oxygen demand, total phosphorus, total kjeldahl nitrogen and color were simultaneously removed. The electrodes used were Fe and Al. The highest removal efficiencies achieved for COD, total kjeldahl nitrogen, total phosphorus and color are 94.0%, 77.51%, 97.0% and 99.0%, respectively. It was noted that these efficiencies were achieved in the neutral or higher pH range, and removal efficiencies were found to be increasing as the current density increased.

The treated water quality met the standards of water pollution regulations in Turkey.

Hence, it was concluded that this combined process of EC and Ef is efficient and sustainable for slaughterhouse wastewater treatment (Akarsu et al., 2021).

1.3.2.4.3   Electrochemical Oxidation

This process for the treatment of wastewater has been used since the early 20th century, but it gained researchers' attention in the last two decades due to its higher efficiency in the removal of ECs, which are difficult to remove by conventional processes (Chen, 2004). Because of its high electrode cost, low oxidation efficiency and high energy consumption, many researchers have been trying to develop an efficient model to treat wastewater containing different types of contaminates and impurities using the electrochemical oxidation process. In another research carried out by Valero et al. (2010), electrochemical oxidation treatment of dye-containing wastewater using photovoltaic panels increases the use of green energy and also decreases the energy cost. The way experiment includes the 40-module photovoltaic (PV) array, electrochemical filter press reactor and their influence on removal efficiency of dye-containing solution were studied. From the above research, a strong influence of PV array on the optimum use of electrical energy generated was concluded, and a proposal was made to adjust the electrooxidation-photovoltaic (EO-PV) system operating conditions to the wastewater treatment. For the treatment of industrial dyeing and finishing wastewater, You et al. (2016) fabricated a monolithic porous magneli phase Ti4O7 electrode for the electrochemical oxidation process. The results demonstrate that the Ti4O7 achieved an efficient and stable removal of organic pollutants without the need for chemicals, and the chemical oxygen demand (COD) and the dissolve oxygen carbon (DOC) were removed with efficiency of 66.5% and 46.7%, respectively, at a current of 8 mA cm² with a retention time of two hours. The Ti4O7 electrodes also exhibit long stability with COD removal capability only declining by 8% after a full 50-cycle operation at a current of 20 mA c/m² (Luu, 2020) used inactive anode in electrochemical oxidation for the treatment of tannery wastewater after activated sludge pre-treatment process. In this study, SnO_2/Ti and PbO_2/Ti inactive anode effects of current, pH, density and reaction time in removal of pollutants were studied using the electrochemical oxidation process. A total removal efficiency of 80% of chemical oxygen demand, total nitrogen and color were achieved for the treatment of tannery wastewater using SnO_2/Ti and PbO_2/Ti at a current density of 66.7 mA/cm² for a reaction time of 90 minutes.

In a study carried out for COD and TOC removal from saline wastewater using the electrochemical oxidation process, Darvishmotevalli et al. (2019) applied response surface methodology (RSM) to find the effect of independent variables such as reaction time, initial pH and salt concentration. The results obtained were COD and TOC removal efficiency of 91.78% and 68.49%, respectively, with optimum conditions of reaction time 30.71 minutes, PH 7.69, voltage of 7.41 V and salt concentration of 30.94 gm/L. It was concluded that electrochemical oxidation is quite successful in removing COD and TOC from the saline wastewater.

1.3.2.4.4   Electrodialysis

The electrodialysis process uses an electric gradient to separate mineral ions from water by using an ion-selective membrane, which produces two different diluted and concentrated streams. Electrodialysis was first introduced around 1960s, but its use for the treatment of wastewater gained popularity in the late 20th century because of its advantage over conventional processes in the treatment of industrial effluent containing toxic and non-biodegradable constituents (Korngold et al., 1977). In search for an efficient method for the treatment of wastewater, studies have been conducted to use the electrodialysis treatment process efficiently for the treatment of wastewater. For instance, Gain et al. (2002) used coupling membrane electrodialysis for the treatment of wastewater containing ammonium nitrate. Throughout the study, the in-situ stripping ammonia was continuously removed and produced gas with a constant current efficiency of about 70%.

In another study, Shiming et al. (2020) used a reverse electrodialysis reactor which was powered by salinity gradient energy (SGE) for the treatment of dye-containing wastewater. Working principles used in the experiment were based on the conversion of low-grade heat (LGH) to the SGE. By reverse electrodialysis, this energy is then converted to the degradation energy of organic pollutants. The results show that some of the characteristics such as the operating parameter of electrode rinse solution and degradation circulation modes have effects on the decolonization efficiency of acid orange 7(AO7) dye. Results demonstrated that after 20 minutes of the experiment, the A07 dye removal efficiency for degradation circulation mode was 99.93% and 96.52% for anodic and cathodic loops, respectively. Several other studies that have been done on the electrochemical treatment of water and wastewater are enlisted in Table 1.5.

1.3.2.5 Adsorption-Based Treatment

The first use of adsorption was recorded in 1862, in which potable water was purified using a carbon-based material that gave rise to the development of commercial

TABLE 1.5
Few Recent Studies on Electrochemical-Based Approaches

S. No.	Technology Used	Type of Impurities Removed	% Removal	Remark	References
1	Electrocoagulation and electroflotation	COD, surfactant, color and microplastic	91%, 94%, 100% and 98%, respectively	pH = 9 Reaction time = 60 minutes	Akarsu and Deniz (2021)
2	Bipolar electrocoagulation and electroflotation	Turbidity, COD, phosphate and surfactants		Pilot scale tests	Ge et al. (2004)
3	Electrocoagulation	COD, total organic carbon and turbidity	93.0%, 83.0% and 99.89%, respectively		Kobya et al. (2011a)
4	Electrocoagulation	Zn, Fe and ammonia nitrogen	93%, 83% and 93%, respectively	120 minutes of retention time	Kasmuri and Tarmizi (2018)
5	Electrochemical oxidation	Noncyanide stripper waste	70%–99.4%	Anode efficiencies were 1,960 and 340 g/hAm^2	Shiming et al. (2020), Vlyssides et al. (1999)
6	Electrochemical oxidation	TOC and N-NH$_3$	90%	Boron-doped diamond anode was used	Anglada et al. (2010)
7	Bipolar membrane electrodialysis	COD and color from textile wastewater	72.02% and 66.9%, respectively		Yuzer and Selcuk (2021)
8	Electrodialysis	Sodium and sulfate	72.02% and 66.9%, respectively	Energy-dispersive X-ray spectroscopy was employed	Berkessa et al. (2019)

use of activated carbon for the treatment of water and wastewater (Çeçen and Aktaş, 2011). With advancements in research for water purification, several new natural as well as synthetic adsorbents were either tried or synthesized. Numerous studies are available with sufficient literature regarding the use of adsorbents for the target treatment of different water contaminants. For instance, Larsen and Schierup (1981) investigated the adsorption capacities of adsorbents, namely, activated carbon and sawdust the uptake of heavy metals. During tests, it was found that 1 g of barley straws absorbed 4.3–15.2 mg of Zn, Cu, Pb, Ni and Cd. On the other hand, activated carbon adsorbed 6.2–19.5 mg and pine sawdust adsorbed in the range from 1.3 to 5.0 mg.

In another study, agricultural and industrial waste in the form of rice husk and fly ash as an adsorbent were tried (Hegazi, 2013). The adsorbent performance was evaluated under various conditions such as pH, heavy metal concentration, mixing speed and adsorbent dose, and the optimum removal conditions were identified for each metal. The removal efficiency increased as the dose increased, and the doses used were 20, 30, 40, 60, and 60 mg/L. Pb showed a maximum removal efficiency of 87.17% for 60 mg/L dose of rice husk, and in case of fly ash with a dosage of 60 mg/L, the efficiency was found to be 76.06%. Cd showed 67.917% for a dose of 60 mg/L of rice husk and 73.54% for fly ash at the same dose. In the case of Cu removal, efficiency was found to be 98.177% for rice husk and 98.545% for fly ash with a dose of 60 mg/L in both. Lastly for Ni, the removal efficiency was found to be 96.954% in the case of rice husk and 96.034% for fly ash. It may be concluded that these low-cost adsorbents can also be potentially used for the removal of heavy metals with moderate initial concentrations. The results showed that rice husk was effective in the simultaneous removal of Fe, Pb and Ni, while on the other hand, fly ash was effective for Cd and Cu. The contact time for maximum adsorption was found to be two hours, and the optimum pH range noted was 6.0–7.0.

In another study carried out considering carbon nanotubes (CNTs), single-walled and multi-walled long carbon cylinders were found to be quite significant in the removal of heavy metals (Burakov et al., 2018). The factors that affected the removal efficiency were initial pH and metal ion concentration. Cu^{2+} removal using NaOCl-modified CNTs showed a maximum removal capacity of 47.39 mg/g at a pH of 6.0, while using as-produced CNTs, the maximum removal capacity was found to be 8.25 mg/g at a pH of 6.0. Ni^{2+} removal using multi-walled CNTs showed the maximum removal capacity of 3.72 mg/g; for NaClo-modified single-walled CNTs, it was found to be 47.86 mg/g, and for NaClo-modified multi-walled CNTs, it was found to be 38.46 mg/g. Pb^{2+} maximum removal capacity using acidified multi-walled CNTs was found to be 49.71 mg/g at a pH of 5.0. CNTs treated with HNO_3 showed a maximum removal capacity of 49.95 gm. Cr^{6+} maximum removal capacity using CNTs was found to be 264.5 mg/g at pH 2.0, and using N-CNTs, it was found to be 638.56 mg/g at a pH of 8.0. Single-walled NaOCl-modified CNTs showed a maximum removal capacity of 43.66 mg/g for Zn^{2+}, and in the case of multi-walled NaOCl-modified CNTs, it was found to be 32.68 mg/g for Zn^{2+}.

In another group of researchers, Ragheb (2013) studied the removal of phosphate using slag and fly ash with an aqueous solution containing 1,000 mg/L of phosphate and real wastewater taken from household products company Procter & Gamble. Various parameters including initial pH, adsorbent dose, initial metal ion concentration and adsorption time were studied. Results showed the sorption process to be fast, and equilibrium was reached at 30-minute contact time; the maximum phosphate removal of 93% and 95% was achieved for slag and fly ash, respectively, at a dosage of 0.5 mg/100 mL; the optimum pH was found to be 5 in the case of slag and 7 for fly ash. In the case of real wastewater, sample removal efficiency for phosphate was found to be 96.15% and 96.9% for slag and fly ash, respectively, for an initial concentration of 40 mg/L of phosphate. Apart from phosphate removal, slag and fly ash, other impurities that adsorbed with efficiency were as follows: 28.5%, 39.2% of COD removal, 42.65%, 17.8% of BOD_5 removal, 26.05%, 12.66% of ammonia removal, 56.1%, 15.6% of nitrate removal, 19.35%, 36.9% of TDS removal, 30.15%, 25.3% if chloride removal, 17.2%, 39.7% of sulfate removal and 63.35%, 66.6% removal of oil and grease.

Similarly, Manikam et al. (2019) targeted organic compounds, namely, ammonia nitrogen, nitrate and phosphorus along with COD removal using palm oil boiler ash composite adsorbent known as composite palm oil fuel ash (POFA). The initial values of COD, ammonia nitrogen, nitrate and phosphorus were 63.39, 6.01, 0.63 and 0.43 mg/L, respectively. The optimum shaking speed was found to be 200 rpm for all the compounds, and the optimum shaking time was found to be 20 minutes for ammonia nitrogen, nitrate and phosphorus and 30 minutes for COD, the optimum pH was obtained at 6 for COD, nitrate and phosphorus and 9 for ammonia nitrogen. And the ideal dosage of composite POFA was 8 g for 100 mL of raw sewage wastewater. The removal % of COD, ammonia nitrogen, nitrate and phosphorus at 8 g dose was found to be 65%, 63%, 100% and 97%, respectively.

The removal of three different types of dyes methylene blue, malachite green and rhodamine B from textile wastewater by Akashkinari coal through the adsorption method was studied by Khan et al. (2004) at different temperatures, PH and concentrations. The results demonstrated that by increasing the concentration in the solution from 5 to 20 mL/L at pH 6.8, 7.2 and 5.8, respectively, the percentage adsorption of methylene blue, malachite green and rhodamin B decreases from 97.18, 89.16 and 78.40 to 83.90, 79.77 and 67.35 respectively. At the temperatures 20°C, 30°C and 40°C, the rate constant of adsorption (kcd) for methylene blue, malachite green and rhodamine B was reported between 4.27 and 3.95×10^{-2}, 4.53 and 4.61×10^{-2} and 4.39 and 4.20×10^{-2}/minutes, respectively. For the adsorption capacity of dyes at different temperatures, the Langmuir constant varies from 2.12, 2.68 and 1.23 to 1.59, 1.15 and 0.78 mg/g. On increasing the pH value from 3.8 to 7.2, the adsorption rate of three dyes increased from 74.39% to 92.80%, 62.16% to 83.42% and 51.57% to 71.78%, respectively; however, on further increment, adsorption rate was found to be decreasing. From the above research, it was concluded that the Akashkinari coal is a good adsorbent and can be used for the treatment of wastewater containing dyes (Khan et al., 2004).

In the constant efforts to produce an efficient adsorptive material for dye removal applications, Bezerra de Araujo et al. (2019) used graphene oxide samples. Graphene oxide has been widely accepted as a removal matter for anionic and cationic dyes, but limited research has been done about its use as an adsorptive material for textile wastewater. Kinetics and equilibrium of adsorption were studied prior to the experiment. The experiment data shows that in 5 minutes, 90% of methylene blue dye was removed with theoretical Qmax at 308.11 mg/g. After 30 minutes of experiment, 60% of color and 85% of turbidity were removed from the wastewater, indicating that graphene oxide can be used as an alternative to treat wastewater (Bezerra de Araujo et al., 2019).

In another study to develop an alternative for the treatment of wastewater produced in aluminum cane production plants, a researcher (Pietrelli, 2005) investigated the adsorption of fluoride onto the metallurgical-grade alumina. Experiment data showed that the best removal condition was obtained at pH 5–6. The experiment showed that the adsorption rate increased by 25% at an appropriate pH, and fluoride was removed through metallurgical-grade alumina with a capacity of 12.21 mg of F per gram. Around 70.6% of the total column length was occupied by the mass transfer zone at 5% of breakthrough (Pietrelli, 2005).

In another research, Kumari et al. (2020) prepared a novel adsorbent (CAZ) by loading zirconium oxide and calcium on activated alumina. Different analytical methods such as SEM/energy dispersive x-ray (EDX) and brunauer-emmett-feller (BET) were used to characterize the properties of the adsorbent, and the results showed a considerable change in loading of calcium and zirconium. As demonstrated by FTIR and x-ray diffraction (XRD) analyses, at an initial pH of 6 and an adsorbent dose of 6 g/L, max de-fluoridation efficiency (97%) of CAZ was achieved. The fluoride adsorption on CAZ was spontaneous and endothermic, and the Q max of CAZ was found to be significantly higher than the parent alumina. The results demonstrated a better 92% de-fluoridation performance of CAZ that is better than the parent alumina (74%) and present CAZ as an alternative adsorbent for de-fluoridation of wastewater (Kumari et al., 2020).

In another research, Kumar et al. (2020) studied the adsorptive applications of nonabsorbent chitosan-coated magnetic nanoparticles (cMNPs) for the treatment of lignocellulosic biorefinery wastewater, which contains 26 phenolic compounds and three heavy metals. The elemental and vibrating sample magnetometer analysis confirmed that the magnetic properties of an adsorbent allow easy separation of particles when an external magnetic field is present. Maximum removal of phenol (46.2%), copper (42.2%), chromium (18.7%) and arsenic (2.44%) was achieved at an optimized adsorbent dosage of 2.0 g/L and 90 minutes contact time at the pH of 6. The adsorption capacity of cMNPs was found to be in order as copper (1.03 mg/g), chromium (0.20 mg/g), arsenic (1.03 mg/g) and phenol (0.56 mg/g). For the use of adsorbent, its reusability was studied by using the same cNMPs for five consecutive rounds; the results showed that cMNPs have retained 20% of their initial adsorption capacity (Kumar et al., 2020). Few other studies that have been done on adsorption are enlisted in Table 1.6.

TABLE 1.6

Few Recently Used Adsorbents and Their Applications in Wastewater Treatment

S. No.	Adsorbent Used	Impurities Removed	Removal Capacity	Remark	References
1	Activated carbon	Pb, Zn, Cu, Cd	21.8, 21.2, 19.5, 15.7 mg/g, respectively	Optimum pH = 6, 7, 7, 8, respectively	Rao et al. (2009)
2	Sawdust	Pb and Cu	94.61% and 99.39%, respectively	Optimum pH = 7 and 6.6, respectively	Ahmad et al. (2009)
3	Oak bark char	As, Cd, Pb	70%, 50% and 100%, respectively	Dosage = 10 g/L pH = 3–4 for As and 4–5 for Cd and Pb	Mohan et al. (2007)
4	Polymeric turmeric powder adsorbent	Methylene blue dye	157.33 mg/g	Optimum pH = 7	Kubra et al. (2021)
5	Green synthesized copper nano adsorbent	Pharmaceutical compounds Ibu, Nab and Dic	74.4%, 86.9% and 91.4%, respectively	pH = 4.5 Temperature = 298 K and 10 mg of Cu NPs	Husein et al. (2019)
6	Adsorption and precipitation	Fluoride	95% with lime precipitation and 90% with $CaCo_3$	pH ranges from 6.5 to 8.5	Ezzeddine et al. (2015)
7	Waste Fe and Cr hydroxide	Cd (II)	70%	pH = 3.80	Namasivayam and Ranganathan (1995)
8	CuO-Wo$_3$ hybrid adsorbent	Industrial dye	38.4% high removal rate than Wo$_3$ nanofibers		Dursun et al. (2020)
9	Multiwall carbon nanotube adsorbent	Anionic and cationic dyes	98.7% and 97.2%, respectively	Contact time = 60 minutes, speed = 240 rpm	Shabaan et al. (2020)

1.4 CONCLUSION

The inception and development of various wastewater treatment techniques have been discussed in this study focusing on how these conventional techniques started and improved over the years to become more efficient and economical. Early stages of use after the evolution of various treatment technologies, their working, factors and parameters that affect the removal process and efficiency along with pros and cons are all included in this study. Membrane technology has an advantage over other conventional technologies because of its ability to remove minor as well as major impurities with low energy consumption; one of the major problems associated with all type of membranes is fouling, which can affect its efficiency and increase energy consumption. Ion exchange has good efficiency in removing inorganic contaminants from wastewater, namely, heavy metals, dyes, fluoride, etc., but it is not as effective for the removal of organic contaminants. Adsorption is effective in removing organic and inorganic contaminants, heavy metals, etc. But the synthesis of new adsorbents on the industrial scale is cumbersome and may not be economical. On the other hand, the biological treatment method of wastewater was mostly reported to be more economical and environment friendly than other physiochemical methods of wastewater treatment; however, disposal issue of excess sewage produced is a major problem for biological methods. This was more of an assessment of how far have we come from where we started in the domain of wastewater treatment; however, there is always room for improvement in almost everything, as we may see more innovations and technological advancements in future.

REFERENCES

Abbasi, T., & Abbasi, S. A. (2012). Formation and impact of granules in fostering clean energy production and wastewater treatment in upflow anaerobic sludge blanket (UASB) reactors. *Renewable and Sustainable Energy Reviews, 16(3).* doi: 10.1016/j.rser.2011.11.017.

Ahmad, A., Rafatullah, M., Sulaiman, O., Ibrahim, M. H., Chii, Y. Y., & Siddique, B. M. (2009). Removal of Cu(II) and Pb(II) ions from aqueous solutions by adsorption on sawdust of Meranti wood. *Desalination, 247*(1–3). doi: 10.1016/j.desal.2009.01.007.

Akarsu, C., & Deniz, F. (2021). Electrocoagulation/electroflotation process for removal of organics and microplastics in laundry wastewater. *CLEAN: Soil Air Water, 49*(1). doi: 10.1002/clen.202000146.

Akarsu, C., Deveci, E. Ü., Gönen, Ç., & Madenli, Ö. (2021). Treatment of slaughterhouse wastewater by electrocoagulation and electroflotation as a combined process: process optimization through response surface methodology. *Environmental Science and Pollution Research, 28*(26). doi: 10.1007/s11356-021-12855-4.

Akbal, F., & Camcidotless, S. (2011). Copper, chromium and nickel removal from metal plating wastewater by electrocoagulation. *Desalination, 269*(1–3). doi: 10.1016/j.desal.2010.11.001.

Alam, R., Khan, S. U., Basheer, F., & Farooqi, I. H. (2021a). Nutrients and organics removal from slaughterhouse wastewater using phytoremediation: A comparative study on different aquatic plant species. *IOP Conference Series: Materials Science and Engineering, 1058*(1), 012068. IOP Publishing, doi: 10.1088/1757-899X/1058/1/012068.

Alam, R., Sheob, M., Saeed, B., Khan, S. U., Shirinkar, M., Frontistis, Z., Basheer, F., & Farooqi, I. H. (2021b). Use of electrocoagulation for treatment of pharmaceutical compounds in water/wastewater: A review exploring opportunities and challenges. *Water, 13*(15), 2105. doi: 10.3390/w13152105.

Ali, A., Wu, Z., Li, M., & Su, J. (2021). Carbon to nitrogen ratios influence the removal performance of calcium, fluoride, and nitrate by Acinetobacter H12 in a quartz sand-filled biofilm reactor. *Bioresource Technology, 333.* doi: 10.1016/j.biortech.2021.125154.

Alleman, J. E., & Prakasam, T. B. S. (1983). Reflections on seven decades of activated sludge history. *Journal of the Water Pollution Control Federation, 55*(5), 436–443.

Alnakeeb, A., & Rasheed, R. M. (2021). Chemical precipitation method for sulphate removal from treated wastewater of Al-Doura refinery. *Engineering and Technology Journal*, *39*(3A). doi: 10.30684/etj.v39i3a.503.

Alonso, J. J. S., El Kori, N., Melián-Martel, N., & del Río-Gamero, B. (2018). Removal of ciprofloxacin from seawater by reverse osmosis. *Journal of Environmental Management*, *217*. doi: 10.1016/j.jenvman.2018.03.108.

Al-Rashdi, B. A. M., Johnson, D. J., & Hilal, N. (2013). Removal of heavy metal ions by nanofiltration. *Desalination, 315*. doi: 10.1016/j.desal.2012.05.022.

Alsentzer, H. A. (1963). Ion exchange in water treatment. *Journal (American Water Works Association)*, *55*(6), 749.

Al-Sheikh, F., Moralejo, C., Pritzker, M., Anderson, W. A., & Elkamel, A. (2021). Batch adsorption study of ammonia removal from synthetic/real wastewater using ion exchange resins and zeolites. *Separation Science and Technology (Philadelphia)*, *56*(3), 462–473. doi: 10.1080/01496395.2020.1718706.

Anglada, A., Ibañez, R., Urtiaga, A., & Ortiz, I. (2010). Electrochemical oxidation of saline industrial wastewaters using boron-doped diamond anodes. *Catalysis Today*, *151*(1–2). doi: 10.1016/j.cattod.2010.01.033.

Bal, A. S., & Dhagat, N. N. (2001). Upflow anaerobic sludge blanket reactor: A review. *Indian Journal of Environmental Health*, *43*(2), 1–83.

Barwal, A., & Chaudhary, R. (2014). To study the performance of biocarriers in moving bed biofilm reactor (MBBR) technology and kinetics of biofilm for retrofitting the existing aerobic treatment systems: A review. *Reviews in Environmental Science and Biotechnology*, *13*(3). doi: 10.1007/s11157-014-9333-7.

Bazrafshan, E., Mohammadi, L., Ansari-Moghaddam, A., & Mahvi, A. H. (2015). Heavy metals removal from aqueous environments by electrocoagulation process: A systematic review. *Journal of Environmental Health Science and Engineering*, *13*(1). BioMed Central Ltd. doi: 10.1186/s40201-015-0233-8.

Behling, E., Diaz, A., Colina, G., Herrera, M., Gutierrez, E., Chacin, E., Fernandez, N., & Forster, C. F. (1997). Domestic wastewater treatment using a UASB reactor. *Bioresource Technology*, *61*(3). doi: 10.1016/S0960-8524(97)00148-X.

Belkacem, M., Khodir, M., & Abdelkrim, S. (2008). Treatment characteristics of textile wastewater and removal of heavy metals using the electroflotation technique. *Desalination*, *228*(1–3). doi: 10.1016/j.desal.2007.10.013.

Benalia, M. C., Youcef, L., Bouaziz, M. G., Achour, S., & Menasra, H. (2021). Removal of heavy metals from industrial wastewater by chemical precipitation: Mechanisms and sludge characterization. *Arabian Journal for Science and Engineering*. doi: 10.1007/s13369-021-05525-7.

Berkessa, Y. W., Lang, Q., Yan, B., Kuang, S., Mao, D., Shu, L., & Zhang, Y. (2019). Anion exchange membrane organic fouling and mitigation in salt valorization process from high salinity textile wastewater by bipolar membrane electrodialysis. *Desalination*, *465*. doi: 10.1016/j.desal.2019.04.027.

Bezerra de Araujo, C. M., Filipe Oliveira do Nascimento, G., Rodrigues Bezerra da Costa, G., Santos da Silva, K., SalgueiroBaptisttella, A. M., Gomes Ghislandi, M., & Alves da Motta Sobrinho, M. (2019). Adsorptive removal of dye from real textile wastewater using graphene oxide produced via modifications of hummers method. *Chemical Engineering Communications*, *206*(11). doi: 10.1080/00986445.2018.1534232.

Boyd, G. (1951). Ion exchange. www.annualreviews.org.

Burakov, A. E., Galunin, E. V., Burakova, I. V., Kucherova, A. E., Agarwal, S., Tkachev, A. G., & Gupta, V. K. (2018). Adsorption of heavy metals on conventional and nanostructured materials for wastewater treatment purposes: A review. *Ecotoxicology and Environmental Safety*, *148*. doi: 10.1016/j.ecoenv.2017.11.034.

Çeçen, F., & Aktaş, Ö. (2011). Water and wastewater treatment: Historical perspective of activated carbon adsorption and its integration with biological processes. In: *Activated Carbon for Water and Wastewater Treatment*. John Wiley & Sons: Hoboken, NJ. doi: 10.1002/9783527639441.ch1.

Chang, M. F., & Liu, J. C. (2007). Precipitation removal of fluoride from semiconductor wastewater. *Journal of Environmental Engineering, 133*(4), 419–425. doi: 10.1061/(asce)0733-9372(2007)133:4(419).

Chen, B. Y., Chen, S. Y., & Chang, J. S. (2005). Immobilized cell fixed-bed bioreactor for wastewater decolorization. *Process Biochemistry, 40*(11). doi: 10.1016/j.procbio.2005.04.002.

Chen, G. (2004). Electrochemical technologies in wastewater treatment. *Separation and Purification Technology, 38*(1). doi: 10.1016/j.seppur.2003.10.006.

Chen, R., Takemura, Y., Liu, Y., Ji, J., Sakuma, S., Kubota, K., Ma, H., & Li, Y. Y. (2019). Using partial nitrification and anammox to remove nitrogen from low-strength wastewater by co-immobilizing biofilm inside a moving bed bioreactor. *ACS Sustainable Chemistry and Engineering, 7*(1). doi: 10.1021/acssuschemeng.8b05055.

Conceição, V., Freire, F. B., & de Carvalho, K. Q. (2013). Treatment of textile effluent containing indigo blue dye by a UASB reactor coupled with pottery clay adsorption. *ActaScientiarum Technology, 35*(1). doi: 10.4025/actascitechnol.v35i1.13091.

Cortés-Lorenzo, C., Rodríguez-Díaz, M., López-Lopez, C., Sánchez-Peinado, M., Rodelas, B., & González-López, J. (2012). Effect of salinity on enzymatic activities in a submerged fixed bed biofilm reactor for municipal sewage treatment. *Bioresource Technology, 121*. doi: 10.1016/j.biortech.2012.06.083.

Dalis, D., Anagnostidis, K., Lopez, A., Letsiou, I., & Hartmann, L. (1996). Anaerobic digestion of total raw olive-oil wastewater in a two-stage pilot-plant (up-flow and fixed-bed bioreactors). *Bioresource Technology, 57*(3). doi: 10.1016/S0960-8524(96)00051-X.

Darvishmotevalli, M., Zarei, A., Moradnia, M., Noorisepehr, M., & Mohammadi, H. (2019). Optimization of saline wastewater treatment using electrochemical oxidation process: Prediction by RSM method. *MethodsX, 6*. doi: 10.1016/j.mex.2019.03.015.

Das, D., & Nandi, B. K. (2020). Simultaneous removal of fluoride and Fe (II) ions from drinking water by electrocoagulation. *Journal of Environmental Chemical Engineering, 8*(1). doi: 10.1016/j.jece.2019.103643.

Daud, M. K., Rizvi, H., Akram, M. F., Ali, S., Rizwan, M., Nafees, M., & Jin, Z. S. (2018). Review of upflow anaerobic sludge blanket reactor technology: Effect of different parameters and developments for domestic wastewater treatment. In Journal of Chemistry (*Vol. 2018*). doi: 10.1155/2018/1596319.

Deniel, R., Himabindu, V., Rao, A. V. S. P., & Anjaneyulu, Y. (2008). Removal of arsenic from wastewaters using electrocoagulation. *Journal of Environmental Science and Engineering, 50*(4), 283–288.

Dursun, S., Koyuncu, S. N., Kaya, İ. C., Kaya, G. G., Kalem, V., & Akyildiz, H. (2020). Production of CuO–WO3 hybrids and their dye removal capacity/performance from wastewater by adsorption/photocatalysis. *Journal of Water Process Engineering, 36*. doi: 10.1016/j.jwpe.2020.101390.

Ezzeddine, A., Bedoui, A., Hannachi, A., & Bensalah, N. (2015). Removal of fluoride from aluminum fluoride manufacturing wastewater by precipitation and adsorption processes. *Desalination and Water Treatment, 54*(8). doi: 10.1080/19443994.2014.899515.

Fane, A. G., Wang, R., & Jia, Y. (2011). Membrane technology: Past, present and future. In: L. K. Wang, J. P. Chen, Y.-T. Hung, & N. K. Shammas (Eds.), *Membrane and Desalination Technologies*. Springer Science & Business Media: Berlin, Germany. doi: 10.1007/978-1-59745-278-6_1.

Gain, E., Laborie, S., Viers, P., Rakib, M., Durand, G., & Hartmann, D. (2002). Ammonium nitrate wastewater treatment by coupled membrane electrolysis and electrodialysis. *Journal of Applied Electrochemistry, 32*(9). doi: 10.1023/A:1020908702406.

Gao, B. Y., Yue, Q. Y., Wang, Y., & Zhou, W. Z. (2007). Color removal from dye-containing wastewater by magnesium chloride. *Journal of Environmental Management, 82*(2), 167–172. doi: 10.1016/J.JENVMAN.2005.12.019.

García-Ruiz, M. J., Castellano-Hinojosa, A., González-López, J., & Osorio, F. (2018). Effects of salinity on the nitrogen removal efficiency and bacterial community structure in fixed-bed biofilm CANON bioreactors. *Chemical Engineering Journal, 347*. doi: 10.1016/j.cej.2018.04.067.

Gashi, S. T., Daci, N. M., Ahmeti, X. M., Selimi, T. J., & Hoxha, E. M. (1988). Removal of heavy metals from industrial wastewaters. *Studies in Environmental Science, 34*(C), 91–97. doi: 10.1016/S0166-1116(08)71281-0.

Ge, J., Qu, J., Lei, P., & Liu, H. (2004). New bipolar electrocoagulation-electroflotation process for the treatment of laundry wastewater. *Separation and Purification Technology, 36*(1), 33–39. doi: 10.1016/S1383-5866(03)00150-3.

Genc, A., & Goc, S. (2018). Electroflotation of oily wastewater using stainless steel sponge electrodes. *Water Science and Technology, 78*(7). doi: 10.2166/wst.2018.422.

Gholami, M., Mirzaei, R., Kalantary, R. R., Sabzali, A., & Gatei, F. (2012). Performance evaluation of reverse osmosis technology for selected antibiotics removal from synthetic pharmaceutical wastewater. *Iranian Journal of Environmental Health Science and Engineering, 9*(19). doi: 10.1186/1735-2746-9-19.

Gould, R. H. (1942). Operating experiences in New York City. *Sewage Works Journal, 14*(1), 70–80.

Gregory, J., & Dhond, R. V. (1972). Wastewater treatment by ion exchange. *Water Research, 6*(6), 681–694. doi: 10.1016/0043-1354(72)90183-2.

Gu, Q., Ng, T. C. A., Zain, I., Liu, X., Zhang, L., Zhang, Z., Lyu, Z., He, Z., Ng, H. Y., & Wang, J. (2020). Chemical-grafting of graphene oxide quantum dots (GOQDs) onto ceramic microfiltration membranes for enhanced water permeability and anti-organic fouling potential. *Applied Surface Science, 502*. doi: 10.1016/j.apsusc.2019.144128.

Haddad, M., Abid, S., Hamdi, M., & Bouallagui, H. (2018). Reduction of adsorbed dyes content in the discharged sludge coming from an industrial textile wastewater treatment plant using aerobic activated sludge process. *Journal of Environmental Management, 223*. doi: 10.1016/j.jenvman.2018.07.009.

Hakami, M. W., Alkhudhiri, A., Al-Batty, S., Zacharof, M. P., Maddy, J., & Hilal, N. (2020). Ceramic microfiltration membranes in wastewater treatment: Filtration behavior, fouling and prevention. *Membranes, 10*(9). doi: 10.3390/membranes10090248.

Harper, T. R., & Kingham, N. W. (1992). Removal of arsenic from wastewater using chemical precipitation methods. *Water Environment Research, 64*(3). doi: 10.2175/wer.64.3.2.

Hassan, M. M., & Carr, C. M. (2018). A critical review on recent advancements of the removal of reactive dyes from dyehouse effluent by ion-exchange adsorbents. *Chemosphere, 209*, 201–219. Elsevier Ltd. doi: 10.1016/j.chemosphere.2018.06.043.

Hegazi, H. A. (2013). Removal of heavy metals from wastewater using agricultural and industrial wastes as adsorbents. *HBRC Journal, 9*(3). doi: 10.1016/j.hbrcj.2013.08.004.

Her, N., Amy, G., Plottu-Pecheux, A., & Yoon, Y. (2007). Identification of nanofiltration membrane foulants. *Water Research, 41*(17). doi: 10.1016/j.watres.2007.05.015.

Huang, H., Liu, J., Zhang, P., Zhang, D., & Gao, F. (2017). Investigation on the simultaneous removal of fluoride, ammonia nitrogen and phosphate from semiconductor wastewater using chemical precipitation. *Chemical Engineering Journal, 307*. doi: 10.1016/j.cej.2016.08.134.

Husein, D. Z., Hassanien, R., & Al-Hakkani, M. F. (2019). Green-synthesized copper nano-adsorbent for the removal of pharmaceutical pollutants from real wastewater samples. *Heliyon, 5*(8). doi: 10.1016/j.heliyon.2019.e02339.

Hussain, S. T., & Khaleefa Ali, S. A. (2020). Removal of heavy metal by ion exchange using bentonite clay. *Journal of Ecological Engineering, 22*(1). doi: 10.12911/22998993/128865.

Ibrahim, Y., Abdulkarem, E., Naddeo, V., Banat, F., & Hasan, S. W. (2019). Synthesis of super hydrophilic cellulose-alpha zirconium phosphate ion exchange membrane via surface coating for the removal of heavy metals from wastewater. *Science of the Total Environment, 690*. doi: 10.1016/j.scitotenv.2019.07.009.

Jia, Y., Zhou, M., Chen, Y., Hu, Y., & Luo, J. (2020). Insight into short-cut of simultaneous nitrification and denitrification process in moving bed biofilm reactor: Effects of carbon to nitrogen ratio. *Chemical Engineering Journal, 400*. doi: 10.1016/j.cej.2020.125905.

Jiang, M., Yang, W., Zhang, Z., Yang, Z., & Wang, Y. (2015). Adsorption of three pharmaceuticals on two magnetic ion-exchange resins. *Journal of Environmental Sciences (China), 31*. doi: 10.1016/j.jes.2014.09.035.

Kasmuri, N., & Tarmizi, N. A. A. (2018). The treatment of landfill leachate by electrocoagulation to reduce heavy metals and ammonia-nitrogen. *International Journal of Engineering and Technology (UAE), 7*(3). doi: 10.14419/ijet.v7i3.11.15940.

Keyser, M., Witthuhn, R. C., Ronquest, L. C., & Britz, T. J. (2003). Treatment of winery effluent with upflow anaerobic sludge blanket (UASB): Granular sludges enriched with Enterobactersakazakii. *Biotechnology Letters, 25*(22). doi: 10.1023/B:BILE.0000003978.72266.96.

Khajouei, M., Najafi, M., & Jafari, S. A. (2019). Development of ultrafiltration membrane via in-situ grafting of nano-GO/PSF with anti-biofouling properties. *Chemical Engineering Research and Design, 142*. doi: 10.1016/j.cherd.2018.11.033.

Khan, S. U., Asif, M., Alam, F., Khan, N. A., & Farooqi, I. H. (2020). Optimizing fluoride removal and energy consumption in a batch reactor using electrocoagulation: a smart treatment technology. In: S. Ahmed, S. M. Abbas, & H. Zia (Eds.), *Smart Cities—Opportunities and Challenges* (pp. 767–778). Springer: Singapore. doi: 10.1007/978-981-15-2545-2_62.

Khan, S. U., Islam, D. T., Farooqi, I. H., Ayub, S., & Basheer, F. (2019). Hexavalent chromium removal in an electrocoagulation column reactor: Process optimization using CCD, adsorption kinetics and pH modulated sludge formation. *Process Safety and Environmental Protection, 122*, 118–130, doi: 10.1016/j.psep.2018.11.024.

Khan, S. U., Noor, A., & Farooqi, I. H. (2015). GIS application for groundwater management and quality mapping in rural areas of District Agra, India. *International Journal Of Water Resources & Arid Environments, 4*(1), 89–96.

Khan, T. A., Singh, V. V., & Kumar, D. (2004). Removal of some basic dyes from artificial textile wastewater by adsorption on AkashKinari coal. *Journal of Scientific and Industrial Research, 63*(4), 355–364.

Kim, S., Ou, R., Hu, Y., Li, X., Zhang, H., Simon, G. P., & Wang, H. (2018). Non-swelling graphene oxide-polymer nanocomposite membrane for reverse osmosis desalination. *Journal of Membrane Science, 562*. doi: 10.1016/j.memsci.2018.05.029.

Kobya, M., Demirbas, E., Bayramoglu, M., & Sensoy, M. T. (2011a). Optimization of electrocoagulation process for the treatment of metal cutting wastewaters with response surface methodology. *Water, Air, and Soil Pollution, 215*(1–4), 399–410. doi: 10.1007/s11270-010-0486-x.

Kobya, M., Ulu, F., Gebologlu, U., Demirbas, E., & Oncel, M. S. (2011b). Treatment of potable water containing low concentration of arsenic with electrocoagulation: Different connection modes and Fe-Al electrodes. *Separation and Purification Technology, 77*(3). doi: 10.1016/j.seppur.2010.12.018.

Korngold, E., Kock, K., & Strathmann, H. (1977). Electrodialysis in advanced waste water treatment. *Desalination, 24*(1–3). doi: 10.1016/S0011-9164(00)88079-0.

Kubra, K. T., Salman, M. S., & Hasan, M. N. (2021). Enhanced toxic dye removal from wastewater using biodegradable polymeric natural adsorbent. *Journal of Molecular Liquids, 328*. doi: 10.1016/j.molliq.2021.115468.

Kuiper, S., van Rijn, C. J. M., Nijdam, W., & Elwenspoek, M. C. (1998). Development and applications of very high flux microfiltration membranes. *Journal of Membrane Science*, *150*(1). doi: 10.1016/S0376-7388(98)00197-5.

Kumar, A. K. R., Saikia, K., Neeraj, G., Cabana, H., & Kumar, V. V. (2020). Remediation of bio-refinery wastewater containing organic and inorganic toxic pollutants by adsorption onto chitosan-based magnetic nanosorbent. *Water Quality Research Journal*, *55*(1). doi: 10.2166/WQRJ.2019.003.

Kumar, M., Kuroda, K., Patel, A. K., Patel, N., Bhattacharya, P., Joshi, M., & Joshi, C. G. (2021). Decay of SARS-CoV-2 RNA along the wastewater treatment outfitted with Upflow Anaerobic Sludge Blanket (UASB) system evaluated through two sample concentration techniques. *Science of the Total Environment*, *754*. doi: 10.1016/j.scitotenv.2020.142329.

Kumar, R., Bhakta, P., Chakraborty, S., & Pal, P. (2011). Separating cyanide from coke wastewater by cross flow nanofiltration. *Separation Science and Technology*, *46*(13). doi: 10.1080/01496395.2011.594479.

Kumar, R., Singh, R. D., & Sharma, K. D. (2005). Water resources of India. *Current Science*, *89*, 794–811.

Kumari, U., Siddiqi, H., Bal, M., & Meikap, B. C. (2020). Calcium and zirconium modified acid activated alumina for adsorptive removal of fluoride: Performance evaluation, kinetics, isotherm, characterization and industrial wastewater treatment. *Advanced Powder Technology*, *31*(5). doi: 10.1016/j.apt.2020.02.035.

Kunst, B., & Sourirajan, S. (1974). An approach to the development of cellulose acetate ultrafiltration membranes. *Journal of Applied Polymer Science*, *18*(11). doi: 10.1002/app.1974.070181121.

Larsen, V. J., & Schierup, H. (1981). The use of straw for removal of heavy metals from waste water. *Journal of Environmental Quality*, *10*(2). doi: 10.2134/jeq1981.00472425001000020013x.

Lee, S., Park, M., Yeon, S., & Park, D. (2016). Optimization of chemical precipitation for phosphate removal from domestic wastewater. *Journal of the Korean Society of Water and Wastewater*, *30*(6), 663–671. doi: 10.11001/jksww.2016.30.6.663.

Leiknes, T. O., & Ødegaard, H. (2007). The development of a biofilm membrane bioreactor. *Desalination*, *202*(1–3). doi: 10.1016/j.desal.2005.12.049.

Li, X. L., Zhu, L. P., Zhu, B. K., & Xu, Y. Y. (2011). High-flux and anti-fouling cellulose nanofiltration membranes prepared via phase inversion with ionic liquid as solvent. *Separation and Purification Technology*, *83*(1). doi: 10.1016/j.seppur.2011.09.012.

Liang, F. B., Song, Y. L., Huang, C. P., Li, Y. X., & Chen, B. H. (2013). Synthesis of novel lignin-based ion-exchange resin and its utilization in heavy metals removal. *Industrial and Engineering Chemistry Research*, *52*(3), 1267–1274. doi: 10.1021/ie301863e.

Liu, S., Daigger, G. T., Liu, B., Zhao, W., & Liu, J. (2020). Enhanced performance of simultaneous carbon, nitrogen and phosphorus removal from municipal wastewater in an anaerobic-aerobic-anoxic sequencing batch reactor (AOA-SBR) system by alternating the cycle times. *Bioresource Technology*, *301*. doi: 10.1016/j.biortech.2020.122750.

Liu, X., Yuan, W., Di, M., Li, Z., & Wang, J. (2019). Transfer and fate of microplastics during the conventional activated sludge process in one wastewater treatment plant of China. *Chemical Engineering Journal*, *362*. doi: 10.1016/j.cej.2019.01.033.

Liu, Y. (2003). Chemically reduced excess sludge production in the activated sludge process. *Chemosphere*, *50*(1). doi: 10.1016/S0045-6535(02)00551-9.

Luu, T. (2020). Tannery wastewater treatment after activated sludge pre-treatment using electro-oxidation on inactive anodes. *Clean Technologies and Environmental Policy*, *22*(8). doi: 10.1007/s10098-020-01907-x.

Manikam, M. K., Halim, A. A., Hanafiah, M. M., & Krishnamoorthy, R. R. (2019). Removal of ammonia nitrogen, nitrate, phosphorus and cod from sewage wastewater using palm oil boiler ash composite adsorbent. *Desalination and Water Treatment*, *149*. doi: 10.5004/dwt.2019.23842.

Matlock, M. M., Howerton, B. S., & Atwood, D. A. (2002). Chemical precipitation of heavy metals from acid mine drainage. *Water Research*, *36*(19), 4757–4764. doi: 10.1016/S0043-1354(02)00149-5.

Maurya, S., & Daverey, A. (2018). Evaluation of plant-based natural coagulants for municipal wastewater treatment. *3 Biotech*, *8*(1). doi: 10.1007/s13205-018-1103-8.

McCarty, P. L. (2001). The development of anaerobic treatment and its future. *Water Science and Technology*, *44*(8), 149–156.

McKinney, R. E. (1965). Research and current development in the activated sludge process. *Journal of the Water Pollution Control Federation*, *37*(12), 1696–1704.

Mella, B., Glanert, A. C. C., & Gutterres, M. (2016). Removal of chromium from tanning wastewater by chemical precipitation and electrocoagulation. *Journal of the Society of Leather Technologists and Chemists*, *100*(2), 55–61.

Modrogan, C., MironRaluca, A., OrbuletOanamari, D., & Costache, C. (2015). Removal of heavy metals ions from wastewater using ion exchange. *International Multidisciplinary Scientific GeoConference Surveying Geology and Mining Ecology Management, SGEM*, *1*(5), 727–734. doi: 10.5593/sgem2015/b51/s20.096.

Mohan, D., Pittman, C. U., Bricka, M., Smith, F., Yancey, B., Mohammad, J., Steele, P. H., Alexandre-Franco, M. F., Gómez-Serrano, V., & Gong, H. (2007). Sorption of arsenic, cadmium, and lead by chars produced from fast pyrolysis of wood and bark during bio-oil production. *Journal of Colloid and Interface Science*, *310*(1). doi: 10.1016/j.jcis.2007.01.020.

Mousazadeh, M., Niaragh, E. K., Usman, M., Khan, S. U., Sandoval, M. A., Al-Qodah, Z., Khalid, Z. B., Gilhotra, V. & Emamjomeh, M. M. (2021). A critical review of state-of-the-art electrocoagulation technique applied to COD-rich industrial wastewaters. *Environmental Science and Pollution Research*, *28*(32), 43143–43172. doi: 10.1007/s11356-021-14631-w.

Naje, A. S., & Abbas, S. A. (2013). Electrocoagulation technology in wastewater treatment: A review of methods and applications. *Civil and Environmental Research*, *3*(11), 29–43.

Namasivayam, C., & Ranganathan, K. (1995). Removal of Cd(II) from wastewater by adsorption on "waste" Fe(III) Cr(III) hydroxide. *Water Research*, *29*(7). doi: 10.1016/0043-1354(94)00320-7.

Nassef, E. (2012). Removal of phosphates from industrial waste water by chemical precipitation. *An International Journal (ESTIJ)*, *2*(3), 409–413.

NavamaniKartic, D., Aditya Narayana, B. C. H., & Arivazhagan, M. (2018). Removal of high concentration of sulfate from pigment industry effluent by chemical precipitation using barium chloride: RSM and ANN modeling approach. *Journal of Environmental Management*, *206*. doi: 10.1016/j.jenvman.2017.10.017.

Nicolaisen, B. (2003). Developments in membrane technology for water treatment. *Desalination*, *153*(1–3). doi: 10.1016/S0011-9164(02)01127-X.

Pan, Z. F., & An, L. (2019). Removal of heavy metal from wastewater using ion exchange membranes. In: Inamuddin, M. I. Ahamed, & A. M. Asiri (Eds), *Applications of Ion Exchange Materials in the Environment*. Springer: Berlin, Germany. doi: 10.1007/978-3-030-10430-6_2.

Patel, N. B., Soni, B. D., & Ruparelia, J. P. (2000). Studies on removal of dyes from wastewater using electro-coagulation process. *Nirma University Journal of Engineering and Technology*, *1*(2), 24–30.

Paździor, K., Bilińska, L., & Ledakowicz, S. (2019). A review of the existing and emerging technologies in the combination of AOPs and biological processes in industrial textile wastewater treatment. *Chemical Engineering Journal, 376*. doi: 10.1016/j. cej.2018.12.057.

Peña, M. R., Rodriguéz, J., Mara, D. D., & Sepulveda, M. (2000). UASBs or anaerobic ponds in warm climates? A preliminary answer from Colombia. *Water Science and Technology, 42*(10–11). doi: 10.2166/wst.2000.0609.

Peters, T. (2010). Membrane technology for water treatment. *Chemical Engineering and Technology, 33*(8). doi: 10.1002/ceat.201000139.

Pietrelli, L. (2005). Fluoride wastewater treatment by adsorption onto metallurgical grade alumina. *Annali Di Chimica, 95*(5). doi: 10.1002/adic.200590035.

Pociecha, M., & Lestan, D. (2010). Using electrocoagulation for metal and chelant separation from washing solution after EDTA leaching of Pb, Zn and Cd contaminated soil. *Journal of Hazardous Materials, 174*(1–3). doi: 10.1016/j.jhazmat.2009.09.103.

Poh, P. E., & Chong, M. F. (2009). Development of anaerobic digestion methods for palm oil mill effluent (POME) treatment. *Bioresource Technology, 100*(1). doi: 10.1016/j. biortech.2008.06.022.

Purkait, M. K., DasGupta, S., & De, S. (2004). Removal of dye from wastewater using micellar-enhanced ultrafiltration and recovery of surfactant. *Separation and Purification Technology, 37*(1). doi: 10.1016/j.seppur.2003.08.005.

Qdais, H. A., & Moussa, H. (2004). Removal of heavy metals from wastewater by membrane processes: A comparative study. *Desalination, 164*(2). doi: 10.1016/S0011-9164(04) 00169-9.

Qiqi, Y., Qiang, H., & Ibrahim, H. T. (2012). Review on moving bed biofilm processes. *Pakistan Journal of Nutrition, 11*(9). doi: 10.3923/pjn.2012.804.811.

Raghav, S., Nair, M., & Kumar, D. (2019). Tetragonal prism shaped Ni-Al bimetallic adsorbent for study of adsorptive removal of fluoride and role of ion-exchange. *Applied Surface Science, 498*. doi: 10.1016/j.apsusc.2019.143785.

Ragheb, S. M. (2013). Phosphate removal from aqueous solution using slag and fly ash. *HBRC Journal, 9*(3). doi: 10.1016/j.hbrcj.2013.08.005.

Rajeshwari, K. V., Balakrishnan, M., Kansal, A., Lata, K., & Kishore, V. V. N. (2000). State-of-the-art of anaerobic digestion technology for industrial wastewater treatment. *Renewable & Sustainable Energy Reviews, 4*(2). doi: 10.1016/S1364-0321(99)00014-3.

Raju, G. B., & Khangaonkar, P. R. (1984). Electroflotation: A critical review. *Transactions of the Indian Institute of Metals, 37*(1), 59–66.

Randall, C. W. (1998). Activated sludge: Latest developments and a look into the future (abridged). *Water and Environment Journal, 12*(5). doi: 10.1111/j.1747-6593.1998. tb00199.x.

Rao, M. M., Ramana, D. K., Seshaiah, K., Wang, M. C., & Chien, S. W. C. (2009). Removal of some metal ions by activated carbon prepared from Phaseolusaureus hulls. *Journal of Hazardous Materials, 166*(2–3). doi: 10.1016/j.jhazmat.2008.12.002.

Sadr Ghayeni, S. B., Beatson, P. J., Schneider, R. P., & Fane, A. G. (1998). Water reclamation from municipal wastewater using combined microfiltration-reverse osmosis (ME-RO): Preliminary performance data and microbiological aspects of system operation. *Desalination, 116*(1). doi: 10.1016/S0011-9164(98)00058-7.

Sahinkaya, E., Sahin, A., Yurtsever, A., & Kitis, M. (2018). Concentrate minimization and water recovery enhancement using pellet precipitator in a reverse osmosis process treating textile wastewater. *Journal of Environmental Management, 222*. doi: 10.1016/j. jenvman.2018.05.057.

Sawyer, C. N. (1965). Milestones in the development of the activated sludge process. *Water Pollution Control Federation, 37*(2), 151–162.

Shabaan, O. A., Jahin, H. S., & Mohamed, G. G. (2020). Removal of anionic and cationic dyes from wastewater by adsorption using multiwall carbon nanotubes. *Arabian Journal of Chemistry*. doi: 10.1016/j.arabjc.2020.01.010.

Shiming, X., Qiang, L., Dongxu, J., Xi, W., Zhijie, X., Ping, W., Debing, W., & Fujiang, D. (2020). Experimental investigation on dye wastewater treatment with reverse electrodialysis reactor powered by salinity gradient energy. *Desalination*, *495*. doi: 10.1016/j. desal.2020.114541.

Showell, E. B. (1951). Ion exchange for water treatment. *Journal: American Water Works Association*, *43*(7). doi: 10.1002/j.1551-8833.1951.tb18991.x.

Singh, G., Kumar, B., Sen, P. K., & Majumdar, J. (1999). Removal of fluoride from spent pot liner leachate using ion exchange. *Water Environment Research*, *71*(1). doi: 10.2175/106143099x121571.

Sonksen, M. G., Sittig, F. M., & Maziarz, E. F. (1978). Treatment of oily wastes by ultrafiltration/reverse osmosis; a case history. *In 33rd Indiana Waste Conference Proceedings*, United States, No. CONF-7805252.

Sonwani, R. K., Swain, G., Giri, B. S., Singh, R. S., & Rai, B. N. (2020). Biodegradation of Congo red dye in a moving bed biofilm reactor: Performance evaluation and kinetic modeling. *Bioresource Technology*, *302*. doi: 10.1016/j.biortech.2020.122811.

Swanckaert, B., Geltmeyer, J., Rabaey, K., de Buysser, K., Bonin, L., & de Clerck, K. (2022). A review on ion-exchange nanofiber membranes: properties, structure and application in electrochemical (waste)water treatment. *Separation and Purification Technology*, *287*, 120529. doi: 10.1016/J.SEPPUR.2022.120529.

Taheran, M., Naghdi, M., Brar, S. K., Verma, M., & Surampalli, R. Y. (2018). Emerging contaminants: Here today, there tomorrow! *Environmental Nanotechnology, Monitoring and Management*, *10*. doi: 10.1016/j.enmm.2018.05.010.

Tripathi, B. D., & Shukla, S. C. (1991). Biological treatment of wastewater by selected aquatic plants. *Environmental Pollution*, *69*(1). doi: 10.1016/0269-7491(91)90164-R.

Usmani, S., Washeem, M., Rajput, B., Mahtab, M. S., Khan, S. U., & Farooqi, I. H. (2021). The fenton-based approaches focusing industrial saline wastewater treatment. In: S. Feroz, & D. W. Bahnemann (Eds.), *Removal of Pollutants from Saline Water* (pp. 375–398). CRC Press: Boca Raton, FL.

Vaaramaa, K., & Lehto, J. (2003). Removal of metals and anions from drinking water by ion exchange. www.elsevier.com/locate/desal.

Valero, D., Ortiz, J. M., Expósito, E., Montiel, V., & Aldaz, A. (2010). Electrochemical wastewater treatment directly powered by photovoltaic panels: Electrooxidation of a dye-containing wastewater. *Environmental Science and Technology*, *44*(13). doi: 10.1021/es100555z.

van der Roest, H. F., de Bruin, L. M. M., Gademan, G., & Coelho, F. (2011). Towards sustainable waste water treatment with Dutch Nereda® technology. *Water Practice and Technology*, *6*(3). doi: 10.2166/wpt.2011.059.

Vasudevan, S., Lakshmi, J., & Sozhan, G. (2011). Effects of alternating and direct current in electrocoagulation process on the removal of cadmium from water. *Journal of Hazardous Materials*, *192*(1). doi: 10.1016/j.jhazmat.2011.04.081.

Verma, S. K., Khandegar, V., & Saroha, Anil. K. (2013). Removal of chromium from electroplating industry effluent using electrocoagulation. *Journal of Hazardous, Toxic, and Radioactive Waste*, *17*(2). doi: 10.1061/(asce)hz.2153-5515.0000170.

Vlyssides, A. G., Karlis, P. K., & Zorpas, A. A. (1999). Electrochemical oxidation of NonCyanide strippers wastes. *Environment International*, *25*(5). doi: 10.1016/S0160-4120(99) 00028-8.

Wang, G., Wang, D., Xu, Y., Li, Z., & Huang, L. (2020). Study on optimization and performance of biological enhanced activated sludge process for pharmaceutical wastewater treatment. *Science of the Total Environment*, *739*. doi: 10.1016/j.scitotenv.2020.140166.

Wang, L., Zhang, Y., Sun, N., Sun, W., Hu, Y., & Tang, H. (2019). Precipitation methods using calcium-containing ores for fluoride removal in wastewater. *Minerals*, *9*(9). doi: 10.3390/min9090511.

Waqas, S., Bilad, M. R., Man, Z., Wibisono, Y., Jaafar, J., IndraMahlia, T. M., Khan, A. L., & Aslam, M. (2020). Recent progress in integrated fixed-film activated sludge process for wastewater treatment: A review. *Journal of Environmental Management*, *268*. doi: 10.1016/j.jenvman.2020.110718.

Watson, B. M., & Hornburg, C. D. (1989). Low-energy membrane nanofiltration for removal of color, organics and hardness from drinking water supplies. *Desalination*, *72*(1–2). doi: 10.1016/0011-9164(89)80024-4.

Wenten, I. G. (2003). Recent development in membrane science and its industrial applications. *Journal of Science and Technology*, *24*, 1010–1024.

Wiesmann, U. (1994). Biological nitrogen removal from wastewater. *Advances in Biochemical Engineering/Biotechnology*, *51*. doi: 10.1007/bfb0008736.

Wu, J., Wang, T., Wang, J., Zhang, Y., & Pan, W. P. (2021). A novel modified method for the efficient removal of Pb and Cd from wastewater by biochar: Enhanced the ion exchange and precipitation capacity. *Science of the Total Environment*, *754*, 142150. doi: 10.1016/J.SCITOTENV.2020.142150.

Wu, R. (2019). Removal of heavy metal ions from industrial wastewater based on chemical precipitation method. *Ekoloji*, *28*(107): 2443–2452.

Yacob, S., Shirai, Y., Hassan, M. A., Wakisaka, M., & Subash, S. (2006). Start-up operation of semi-commercial closed anaerobic digester for palm oil mill effluent treatment. *Process Biochemistry*, *41*(4). doi: 10.1016/j.procbio.2005.10.021.

Yasui, H., & Shibata, M. (1994). An innovative approach to reduce excess sludge production in the activated sludge process. *Water Science and Technology*, *30*(9 pt 9). doi: 10.2166/wst.1994.0434.

You, S., Liu, B., Gao, Y., Wang, Y., Tang, C. Y., Huang, Y., & Ren, N. (2016). Monolithic Porous Magnéli-phase Ti4O7 for electro-oxidation treatment of industrial wastewater. *ElectrochimicaActa*, *214*. doi: 10.1016/j.electacta.2016.08.037.

Yuzer, B., & Selcuk, H. (2021). Recovery of biologically treated textile wastewater by ozonation and subsequent bipolar membrane electrodialysis process. *Membranes*, *11*(11). doi: 10.3390/membranes11110900.

Zhang, X., Ye, C., Pi, K., Huang, J., Xia, M., & Gerson, A. R. (2019). Sustainable treatment of desulfurization wastewater by ion exchange and bipolar membrane electrodialysis hybrid technology. *Separation and Purification Technology*, *211*, 330–339. doi: 10.1016/J.SEPPUR.2018.10.003.

Zhang, Y., & Duan, X. (2020). Chemical precipitation of heavy metals from wastewater by using the synthetical magnesium hydroxy carbonate. *Water Science and Technology*, *81*(6). doi: 10.2166/wst.2020.208.

Zhu, L., Guo, Z., Hua, X., Dong, D., Liang, D., & Sun, Y. (2012). Ammonia nitrogen removal from chlor-alkali chemical industry wastewater by magnesium ammonium phosphate precipitation method. *Advanced Materials Research*, *573–574*, 1096–1100. doi: 10.4028/www.scientific.net/AMR.573-574.1096.

2 Introduction to Aerobic Granulation Technology

A Breakthrough in Wastewater Treatment System

Ghazal Srivastava and Absar Ahmad Kazmi

Indian Institute of Technology Roorkee

CONTENTS

2.1 Introduction 38
2.2 Difference between Aerobic and Anaerobic Granulation 40
2.3 Applications of Aerobic Granulation Technology as a Breakthrough in
Wastewater Treatment System 41
    2.3.1 Lab-Scale Applications..41
        2.3.1.1 Biological Treatment of Organics from Wastewater............41
        2.3.1.2 Biological Nutrient (Nitrogen and Phosphorus)
                Removal from Wastewater 42
        2.3.1.3 Degradation of Toxic Substances...42
        2.3.1.4 Biosorption of Dyes and Heavy Metals 42
        2.3.1.5 Mathematical Modeling Practices 44
    2.3.2 Full-Scale Applications 44
2.4 Characteristics of Aerobic Granular Sludge 44
2.5 Technologies Associated with Aerobic Granulation Process 45
2.6 Biochemical Processes Undergo during Aerobic Granulation and
Microbiology Involved..45
    2.6.1 Biochemical Processes 45
    2.6.2 Microbiology...46
2.7 A Case Study of Aerobic Granulation in Pre-Anoxic Selector Attached
SBR in Roorkee, India..47
2.8 Future Scope and Objectives 47
2.9 Conclusion 49
References..49

DOI: 10.1201/9781003202431-2

2.1   INTRODUCTION

The granular sludge is a self-immobilized microbial conglomerate with an elevated density and is utilized in wastewater treating reactors/sewage treatment plants. Granular sludge is a type of consortium formed by microorganisms and has a granule appearance; its shape does not rely on the presence of water, and it can withstand a certain amount of pressure (Gao et al., 2010). According to the International Water Association (IWA) in 2005, granules making up aerobic granular activated sludge are analyzed as aggregates of microbial basis, which do not coagulate under less hydrodynamic shear, and which settle at appreciably rapid rates than activated sludge flocs (de Kreuk et al., 2005; Gao et al., 2010).

Mishima and Nakamura (1991) first noticed aerobic granular sludge (AGS) that was taking place in an aerobic up-flow sludge blanket reactor. Morgenroth and Sherden (1997) used a sequencing batch reactor (SBR) to produce Gasman the granular and flocculated biomass appeared after 40 days of incubation. SBR has been observed as applicable for aerobic granulation and recognized as an appropriate reactor configuration (Liu and Tay, 2004). The research endeavors have concentrated on the cultivation conditions, factors controlling granulation, and the microbial community prevalence of the granular sludge. Conventional activated sludge processes have typical drawbacks such as slow settling flocculent biomass necessitating large clarifiers, lesser effectiveness for biological nutrient removal, lesser reactor biomass concentrations, huge treatment system footprints, and relatively high system energy practices. It is known through lab and the full-scale systems that AGS has diverse advantages over CAS systems, together with improved settling features. These characteristics consecutively allow for greater biomass concentrations and more compressed treatment systems.

The main features of AGS technology are as follows:

1. Aerobic granular sludge has high biomass withholding capacity which permits a reactor to gain a high biomass concentration and treats wastewater at a comparatively high volumetric loading rate.
2. Aerobic granular sludge has excellent settling property. The size and density of the granules determine the settling velocity of an AGS, and it can reach up to 90 m/hour.
3. An elongated sludge retention time of the denser sludge can be attained, which is especially helpful to nitrifiers and ANAMMOX bacteria. Aerobic and anoxic zones are present inside the granules that can simultaneously perform different biological processes in the same system, i.e., simultaneous nitrification-denitrification (SND) could occur (Lei and Liu, 2006). Partial nitritation of ammonium to nitrite and the ANAMMOX reaction were also examined in the sludge granules developed in the SBR (Shi et al., 2009).
4. Aerobic granular sludge can endure shock loading due to its peculiar granule structure and high biomass concentration in the reactor. Moreover, no sludge bulking was observed in this technology (Moy et al., 2002).

The Nereda system is an organized research partnership in the Netherlands, considered to be the first AGS technology applied at full-scale, and more than 40 municipal and industrial plants are now in function or underproduction worldwide. The new plants are in the planning and design phase, including plants with a capability exceeding 1 million population equivalent (PE). Results from operational plants substantiate the advantages of the system concerning the plant's performance, energy and cost efficiency (Pronk et al., 2017). Almost1/4th to 3/4th reduction is observed in treatment system footprints as an outcome of greater reactor biomass concentrations and the non-use of secondary settling tanks; 20%–50% energy usage minimization and associated capital and operational cost savings are also observed with this technology.

The development of aerobic granules is a complex process managed by several parameters. The essential constraints are summarized as follows:

a. *Divalent metal ions* have been confirmed to have an affirmative result on granulation. Metal ions such as calcium and magnesium enhance sludge granulation in SBR (Li et al., 2009; Ren et al., 2008). The mechanisms of improvement of granulation may engage several functions: first, inorganic ions neutralize the negative charge on the faces of bacteria and improve aggregation/ flocculation (Morgan et al.,1990); second, the development of precipitates by inorganic ions provides nuclei to hasten the microbial aggregation, especially in the case of Ca^{2+}; and third, inorganic ions form an ionic bond on the particles' surface and proceed as a cation bridge to enhance granulation. Following granulation, calcium binds to extracellular polymeric substance (EPS) and thereby produces the EPS-Ca^{2+}-EPS cross-linkages to reinforce the granule structure (Ren et al., 2008). The effect of Ca^{2+} augmentation on aerobic granulation in sequential aerobic sludge blanket reactors was tested by Jiang et al. (2003). The augmentation with Ca^{2+} (100 mg/L) and Mg^{2+} (10 mg/L) considerably decreased the granulation period from 32 to 16 days and 32 to 18 days, respectively, using glucose-containing synthetic wastewater (Li et al., 2009).

b. The management of the *aerobic starvation* phase in SBR can increase the speed of the granulation process. An extended aerobic starvation period could enhance the EPS generation in the sludge. This suggests that the elongated starvation period chooses microorganisms that secrete more EPS, which would assist aerobic granulation (Li et al., 2006). Li et al. (2006) suggested that the starvation period provided an anaerobic atmosphere, which supports the anaerobic metabolism of facultative microorganisms, which in turn facilitates the granulation process.

c. The *selection pressure* comprises hydraulic shear force and settling time. Hydraulic shear force was observed to have a noteworthy impact on the creation and structure of AGS floc, and the high shearing force is one of the key factors for the formation of aerobic granules (Dulekgurgen et al., 2008). Studies have demonstrated that AGS is formed with a little settling time, and granules with a larger diameter increased as shorter settling times were pertained (Qin et al., 2004).

d. Li et al. (2008) specified that the aerobic granules developed at different *organic loading rates* had variable morphologies, structures, and bacterial varieties. A higher organic loading rate affects the formation of larger and looser granules within a lesser duration, while a lesser organic loading rate leads to the creation of smaller and compacted granules over a lengthy period. Further results demonstrated that the optimum Chemical Oxygen Demand (COD) loading rate for aerobic granulation in the SBR was 2.52 kg/m^{3}day (Kim et al., 2008).

e. It was monitored that lesser *dissolved oxygen* (DO) levels in anoxic selector compartments and high substrate levels (organics) are accompanied by return-activated sludge boost floc formers and suppressed filaments (Srivastava et al., 2021). This condition further improves the floc/granule size and promotes simultaneous nitrification and denitrification in the large sludge flocs of aeration tanks at a controlled DO concentration of 0–2.5 mg/L (Srivastava et al., 2021; Viaeminck et al., 2008).

f. The *temperature* also affects aerobic granulation. Song et al. (2009) examined that 30°C was the optimal temperature for aerobic granulation. The granules that appeared at 30°C temperature had a very dense structure, settling tendency, and higher bioactivity than other temperatures.

g. According to Gao et al. (2010), reactor configuration influences the sludge granulation process. Similarly, Rollemberg et al. (2019) assessed the effect of two SBR configurations, conventional and constant volume-based SBR for AGS cultivation on the physical and microbiological distinctiveness of the granules. The granules were formed within a month with a larger size of >1 mm and improved settleability of SVI$_{30}$~44.8 mL/g in the conventional SBR. On the other hand, constant volume SBR showed a slower formation of granules for more than one-and-a-half months with a minor diameter of 0.8 mm and an inferior settleability of Sludge Volume Index (SVI$_{30}$) ~70.7 mL/g.

Therefore, the objective of this study is to analyze some basic aspects of aerobic granulation technology and the way it can be utilized to promote nutrient removal performances of wastewater treatment systems (especially SBRs). We need to comprehend the modification and upgradation of existing sewage treatment plants to aerobic granulation technology for the improvement of the systems in terms of nutrient (N and P) removal, organic matter (COD, Biochemical Oxygen Demand (BOD), and Total Suspended Solids (TSS)) removal, better sludge settling characteristics and reduction in sludge production.

2.2 DIFFERENCE BETWEEN AEROBIC AND ANAEROBIC GRANULATION

Some analyses observed that both aerobic and anaerobic bio-granules have special characteristics and have particular advantages and disadvantages (Figure 2.1). Granulation technology for treating wastewater proposes more benefits than conventional wastewater treatment. Since the beginning of 2000, the AGS system has become very attractive for the treatment of wastewater shooting from its feature

AEROBIC GRANULATION	ANAEROBIC GRANULATION
• Bacteria require oxygen to survive and grow	• Bacteria do not rely on oxygen for metabolic process and survival
• Operating-Anaerobic and aeorbic (saturated DO concentration)	• Operation-Anaerobic (0 mg/L DO concentration)
• Upflow velocity-Higher than 43 m/h (1.2 cm/s)	• Upflow velocity-0.6-2.0 m/h
• Biomass concentration-5-15 g/L	• Biomass concentration-5-40 g/L (top) and 50-100 g/L (bottom)
• Complete degrade substrate to the end products. e.g. textile wastewater	• Not completely degrade the influent waste. e.g. textile wastewater
• Formation occurs mostly on SBR reactor	• Formation occurs mostly on UASB reactor
• Reactor start-up takes about one month	• Reactor start-up takes about three months
• Operating temperature- Stable at 8-12 °C	• Opetating temperature- Preferable at mesophillic and thermophillic temperature range
• Effluent suspended solids-80-100 mg/L	• Effluent suspended solids-30-150 mg/L
• Wastewater strength-Low to high strength wastewater	• Wastewater strength-High strength wastewater
• Nutrient removal-High	• Nutrient removal-Low
• Simultaneous nitrification denitrification-High	• Simultaneous nitrification denitrification-Impossible

FIGURE 2.1   Comparison between two types of granulation processes (Affam et al., 2018).

dense and compact microbial formation, excellent settleability, high biomass retention, ability to withstand high organic loading rates, SND, little investment, etc. (Affam et al., 2018). Due to this advantage, aerobic granulation has been used in various types of wastewater treatment including organics removal, phosphorus removal, nitrogen removal, and toxic substances (like heavy metals, carcinogenic pollutants, certain harmful chemicals, and dyes) removal (Affam et al., 2018).

2.3   APPLICATIONS OF AEROBIC GRANULATION TECHNOLOGY AS A BREAKTHROUGH IN WASTEWATER TREATMENT SYSTEM

2.3.1   Lab-Scale Applications

Currently, aerobic granulation technology has been occurring at the laboratory scale for treating different wastewaters, and for nutrient removal, organic toxic compounds degradation, dyestuffs removal, metal removal, etc.

2.3.1.1   Biological Treatment of Organics from Wastewater

Aerobic granules can be developed and used to treat various organic wastewaters including dairy, brewery, fish can, municipal, and landfill leachates. Studies have shown that the COD removal efficiency was stabilized to treat high-strength organic wastewater with aerobic granules by a step-up increase in organic loading from 6 to 15 kg COD/m^3·day (Moy et al., 2002). The granules can be satisfactorily

TABLE 2.1
COD Removal in Different Types of Wastewater

Type of Wastewater	COD Removal (%)	References
Dairy wastewater	90	Schwarzenbeck et al. (2005)
Brewery wastewater	89	Wang et al. (2007a)
Raw sewage	97	Song et al. (2009)
Slaughterhouse wastewater	98	Cassidy and Belia (2005)
Synthetic wastewater	95	Wang et al. (2007b)
Synthetic wastewater	95	You et al. (2008)

formed at influent COD concentrations of 500–3,000 mg/L (Liu et al., 2003a). Arrojo et al. (2004) increased AGS using industrial wastewater from dairy products with a total COD of 1.5–3.0 g/L, soluble COD of 0.3–1.5 mg/L, and total nitrogen of 0.05–0.2 g/L in two SBRs. The COD removal through the aerobic granulation process in literature is depicted in Table 2.1.

2.3.1.2 Biological Nutrient (Nitrogen and Phosphorus) Removal from Wastewater

Aerobic granules have an excellent potential for the removal of nitrogen. In an aerobic granule, the DO gradient subsists and the redox profile of the granules can be segregated into three phases, i.e., the aerobic zone followed by a micro-oxygen zone and an anoxic (or anaerobic) zone with a broad range of microbial surrounding. The atmosphere allows the growth of different trophic bacteria inside the granules with several metabolic functions consisting of nitrifiers, denitrifiers, and anaerobes (ANAMMOX and even methanogens). Simultaneous nitrification and denitrification are found to occur in large granules due to DO gradients. On the periphery of the floc, nitrification exists, and inside the floc, anoxic zones form, which is the shelter to denitrifiers like Zooglea, Paracoccus, Rhodococcus, etc. contributing to denitrification. Table 2.2 shows N and P removal during aerobic granulation in different types of wastewater.

2.3.1.3 Degradation of Toxic Substances

The aerobic granules are applicable for the degradation of toxic phenolic compounds (Gao et al., 2010). Aerobic granules were also developed for the biological degradation of 2, 4-dichlorophenol (2, 4-DCP) with glucose as a co-substrate (Wang et al., 2007b). Table 2.3 shows the removal of toxic compounds from different wastewater by applying aerobic granulation technology.

2.3.1.4 Biosorption of Dyes and Heavy Metals

Aerobic granules have a large surface area and high porosity and are considered to be used as biosorbent for dye-stuff and heavy metal removal. Several dyes and coloring pigments are used for coloring textiles, clothes, leather, food, paper and pulp, printing, carpet, and mineral processing industries or factories. They can be severe pollutants in the environment if discharged without appropriate treatment. Table 2.4 shows the biosorption capacity of dyes and heavy metals using aerobic granulation from synthetic wastewater.

TABLE 2.2
N and P Removal in Different Types of Wastewater

Type of Wastewater	N Removal	P Removal	References
Fish canning industry wastewater	Ammonium-N removal~40%	-	Figueroa et al. (2008)
Synthetic wastewater	Ammonium-N removal~85.4%–99.7% and total nitrogenremoval~41.7%–78.4%	-	Wang et al. (2008)
Synthetic wastewater	-	99.6%	Dulekgurgen et al. (2003)
Slaughterhouse wastewater	97%	98%	Cassidy and Belia (2005)
Synthetic wastewater	-	100%	You et al. (2008)

TABLE 2.3
Toxic Compounds Removal in Different Types of Wastewater Using Aerobic Granulation

Type of Wastewater	Toxic Compound Removal	References
Synthetic wastewater	The reduction rate of 0.53 gphenol/ g VSS day	Tay et al. (2004)
Saline wastewater	Phenol removal~99%	Moussavi et al. (2010)
Synthetic wastewater	2,4-DCP removal ~94%	Wang et al. (2007b)
Synthetic wastewater	Phenol removal ~97%	Shams et al. (2008)
Synthetic wastewater	Methyl tert-butyl ether (MTBE) removal 99.9%	Zhang et al. (2008)

TABLE 2.4
Biosorption Capacity of Aerobic Granulation from Different Types of Wastewater

Type of Wastewater	Biosorption Capacity of Dyestuffs and Heavy Metals	References
Synthetic wastewater	56.8 mg of malachite green per gram of aerobic granules at 30°C	Sun et al. (2008)
Synthetic wastewater	The maximum adsorption capacity for Cd^{2+} and Zn^{2+} were 566 and 270 mg/g, respectively	Liu et al. (2002, 2003b)
Synthetic wastewater	The adsorption power of the flocs was 55.3 mg Co/g at pH 7 and 62.5 mg Zn/g at pH 5 in single systems	Sun et al. (2008)

2.3.1.5 Mathematical Modeling Practices

Researchers have applied and created various mathematical models to suggest the bio-reactions undergone in AGS and granular sludge reactor systems. The model provided superior insights into aerobic granule-based absorbers for the elimination of heavy metals, i.e., biosorption capacity of Cd^{2+}, Cu^{2+}, and Zn^{2+}. A model of the concurrent storage and growth systems in an aerobic granular SBR fed with soybean processing wastewater was developed with the adsorption mechanism, microbial safeguarding, and substrate dispersion (Ni and Yu, 2008). An additional model was suggested to simulate the synthesis of EPS, soluble microbial products, and internal storage products in AGS. Multi scale models were developed by researchers at Universidade Nova de Lisboa in Portugal, Waseda University in Japan, and TU Delft in the Netherlands. Wichern et al. (2008) developed a dynamic mathematical model describing COD and nitrogen removal from dairy wastewater in a granular SBR. The modeling results were based on granules with an average diameter of 2.5 mm and a density of 40 g VSS/L (Gao et al., 2010).

2.3.2 FULL-SCALE APPLICATIONS

There are lesser studies available to demonstrate aerobic sludge granulation (AGS) in full-scale treatment plants. However, studies are ongoing as several full-scale plants have been built in the Netherlands, Portugal, and South Africa (Gao et al., 2010). The Nereda scheme is regarded as the earliest AGS technology established at fullscale, and >40 municipal and industrial plants are at present in process or construction globally (Pronk et al., 2017). Additional plants are in the setting up and design phase, together with plants with capacities above 1 million PE (Pronk et al., 2017). Some studies on SBR at fullscale gave evidence of larger floc formation (aerobic granules) and SND in India due to the consequences of selector phase biology (Magdum et al., 2015; Srivastava et al., 2021).

2.4 CHARACTERISTICS OF AEROBIC GRANULAR SLUDGE

This section describes the essential characteristics of AGS, which consists of physical, chemical, and microbiological properties. Physical properties include settling velocity, specific gravity, water content, strength, floc/ granule morphology, granule size, storage stability, and porosity. Chemical properties include cell-surface hydrophobicity, cell-surface charge, bridging actions, and EPS. Microbiological parameters combine microbial species identification including filamentous species, microbial composition (intracellular polymeric substances like poly-β-hydroxybutyrate (PHB), polyphosphates, glycogen, etc.), and factors affecting microbial diversity (granule size, DO, temperatures, pH, etc.). A polymerase chain reaction-denaturing gradient gel electrophoresis (PCR-DGGE) analysis demonstrated an obvious shift of the microbial populations occurring after the sludge in the SBR was developed from a flocculent to the granular form. The microbial population of aerobic granules was completely different from that of the seed sludge (Li et al., 2008).

2.5 TECHNOLOGIES ASSOCIATED WITH AEROBIC GRANULATION PROCESS

SBR technologies are generally observed to be associated with the aerobic granulation process (Affam et al., 2018). Almost all aerobic granules can form only in SBR, but the reason is not well understood. It has been observed that after attaching the anaerobic zone in the form of selectors, the sludge EPS concentration was increased from 84.4 to 104.0 mg. (gMLSS)$^{-1}$ (Yao et al., 2019). Thus, the addition of an anaerobic step can inhibit the growth of filamentous bacteria, increasing the sludge EPS concentration and promoting the precipitation of activated sludge, i.e., sludge aerobic granulation (Yao et al., 2019). The development of aerobic granule was also noticed in an aerobic up-flow sludge blanket reactor (AUSB) (Affam et al., 2018). It was assumed that the formation of aerobic granules is analogous to the anaerobic granule as filamentous bacteria approach together to develop a structure of granules (Figueroa et al., 2008). There are many processes like aerobic sludge granulation-continuous flow reactor (AGS-CFR), aerobic sludge granulation-membrane bioreactor (AGS-MBR), aerobic granular sludge membrane bioreactor (GMBR), and mainly SBR systems performed for AGS under hydraulic selection pressure.

2.6 BIOCHEMICAL PROCESSES UNDERGO DURING AEROBIC GRANULATION AND MICROBIOLOGY INVOLVED

During aerobic granulation, several biochemical processes occur within the bacterial cells and undergoing the treatment plants. Also, many research works have focused on microbial community dynamics during aerobic granulation in SBR and other bioreactors.

2.6.1 BIOCHEMICAL PROCESSES

It has been observed that organic matter removal, nutrient removal (N and P), xenobiotics, and dyes removal using biosorption mechanisms are associated with aerobic granulation. For nitrogen removal, simultaneous nitrification and denitrification occur in large granules. During the enhanced biological phosphorus removal process and during the anaerobic phase, PHB is formed/ observed inside the filaments and flocs of the sludge/granules, and in the aeration phase, polyphosphate globules are identified in big sludge granules. Aerobic granular sludge is appeared by microbial self-aggregation and has benefits such as excellent settling ability, dense and strong microbial structure, high biomass retention, ability to survive a high organic loading rate, and tolerance to toxicity compared with conventional activated sludge (Li et al., 2014). The evolution of aerobic granulation can be monitored using optical microscopy, an image analysis technique, and scanning electron microscope during the operation of bioreactors fed with glucose and acetate, respectively. The substantial characteristics and microbial activity of sludge are in terms of SVI, settling velocity, size, hydrophobicity, SOUR, strength, and roundness were also determined by several researchers (Liu and Tay, 2004). Figure 2.2, according to Wilen et al. (2018), shows the parameters involve in the input, granule, and output stage of the aerobic granulation process.

INPUT	GRANULES	OUTPUT
Parameters controlled by engineer: (a) Substrate loading rate (b) F/M ratio (c) Feast/famine conditions (d) DO concentration (e) Shear forces (f) Settling time (g) Reactor volume and configuration (h) SRT Parameters limited controlled by the engineer: (a) Temperature (b) Composition of seed (c) Influent microorganisms (d) Influent chemical composition (e) Influent particulate matter	(a) Cell-cell communication (b) Cell surface properties (c) Microbial community structure (d) EPS content (e) Chemical gradients (f) Stratification of microorganisms (g) Substrate storage (h) Size, structure and density (i) Predation of bacteria	Degree of removal of organic carbon, N, and P Suspended sludge concentrations in effluent Mechanical stability of granules in short- and long-term Resilience/resistance towards temporal variations and variations in influent parameters Biosorption of xenobiotics and dyes

FIGURE 2.2 Parameters involve in the input, granule, and output stages of the aerobic granulation process.

2.6.2 Microbiology

Several types of research by Gomez-Basurto et al. (2019), Li et al. (2014), Świątczak and Cydzik-Kwiatkowska (2018), and Yao et al. (2019) have shown and revealed the different microbial species found and are responsible for AGS. According to the analysis of Gomez-Basurto et al. (2019), at the beginning of the granulation stage, the relative abundance of *Agrobacterium* attained 36% and *Dipodascus* achieved 90% during the grown-up granule phase. Families of the *Gammaproteobacteria* were the most abundant bacterial classes with most phylotypes belonging to Acinetobacter. As the granulation process ended, the SVI achieved a minimum with intense and stable granules/ flocs. Li et al. (2014) used DGGE analysis and indicated that *Flavobacterium* sp., uncultured beta proteobacterium, uncultured *Aquabacterium* sp., and uncultured *Leptothrix* sp. were presently dominant in SBR, whereas uncultured bacteroidetes were only originated in A/O and oxidation ditch. Yao et al. (2019) observed that the propagation of *norank_o_Sphingobacteriales*, *Thiothrix*, and *Trichosporon* was the main reason for the detected sludge bulking in the reactor. However, as aerobic granulation started, the number of species got reduced considerably. Świątczak and Cydzik-Kwiatkowska (2018) detected that in old and mature granules, the quantity of ammonium-oxidizing bacteria was extremely low, while the abundance of the nitrite-oxidizing bacteria *Nitrospira* sp. was $0.5\% \pm 0.1\%$. The major genera found in the AGS process were *Dechloromonas*,

Tetrasphaera, *Flavobacterium*, *Ohtaekwangia*, and *Sphingopyxis*. Bacteria related to these genera make EPS, helping in stable granule structure formation and phosphorus accumulation. The results of their study could have been helpful for designers of aerobic granulation in wastewater treatment plants. Also, molecular data provided insights into the ecology of grown-up aerobic granules from a full-scale treatment plant (Świątczak and Cydzik-Kwiatkowska, 2018).

2.7 A CASE STUDY OF AEROBIC GRANULATION IN PRE-ANOXIC SELECTOR ATTACHED SBR IN ROORKEE, INDIA

The 3-MLD SBR plant installed near IIT Roorkee was a cyclic technology-based plant attached to anoxic selectors. The plant was analyzed in the study for sludge granulation (denser and stronger microbial structure) and biochemical processes undergoing in the plant. The sludge granulation mechanism helps remove carbon, nitrogen, and phosphorous as well as other pollutants in a single bioreactor under the same operational conditions. The sludge biomass was remarkably diverse in the plant; having various morphotypes being present (rods, filaments, tetrads/ sarcina-like cells, coccoid-clusters, diplococci, and elongated rod-shaped cells unique for the system) can be observed in Srivastava and Kazmi (2021). Phenotypic depiction through chemical staining and conventional light microscopy revealed the occurrence of the polyphosphate accumulating organisms (PAOs) cycling their intracellular poly-P and PHB inclusions between the anaerobic and aerobic zones (Figure 2.3). The plant's performance is shown in Table 2.5.

2.8 FUTURE SCOPE AND OBJECTIVES

Currently, although the studies on AGS have made noteworthy progress, future study is still required to fundamentally realize the granulation and also for the advancement as a cost-effective technology. Future study necessities can be summarized according to the following features (Gao et al., 2010):

a. Studies at full-scale treatment plants are needed at raw sewage/ industrial effluents/ mixed wastewaters with a convenient start-up period.
b. Also, there is a need for modification of existing SBR plants to aerobic granulation technology for the advancement in the treatment performance specifically nutrient removal, using the cost-effective approach in developing countries.
c. More research is still needed to understand the stability of its long-term operation at a large scale.
d. The study of the mechanisms concerning microbial community structures or dynamics should also be well addressed in detail. Even the microbial community of AGS using molecular biotechnology as a tool is still limited which should be studied in detail.

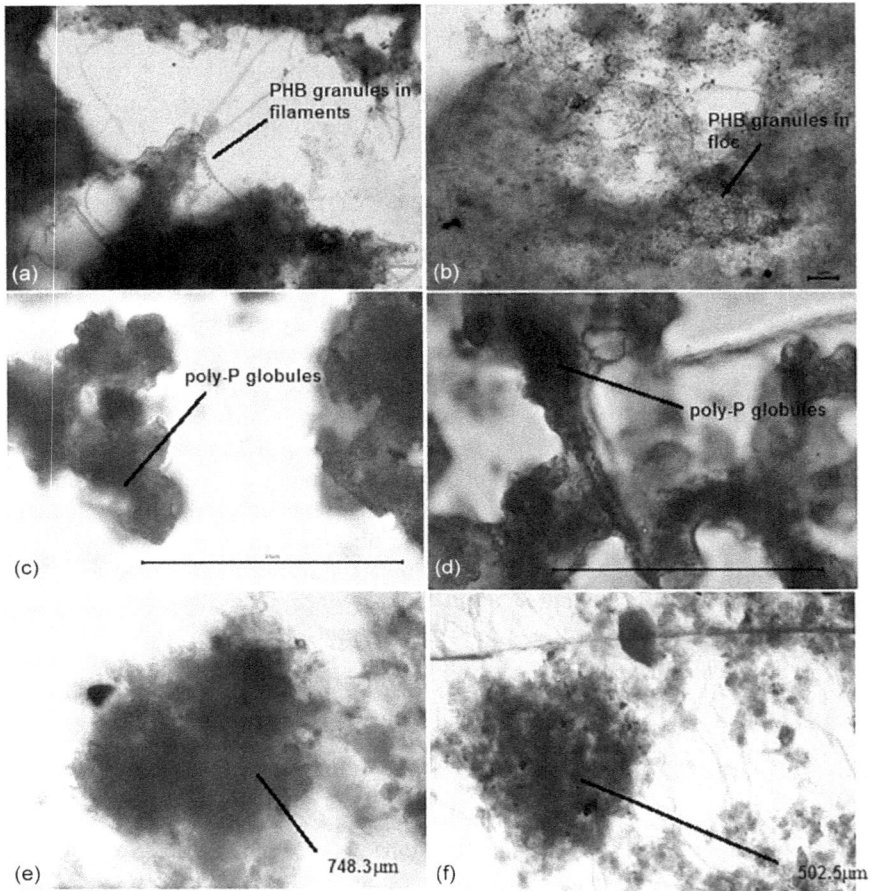

FIGURE 2.3 Bright-field and phase-contrast images of PHB granules (blue-black as PHB (+)), poly-P globules (violet-blue-black as poly-P (+)) at 100× magnification with immersion oil, and matured sludge flocs at 10× magnification under a light microscope (equivalent diameter >500 μm) of the anoxic selector and aeration tank's sludge samples of 3-MLD SBR plant (Srivastava et al., 2022).

e. The microbial ecosystem of the granules should be addressed to improve specifically nitrogen and phosphorus removal.

f. A relationship between sludge granulation, biochemical parameters, and microbial species prevalence should be figured out in full-scale and small-scale STPs to understand the influence of microbial activity on granulation and stability.

TABLE 2.5
3-MLD SBR Plant's Performance

Treatment	Values	Influent Parameters
COD removal	95%	COD~ 400–500 mg/L
BOD removal	96%	BOD~ 150–200 mg/L
TSS removal	95%	TSS~ 200–300 mg/L
TN removal	69%	TN~ 35–50 mg/L
Ammonia-N removal	97%	NH_4-N~ 15–25 mg/L
PO_4-P removal	31%	PO_4-P~ 2–5 mg/L
TP removal	42%	TP~ 5–9 mg/L
SND	78%	-
Average MLSS (mg/L)	7,177	-
SVI (mL/g)	41	-
VSS: TSS	0.5–0.6	0.5–0.6

2.9 CONCLUSION

As per the literature, it is clear and comprehendible that aerobic granulation is a breakthrough technology in wastewater treatment systems, but it needs to be further studied at full-scale and small-scale treatment plants. Moreover, a great focus is needed on intracellular polymeric substances production by the bacteria prevailing in these kinds of treatment systems useful for nitrogen and phosphorus removal. Aerobic granulation seems to be the cheapest technology for effective organic matter, nutrients, and toxic substance removal if applied by modifying existing SBR-based treatment plants in developing countries like India. However, thorough research regarding the relationship between sludge granulation, biochemical parameters, and microbial species prevalence is required at full-scale and small-scale STPs to understand the influence of microbial activity on granulation and stability.

REFERENCES

Affam, AC, Chung, WC, Swee, WC. 2018. Can induced magnetic field enhance bioprocesses: Review. *MATEC Web Conf* 203: 03007.

Arrojo B, Mosquera Corral A, Garrido JM, Méndez R. 2004. Aerobic granulation with industrial wastewater in sequencing batch reactors. *Water Res* 38: 3389–3399.

Cassidy DP, Belia E. 2005. Nitrogen and phosphorus removal from an abattoir wastewater in a SBR with aerobic granular sludge. *Water Res* 39: 4817–4823.

de Kreuk MK, McSwain BS, Bathe S, Tay J, Schwarzenbeck STL, Wilderer PA. 2005. Discussion outcomes. In: Bathe S, de Kreuk MK, McSwain BS, Schwarzenbeck N (Eds.), Aerobic Granular Sludge (Water and Environmental Management Series (WEMS)). Munich: IWA Publishing, pp. 165–169.

Dulekgurgen E, Ovez S, Artan N. 2003. Enhanced biological phosphate removal by granular sludge in a sequencing batch reactor. *Biotechnol Lett* 25: 687–693.

Dulekgurgen E, Yilmaz M, Wilderer PA. 2008. Shape and surface topology of anaerobic/aerobic granules influenced by shearing conditions. *4th IWA Specialized Conference on Sequencing Batch Reactor Technology*, Roma, Italy, pp. 311–320.

Figueroa M, Mosquera Corral A, Campos JL, Méndez R. 2008. Treatment of saline wastewater in SBR aerobic granular reactors. *Water Sci Technol* 58: 479–485.

Gao D, Liu L, Liang H, Wu W. 2010. Aerobic granular sludge: Characterization, mechanism of granulation and application to wastewater treatment. *Crit Rev Biotechnol* 31(2): 137–152.

Gomez-Basurto F, Vital-Jacome M, Gomez-Acata ES, Thalasso F, Luna-Guido M, Dendooven L. 2019. Microbial community dynamics during aerobic granulation in a sequencing batch reactor (SBR). *PeerJ* 7: e7152. doi: 10.7717/peerj.7152.

Jiang HL, Tay JH, Liu Y. 2003. Ca^{2+} augmentation for enhancement of aerobically grown microbial granules in sludge blanket reactor. *Biotechnol Lett* 25: 95–103.

Kim IS, Kim SM, Jang A. 2008. Characterization of aerobic granules by microbial density at different COD loading rates. *Bioresour Technol* 99: 18–25.

Lei Q, Liu Y. 2006. Aerobic granulation for organic carbon and nitrogen removal in alternating aerobic-anaerobic sequencing batch reactor. *Chemosphere* 63: 926–933.

Li AJ, Yang SF, Li XY, Gu JD. 2008. Microbial population dynamics during aerobic sludge granulation at different organic loading rates. *Water Res* 42: 3552–3560.

Li J, Ding L, Cai A, Huang G, Horn H. 2014. Aerobic sludge granulation in a full-scale sequencing batch reactor. *Biomed Res Int* 2014: 1–12.

Li XM, Liu QQ, Yang Q, Guo L, Zeng GM, Hu JM, Zheng W. 2009. Enhanced aerobic sludge granulation in sequencing batch reactor by Mg2+ augmentation. *Bioresour Technol* 100: 64–67.

Li ZH, Kuba T, Kusuda T. 2006. The influence of starvation phase on the properties and the development of aerobic granules. *Enzyme Microb Technol* 38: 670–674.

Liu QS, Tay JH, Liu Y. 2003a. Substrate concentration-independent aerobic granulation in sequential aerobic sludge blanket reactor. *Environ Technol* 24: 1235–1242.

Liu Y, Shu FY, Tay JH. 2002. Aerobic granules novel zinc biosorbent. *Lett Appl Microbiol* 35: 548–551.

Liu Y, Shu FY, Tay JH. 2003b. Biosorption kinetics of Cadmium (II) on aerobic granular sludge. *Process Biochem* 38: 997–1001.

Liu Y, Tay JH. 2004. State of the art of biogranulation technology for wastewater treatment. *Biotechnol Adv* 22: 533–563.

Magdum, SS, Varigala, SK, Minde, GP, Bornare, JB, Kalyanraman, V. 2015. Evaluation of sequential batch reactor (SBR) cycle design to observe the advantages of selector phase biology to achieve maximum nutrient removal. *Int J Sci Res Environ Sci* 3(6): 0234–0238.

Mishima K, Nakamura M. 1991. Self-immobilization of aerobic activated sludge: A pilot study of the process in municipal sewage treatment. *Water Sci Technol* 23: 981–990.

Morgan JM, Forster CF, Evison L. 1990. A comparative study of the nature of biopolymers extracted from anaerobic and activated sludges. *Water Res* 24: 743–750.

Morgenroth E, Sherden T. 1997. Aerobic granular sludge in a sequencing batch reactor. *Water Res* 31: 3191–3194.

Moussavi G, Barikbin B, Mahmoudi M. 2010. The removal of high concentrations of phenol from saline wastewater using aerobic granular SBR. *Chem Eng J.* 158: 498–504.

Moy BYP, Tay JH, Toh SK, Liu Y, Tay STL. 2002. High organic loading influences the physical characteristics of aerobic sludge granules. *Lett Appl Microbiol* 34: 407–412.

Ni BJ, Yu HQ. 2008. Growth and storage processes in aerobic granules grown on soybean wastewater. *Biotechnol Bioeng* 100: 664–672.

Pronk, M, Giesen, A, Thompson, A, Robertson, S, van Loosdrecht, M. 2017. Aerobic granular biomass technology: Advancements in design, applications and further developments. *Water Pract Technol* 12(4): 987–996.

Qin L, Tay JH, Liu Y. 2004. Selection pressure is a driving force of aerobic granulation in sequencing batch reactors. *Process Biochem* 39: 579–584.

Ren TT, Liu L, Sheng GP, Liu XW, Yu HQ. 2008. Calcium spatial distribution in aerobic granules and its effects on granule structure, strength and bioactivity. *Water Res* 42: 3343–3352.

Rollemberg, S, Barros, ARM, de Lima, JPM, Santos, AF, Firmino, PIM, Santos, AB. 2019. Influence of sequencing batch reactor configuration on aerobic granules growth: Engineering and microbiological aspects. *J Cleaner Prod* 238: 117906.

Schwarzenbeck N, Borges JM, Wilderer PA. 2005. Treatment of dairy effluents in an aerobic granular sludge sequencing batch reactor. *Appl Microbiol Biotechnol* 66: 711–718.

Shams QU, Suhail S, Izharul HF, Ahmad A. 2008. Biodegradation of phenols and p-cresol by sequential batch reactor. *International Conference on Environmental Research and Technology, Cleaner Tech Control Treatment & Remediation Technique*, pp. 906–910.

Shi XY, Yu HQ, Sun YJ, Huang X. 2009. Characteristics of aerobic granules rich in autotrophic ammonium-oxidizing bacteria in a sequencing batch reactor. *Chem Eng J* 147: 102–109.

Song ZW, Ren NQ, Zhang K, Tong LY. 2009. Influence of temperature on the characteristics of aerobic granulation in sequencing batch airlift reactors. *J Environ Sci* 21: 273–278.

Srivastava, G, Kazmi, AA. 2021. Perspectives of intracellular polymers in functional evaluation of the microbes for EBPR. *Water Pract Technol*. doi: 10.2166/wpt.2021.112.

Srivastava G, Rajpal A, Kazmi, AA. 2021. A comparative analysis of simultaneous nutrient removal in two full-scale advanced SBR-based sewage treatment plants. *Int J Sci Res (IJSR)* 10(2): 1407–1414.

Srivastava, G, Rajpal, A, Khursheed, A, Nadda, AK, Tyagi, VK, Kazmi, AA. 2022. Influence of variations in wastewater on simultaneous nutrient removal in pre-anoxic selector attached full-scale sequencing batch reactor. *Int J Environ Sci Technol*. 1–18. https://doi.org/10.1007/s13762-022-04052-8

Sun XF, Wang SG, Liu XW, Gong WX, Bao N, Gao BY, Zhang HY. 2008. Biosorption of malachite green from aqueous solutions onto aerobic granules: Kinetic and equilibrium studies. *Bioresour Technol* 99: 3475–3483.

Świątczak P, Cydzik-Kwiatkowska A. 2018. Performance and microbial characteristics of biomass in a full-scale aerobic granular sludge wastewater treatment plant. *Environ Sci Pollut Res* 25: 1655–1669.

Tay JH, Jiang HL, Tay STL. 2004. High-rate biodegradation of phenol by aerobically grown microbial granules. *J Environ Eng* 130: 1415–1423.

Viaeminck SE, Cloetens LFF, Carballa M, Boon N, Verstraete W. 2008. Granular biomass capable of partial nitritation and anammox. *Water Sci Technol* 58: 1113–1120.

Wang JF, Wang X, Zhao ZG, Li JW. 2008. Organics and nitrogen removal and sludge stability in aerobic granular sludge membrane bioreactor. *Appl Microbiol Biotechnol* 79: 679–685.

Wang SG, Liu XW, Gong WX, Gao BY, Zhang DH, Yu HQ. 2007a. Aerobic granulation with brewery wastewater in a sequencing batch reactor. *Bioresour Technol* 98: 2142–2147.

Wang SG, Liu XW, Zhang HY, Gong WX, Sun XF. 2007b. Aerobic granulation for 2, 4-dichlorophenol biodegradation in a sequencing batch reactor. *Chemosphere* 6: 769–775.

Wichern M, Lübken M, Horn H. 2008. Optimizing sequencing batch reactor (SBR) reactor operation for treatment of dairy wastewater with aerobic granular sludge. *Water Sci Technol* 58: 1199–1206.

Wilen B, Liébana R, Persson F, Modin O, Hermansson M. 2018. The mechanisms of granulation of activated sludge in wastewater treatment, its optimization, and impact on effluent quality. *Appl Microbiol Biotechnol* 102: 5005–5020. doi: 10.1007/s00253-018-8990-9.

Yao J, Liu J, Zhang Y, Xu S, Hong Y, Chen Y. 2019. Adding an anaerobic step can rapidly inhibit sludge bulking in SBR reactor. *Sci Rep* 9:10843.

You Y, Peng Y, Yuan ZG, Li XY, Peng YZ. 2008. Cultivation and characteristic of aerobic granular sludge enriched by phosphorus accumulating organisms. *Environ Sci* 29: 2242–2248.

Zhang LL, Chen JM, Fang F. 2008. Biodegradation of methyl t-butyl ether by aerobic granules under a cosubstrate condition. *Appl Microbiol Biotechnol* 78: 543–550.

3 Modified Sequencing Batch Reactors for Wastewater Treatment

Sakina Bombaywala and Vinay Pratap
CSIR-National Environmental Engineering
Research Institute (CSIR-NEERI)
Academy of Scientific and Innovative Research (AcSIR)

CONTENTS

3.1 Introduction 54
3.2 Application of Sequencing Batch Reactor 55
    3.2.1 Biological Nitrogen Removal Process 55
    3.2.2 Simultaneous Nitrification-Denitrification (SND) Process 57
    3.2.3 Short-Cut Nitrogen Removal Process 58
    3.2.4 Anammox Process 58
    3.2.5 Enhanced Biological Phosphorus Removal (EBPR) 59
    3.2.6 Simultaneous Removal of Nitrogen and Phosphorus in an
        SBR Process 60
3.3 Different Variants of SBR Technology 61
    3.3.1 Cyclic Activated Sludge System 61
    3.3.2 UNITANK Technology 61
    3.3.3 Intermediate Cycle Extended Aeration System (ICEAS) 62
3.4 Recent Developments in the Application of SBR 62
    3.4.1 Algae-Based Sequencing Batch Suspended Biofilm
        Reactor (A-SBSBR) 66
    3.4.2 An Airlift Loop Sequencing Batch Biofilm Reactor 66
    3.4.3 Pressurized Sequencing Batch Reactor 66
    3.4.4 Micro-Electrolysis in Sequencing Batch Reactor 66
    3.4.5 Granular Sequencing Batch Reactor 67
    3.4.6 Fixed Bed Sequencing Batch Reactor (FBSBR) 68
    3.4.7 Moving Bed Sequencing Batch Reactor (MBSBR) 68
    3.4.8 Integrated Fixed-Film-Activated Sludge Sequencing Batch
        Reactor (IFAS-SBR) 68
    3.4.9 Membrane-Coupled Sequencing Batch Reactor 69
    3.4.10 Ultrasound-Induced Sequencing Batch Reactor 69
    3.4.11 Photo-Sequencing Batch Reactors (PSBRs) 69
    3.4.12 Photo-Fermentative Sequencing Batch Reactor (PFSBR) 70
    3.4.13 Photocatalytic Hybrid Sequencing Batch Reactor (PHSBR) 70

DOI: 10.1201/9781003202431-3

3.5   Conclusion 71
References... 71

3.1   INTRODUCTION

The sequential batch reactor (SBR) is divided into a cycle of four phases that are filling, reaction, settling and, finally, drawing of effluent (as shown in Figure 3.1). Each phase of SBR has a fixed time. The SBR starts with filling of the reactor tank with wastewater followed by application of stirring, aeration or oxygenation leading up to the second phase. The second phase can be subdivided into aerobic and anoxic phases. In the next phase, the waste sludge is allowed to settle by discontinuing stirring. In the last phase, the effluent is drawn from the reactor tank. After the tank is completely emptied, the next cycle begins with filling the tank with wastewater. The duration of aerobic/anoxic may vary for each cycle depending on the physiochemical properties of the influent water. The operational parameters of the SBR wastewater treatment plant (WWTP) are changed to comply with environmental standard policies for different types of influent wastewater and the quality of effluent required.

It has been found that the uncertainty of suspended biomass severely affects the functioning of the SBR system. The uncertainties are referred to as kinematic, and non-linear dynamic changes in the microbial diversity and nutritional behavior (heterotrophy and autotrophy) vary with every batch of SBR. Apart from microbial disturbances, other factors that vary from batch to batch are hydraulic loading, inlet flow rate, inflow wastewater properties, instrument limitations or failure (inlet pump, mixer, blower, nitrogen and pH sensors) and the dilapidation of dissolved oxygen (DO)

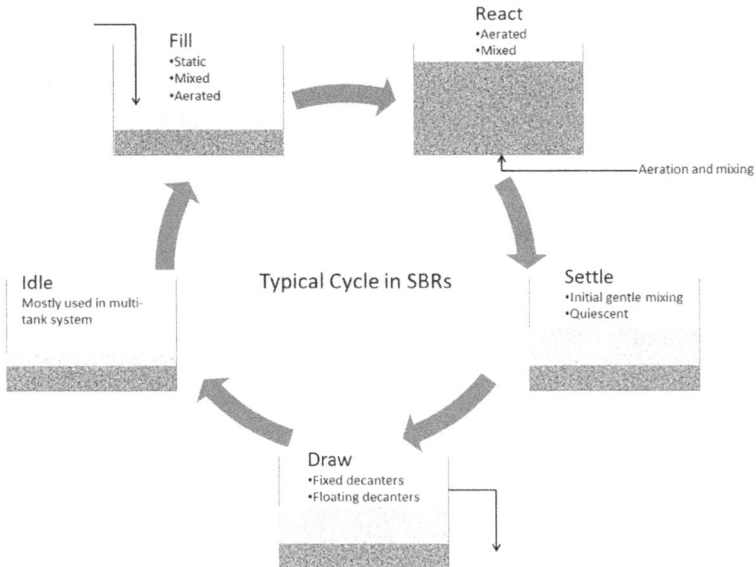

FIGURE 3.1   A typical diagram of sequential batch reactor showing cycle in four phases that are filling, reaction, settling and, finally drawing of effluent.

closed-loop control performance. Additionally, variation in inflow concentration or flow rate can change the dynamics of non-linear biological processes, in turn affecting the growth of sludge biomass and effluent quality. Consequently, the deterioration of SBR performance is unavoidable over a long period of time (Dutta and Sarkar, 2015).

Interestingly, SBR processes can be automated for controlling the mixing rate and aeration time, therefore making the operation easily adjustable in accordance with wastewater quality measures. The automation control can also be used during the aerobic reaction phase to swing between continuous and intermittent aeration controls. Other more complex operational modes can also be entailed using high-level automation machinery. Due to automation, less energy is needed in the treatment process. However, the major advantage is the improved efficiency of the wastewater treatment process.

The duration of a cycle in conventional SBR is fixed despite changes in the characteristics of influent or other processes, which brings ease to the operation of WWTP. However, the conventional approach requires more consumption of energy and is also not very efficient. Thus, controlling and monitoring mechanics are installed in SBR to readily adapt the reaction phase duration to the effluent requirements. Correct estimation of phase duration is necessary due to non-linear kinematics of microbial growth and inherent uncertainties of the treatment process. It means that an advanced controlling mechanic should be able to readily adapt to avoid above-mentioned drawbacks. Less phase duration can result in incomplete removal of nitrogen and longer phase duration can lead to increase in the operational cost (aeration time and consumption of external biomass). Therefore, the accuracy in estimating the phase time is critical, as both less and longer time durations will affect the ammonia-oxidizing bacteria (AOB)/nitrite-oxidizing bacteria (NOB) ratio, resulting in incomplete nitrification over the long term (Jaramillo et al., 2018).

3.2 APPLICATION OF SEQUENCING BATCH REACTOR

The increase in the demand for clean water has inevitably led to installation of tertiary treatment for enhanced removal of nutrients from wastewater. The tertiary water treatment should be able to remove pollutants other than usual pollutants such as chemical oxygen demand (COD), biochemical oxygen demand (BOD), pathogens and suspended solids (Table 3.1). The SBR process can be easily upgraded for enhanced nutrient removal by optimizing the anoxic, aerobic and anaerobic sequence without the need for additional infrastructure. Thus, SBR-based water treatment is comparable to tertiary treatment (as shown in Figure 3.2).

3.2.1 BIOLOGICAL NITROGEN REMOVAL PROCESS

The SBR cycles are operated for time durations rather than in space. This provides more flexibility for removal of nutrients through adjustments to mixing and aeration systems in order to create an alternating anoxic and aerobic environment in the reaction stage. The biological treatment of wastewater involves the removal of nitrogen through nitrification and denitrification processes. Nitrification involves the oxidation of organic nitrogen or COD. The conversion of organic nitrogen to ammonical nitrogen is the first step that is followed by its conversion to nitrate-nitrogen by action of chemoautotrophs. Thus, the two reaction of nitrification is nitritation (conversion of

TABLE 3.1

Different Operational Modes in SBR

Objective			Operational Mode	Function
Carbon removal	Low-load wastewater (municipal sewerage, domestic wastewater)		Static fill, aeration, settle, draw	Increase the reaction driving force
	Inhibitive sewerage, high-load organic wastewater		Aerated fill, aeration, settle, draw	Strong resist impact load capability
Nitrogen removal	Non-advanced nitrogen removal	Pre-denitrification	Mixed fill, aeration, settle, draw	Make the best of carbon source in raw water, low nitrogen removal efficiency
		Post-denitrification	Static fill, aeration, mixed, settle, draw	Utilize carbon source stored in microorganism, low nitrogen removal efficiency
		Alternate aerobic-anoxic	Fill, aeration, mixed, aeration, mixed, aeration, mixed, n times, settle, draw	Nitrogen removal efficiency is improved to some extent, complex operation
	Advanced nitrogen removal	Advanced nitrogen removal add carbon source	Fill, aeration, adding carbon source mixed, settle, draw	High nitrogen removal efficiency, but the high cost
		Step feed	Fill, aeration, fill and mixed, aeration, fill and mixed, n times, settle, draw	High nitrogen removal efficiency makes the best carbon source in raw water, complex operation and low space usage
Phosphorous removal			Mixed fill, aeration, settle, draw	Enhanced biologic phosphorous removal

ammonia to nitrite) and nitratation (conversion of nitrite to nitrate). The two reaction steps are catalyzed by two bacterial groups, namely, AOB and NOB. Both AOB and NOB bacteria use CO_2 as a carbon source and oxygen as the final electron acceptor during respiration. Thus, nitrification is strictly an aerobic process, thereby requiring continuous aeration in the reactor (Blackburne et al., 2008). On the other hand, heterotrophic bacteria are responsible for the denitrification step. The heterotrophs are facultative anaerobes, that is, utilize organic substance as a carbon source and nitrate as the terminal electron acceptor. The denitrification step takes place in the anaerobic or anoxic conditions in the reactor. Nitrogen gas is liberated during the denitrification process. Important parameters for nitrification are DO level, solid retention time (SRT), temperature and pH. The favorable temperature ranges from 25°C to 35°C, whereas the nitrification rate decreases by 50% if the pH is not between 7.5 and 9.8. The nitrification reaction tends to lower the pH of the system. The maximal nitrification rate was achieved at DO levels of more than 2.0 mg/L. Moreover, the rate of nitrification increases with increase in the concentration of nitrifying bacteria. This is attained by increasing the SRT that ensures the increase in the concentration of

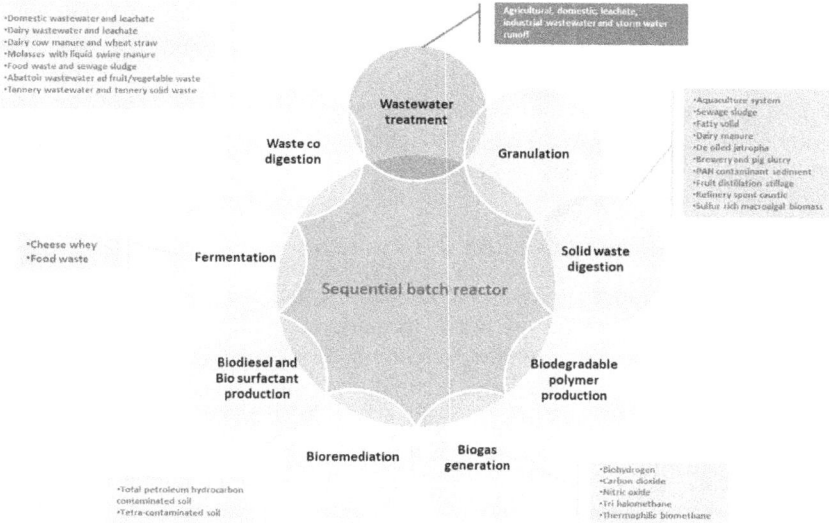

FIGURE 3.2 A diagram showing the comparison of SBR-based water treatment with respect to tertiary treatment.

mixed liquor suspended solids (MLSS) in the reactor tank. SRT is increased when the flow rate is lowered in order to retain waste-activated sludge (WAS). Unlike nitrification, the rate of denitrification is maximum when anoxic conditions (DO level <0.5 mg/L) are achieved in the reactor vessel. Denitrification reactions are suppressed at DO levels above 1 mg/L. The optimum pH range for denitrification reactions ranges from 6.5 to 9 (Qi et al., 2020).

3.2.2 Simultaneous Nitrification-Denitrification (SND) Process

The SBR treatment method is a feasible alternative choice to the continuous flow system for the removal of nutrients as it involves aerobic, anaerobic and anoxic reactions inside a single reactor tank in every cyclic treatment. Therefore, SBR is able to achieve concurrent nitrification-denitrification when the removal of nitrogen is the only target. The carbon to nitrogen ratio (C/N) is the governing factor for the SND reactions. It is reported that when the C/N was adjusted to 11:1, total COD and NH_4^+-N removal was achieved without any NO_2-N residues in the outflow water from the SND-SBR reactor. A low COD to NH_4^+-N ratio will cause imbalances in SND, therefore resulting in no denitrification process (Bonassa et al., 2021). Comparatively, SND utilizes ammonia and nitrate at a slower rate than in separate tanks because only a proportion of the total microbial population is available either for nitrification or denitrification process. Under the optimum operation of SND-SBR, the nitrification efficiency is about 100%, although total nitrogen removal is between 90% and 95%. The efficient performance of SBR in terms of nitrogen removal has resulted in the installation of numerous plants for treatment of landfill leachate. Theoretically, the addition of an anoxic phase following the aerobic phase will improve the efficiency of nitrogen removal, although for low-strength influent, water addition of carbon supplements is necessary. The need for additional carbon is reduced to less than 5% with high-strength wastewater. The advancement in monitoring mechanics will result in a step feed system optimized for oxidation reduction potential (ORP) and DO through an intelligent control regime. This is thought to reduce or completely eliminate the need for carbon supplements (MENG et al., 2008).

3.2.3 SHORT-CUT NITROGEN REMOVAL PROCESS

Another viable treatment option is incomplete nitrification followed by denitrification. Partial nitrification takes up nitrite formation (nitrification) followed by direct conversion of nitrite to nitrogen (denitrification). This process is referred to as short-cut nitrogen removal. This alternative technique is beneficial as the aeration requirement is cut down by 25% and carbon supplements by 40%. Thus, energy expenditure in short-cut nitrogen removal process is also less compared to that in conventional SND treatment. Additionally, this process can achieve a high rate of denitrification and decrease in the amount of waste sludge generated. In the case of a low C/N ratio or nitrogen-containing wastewater, promising results for nitrogen removal can be obtained using this process. Examples of strong nitrogen-containing wastewater are biowaste anaerobic digestion effluents, landfill leachate and sewage sludge digestate (Dutta and Sarkar, 2015).

3.2.4 ANAMMOX PROCESS

The anammox process involves oxidation of ammonia-N to nitrite that is further used by anaerobic ammonia-oxidizing (anammox) bacteria as a terminal electron acceptor in the reaction that converts ammonia-N to nitrate and nitrogen gas. The major advantage is the complete elimination of need for external carbon supplementation and aeration. Nevertheless, the toxicity problem and slow growth of anammox bacteria have limited the initial development of the anammox process using biofilms, suspended biomass or granules (Choi et al., 2019). The anammox process is composed of two reactions, namely, nitritation and anammox reaction. The two reactions can be separated either spatially or temporally. Space separation can be done using a two-stage reactor, whereas strict operational regime and control strategies are required to perform temporal separation. It was observed that the ORP is correlated with different levels of functional activity during different aeration rates and feeding intervals. Thus, intermittent feeding together with intermittent aeration is the best strategy for optimal performance of the anammox process in terms of ammonia-N elimination and nitrate-N generation, along with pH stability throughout the process (Li et al., 2018). In conventional SBR, effluent constitutes only a fraction of the total liquid in the reactor; therefore, there is an increase in the concentration of nitrate and total nitrogen in the SBR reactor over a long period. Logically, it is expected to include denitrification in the anammox process to resolve the issue. Denitrification reaction requires the inclusion of exogenous carbon sources. It is reported that the concentration of COD controls whether an anammox or denitrification reaction will take place, where the anammox reaction is suppressed in high COD conditions. However, at conditions of low COD, both anammox and denitrification activities are not sup-pressed, resulting in a rather improved nitrogen removal efficacy. It is observed that optimum nitrogen removal efficiency is achieved with a combination of denitrifica-tion and anammox (Du et al., 2017). The combined SBR operation is feasible for wastewaters with high nitrogen concentrations and low carbon content. Despite this, the suppression of anammox activity at high carbon concentrations is reversed when the carbon load is removed. In a study online tool to sense ammonia and continuous aeration was given to attain DO level of <1 mgO_2/L. The study reported a maxi-mum conversion of ammonia (90%) to nitrite with 500 gN/m^3 of ammonia oxidized per day (Xu et al., 2015). The anammox and nitrite oxidation occur simultaneously, allowing for an increase in the process performance and easy monitoring compared to when anaerobic and aerobic stages take place in two separate reactors.

3.2.5 ENHANCED BIOLOGICAL PHOSPHORUS REMOVAL (EBPR)

The removal of phosphorus is a vital function of WWTPs. It is carried out by a unique class of microorganisms called polyphosphate accumulating organisms (PAO). As the name suggests, these bacteria accumulate P in their cells in an alter-nate cycle of aerobic and anaerobic conditions. Subsequently, the sludge is discarded from the reactor. During the anaerobic stage, PAOs utilize volatile fatty acids (VFA) as a carbon source, and deposits of polyhydroxyalkanoates (PHAs) are found in bac-terial cells. The hydrolysis of poly-P stores in the cells of PAOs provides energy for carrying out reactions in anaerobic conditions and also the subsequent discharge of orthophosphates in the water. During anoxic or normoxic conditions, PAOs have the ability to accumulate excess phosphorus intracellularly in the form of poly-P. Further, they use PHA stores for the growth of biomass and replenishment of glycogen. Net phosphorus released during the anaerobic phase is less than accumulated during the anoxic or normoxic phase; therefore, the net amount of phosphorus can be eliminated in the sludge wasting stage, which is rich in poly-P. In SBR, parameters can be con-trolled in order to achieve repeating aerobic and anaerobic conditions and therefore SBRs have blooming prospects for biological removal of phosphorus (Rajab et al., 2022). In SBR, highest phosphorus removal efficiency reported is 90% compared to only 10%–20% removal efficiency in the conventional activated sludge process. Another special class of microorganisms called glycogen-accumulating organisms (GAOs) is closely related to PAOs and competes with them. Due to metabolic com-petition, GAOs are responsible for decreasing the efficacy of P removal. Therefore, optimal conditions are important for favoring the growth of PAOs over GAOs. The parameters that need to be monitored are temperature, pH and type of substrate. Cold temperature ranges are more favorable for the growth of PAOs. Increasing pH favors PAOs, and the optimum pH range is 7.2–8, which also keeps GAO growth under control. Both GAO and PAO intracellar deposits are synthesized using carbon alone (Ni et al., 2021). Treatment of wastewater having COD to phosphorus ratio (mg/mg) of more than 40 such as raw water can generate effluent with low phosphorus con-centration. Additionally, the process is highly stable when such wastewaters are used as substrates in the SBR treatment process. The enrichment of PAOs is dependent on the type of carbon substrate. If the incoming water has organic carbon load in the form of VFAs or COD that can get readily degraded into VFAs, the growth of PAOs transcends those of GAOs, resulting in high phosphorus removal. In a recent study, wastewater treatment in EBPR has reported the possibility of attaining large Bio-P discharge at short SRT (2–4 days) (Ge et al., 2015). At lower SRT, complete denitrification/nitrification cannot be achieved; thus, there is need for post treatment before final effluent discharge. It is demonstrated that the enrichment of PAOs is favorable in high organic carbon load conditions; therefore, short SRT and low COD of wastewater can be challenging for effective removal of phosphorus. The non-working time during SBR operation has a significant effect on Bio-P removal rate and is especially true for incoming water with high P concentration. The drawback of extended idle time in EBPR much lengthier than in anaerobic/oxic (A/O) SBR can be addressed by using a novel SBR configuration called sludge tank halves (STH). This design has successfully demonstrated high removal of P compared to the conventional A/O regime (Wang et al., 2014).

3.2.6 Simultaneous Removal of Nitrogen and Phosphorus in an SBR Process

Theoretically, in an SBR, nitrification, denitrification and EBPR reactions should occur concurrently for the removal of phosphorus and nitrogen together. However, this will require optimal control of interaction between the three processes so as to prevent failure of SBR operation. The nitrite and nitrous acid from denitrification/nitrification reactions make the environment acidic, which is known to favor the growth of PAOs more than GAOs. It is reported that PAO growth and activity are severely inhibited at a nitrite-N concentration of 2 mg/L, whereas it is completely inhibited at 6 mg/L concentration of nitrite-N (Hu et al., 2018). Previous studies have indicated a reduction in the amount of phosphors removed in high nitrate levels present in the anaerobic phase. The nitrate molecules have their effect by disrupting the anaerobic environment. In addition to this, fatty acids are utilized by nonpolyphosphate heterotrophs and an incomplete denitrification results in inhibition of PAOs activity by the nitrite. Both denitrifying microbes and PAOs require organic carbon, and natural competition exists between them at low oxygen levels. Exposure of nitrate to anaerobic conditions for extended periods was reported to decrease the diversity of PAO, therefore inhibiting PAO activity or stimulating competition between denitrifying organisms and PAOs. A significant fraction of PAOs has denitrifying activity denitrifying phosphorus accumulating organisms (DPAO) in anoxic conditions. The bacteria use intracellular PHAs for energy and use nitrate as an electro acceptor during the uptake of phosphorus (Lv et al., 2014). The DPAOs are important because they have the ability to remove phosphorus and nitrogen at the same time, have low biomass production and therefore less amount of sludge waste generation. The nitrite-utilizing PAOs have been found but not nitrate utilizing. Hence, three types of PAOs have been found on the basis of who is used as a terminal electron acceptor: (1) oxygen only, (2) oxygen and nitrate and (3) oxygen, nitrate and nitrite (Nielsen et al., 2019). A full-scale WWTP working on a conventional A/O SBR process has been employed for simultaneous removal of phosphorus and nitrogen. Using the single sludge method, a significant removal of P was achieved by the activities of nitrifies and DPAOs. Several studies have been conducted on the anaerobic-aerobic-anoxic-aerobic system for N and P removal, and it is suggested that DPAO proliferation can be attained by employing advanced control strategies. Another scheme that can be employed for the simultaneous removal of P and N is the anaerobic/aerobic-anoxic process (Soejima et al., 2008). A novel strategy was suggested that employed an organic fraction of MSW and activated sludge waste for anaerobic co-digestion in an SBR plant. The phosphorus removal was reported due to the activity of denitrifiers via nitrite. Therefore, this strategy is suggested to have successful future implementation for treating the side stream sludge of rejected water, for improving nutrient removal and reducing energy costs and carbon footprints of the treatment process (Frison et al., 2016).

3.3 DIFFERENT VARIANTS OF SBR TECHNOLOGY

There are different types of SBR technologies available which are discussed below.

3.3.1 CYCLIC ACTIVATED SLUDGE SYSTEM

The cyclic activated sludge system (CASS) comprises a single reactor tank with varying operational volumes in an alternative manner. A single reactor basin consists of a plug-flow at the base zone followed by a complete-mix, main aeration and secondary zones. The activated sludge is recirculated from the main aeration zone in the selector zone present above the completely mixed zone. Here, it gets mixed with inflow wastewater. The incorporation of a feasible design for high-rate plug-flow assists in the stable and uniformity of sludge metabolic activities present in the complete-mix zone resulting in fast organic content biodegradation and enhanced floc settlability. Therefore, the operation is indifferent to any alterations in the COD and slow rate of the raw wastewater. Additional advantages of the process are its superior design for simultaneous N and P removal in CASS than in conventional SBR systems. Also, this process can be used for treating both municipal and industrial wastewater.

3.3.2 UNITANK TECHNOLOGY

The UNITANK SBR system accommodates the advantages of SBR. It comprises three reactor ditches for oxidation reactions and an aeration unit. A simple configuration of UNITANK comprises a series of compartments connected hydraulically in a single reactor tank. The aeration system is present in each compartment with no provision for sludge recirculation. The outer compartments act as an aeration unit for sludge settling in alternate cycles, whereas the middle is only an aeration unit. One operational cycle is divided into two major phases which have three primary steps. Each phase occurs in a symmetric fashion starting from the outer compartments toward the middle ones. As there is no separate sludge-settling unit with scraper, outer compartments are deployed with slots for sludge sedimentation and weirs for effluent discharge. Advanced configurations of UNITANK consisting of additional compartments having anoxic/anaerobic conditions and equipped with mechanics for mixed liquor recirculation have been employed for the removal of N and P. The UNITANK–SBRs are cost effective, reliable, needs less land and have a simple design. Therefore, it is more apt for small to middle-scale WWTPs. The UNITANK SBR plants are installed in many countries like Argentina, Brazil, Mexico, Vietnam, China, etc.

3.3.3 INTERMEDIATE CYCLE EXTENDED AERATION SYSTEM (ICEAS)

The ICEAS is advancement over conventional SBR treatment. Unlike SBR, it is a continuous flow reactor. The variation in flow rate is controlled by a distribution box that spreads inflow wastewater uniformly in reactor tanks so as to prevent overloading in a single reactor tank. A high F/M in the pre-react zone affects the selection of microbial activity. Therefore, enhanced sludge settlability and filamentous microbial growth can be achieved. After the pre-react zone, the main react zone starts which is operated in three sequential modes: aeration, settle and draw. The uniform loading and flow in all the basins at any given time point simplifies the complex control of processes, their operation and maintenance. Because only a single rector tank is required, there is a significant reduction in the cost of capital compared to the usual SBR process. The ICEAS plants are being recognized in the UK, the US, China, Qatar, Peru, etc. The ICEAS are suitable where there is space limitation or requirements of effluent quality (Dutta and Sarkar, 2015).

3.4   RECENT DEVELOPMENTS IN THE APPLICATION OF SBR

Major advantages of SBR technology are its operational flexibility. This owes for its application in numerous treatment plants. The SBR is a versatile technology because of its ability to carry out equalization of flow, biological treatment and secondary clarification in a single set of tanks. This is conferred by changing the time duration for aeration and each phase of the process. Recently, assessments of various combinations of wastewater treatment technologies at a lab scale have been done that further extended the scope of SBR technology (Table 3.2).

The sequencing batch biofilm reactor (SBBR) is a variation of conventional SBR. It is composed of suspended and attached growth (CSAG) system. The growth of biofilm takes place on a support material, which is a solid-liquid interface. The biofilm allows for the proliferation of slow-growing microbes regardless of the bio-aggregate settling characteristics and hydraulic retention time (HRT). The type and size of support material is dependent on the characteristics of the influent water and effluent requirements. The support material can be suspended or packed inside the reactor. The SBBR treatment cycle has only three phases fill, react and draw. Similar to CASS, SBBR consists of a plug-flow system. The sludge-settling time in SBR is equivalent to support media washing time in SBBR. In case of high total suspended solids (TSS) of raw wastewater and undue surge in microbial growth, the SBBR for wastewater treatment is not a suitable option due to unnecessary sloughing off and head losses. The immobilized microbe on a support media prevents microbial washout and decreases the effect of pH, temperature and toxic compounds on microbial growth. The landfill leachate from Georgswerder, Germany was treated

TABLE 3.2
Advantages of Different SBR Configurations

Configuration	Advantages
Algae-based sequencing batch suspended biofilm reactor (A-SBSBR)	• Suspended carriers provide an enabling environment for algae enrichment • Lower HRT and SRT than in the traditional biological systems • Independent sludge discharge and carrier's replacement could be used to separate sludge and algae SRT • Carriers replacement reduces pollution caused by algae loss or death
An airlift loop sequencing batch biofilm reactor	• Integrating nitrification and denitrifying dephosphatation in one reactor for simultaneous phosphorus and nitrogen removal • Competition between nitrifiers and denitrifying phosphorus removal bacteria in biofilm could be avoided by the reactor
Micro-electrolysis in sequencing batch reactor	• Simple convenient and centralized automated operating system • Reduced safety risks • Steady treatment effect • Less area requirement alongside construction, operating and maintenance cost
Pressurized sequencing batch reactor	• Improves aeration efficiency standard and decreases sludge generation resulting in lower sludge disposal cost • Increases DO with increased contact time between air flashes and wastewater threefold
Granular sequencing batch reactor	• Lower energy consumption, smaller footprint, good settling ability • Diverse microbial species and high biomass retention • High-rate simultaneous nitrification, denitrification and phosphorus removal (SNDPR)
Fixed bed sequencing batch reactor (FBSBR)	• High SND • Less excess sludge generation
Integrated fixed-film activated sludge sequencing batch reactor (IFAS-SBR)	• Resistance to adverse shock load and reduced capital cost of upgrading existing reaction tanks • Reduces the risk of active biomass loss • Improves process capacity while providing system stability
Moving bed sequencing batch reactor (MBSBR)	• Flexible operation, discharge control, lower footprint and tolerance to organic shock and toxic loads • The use of inexpensive porous media, robustness against starvation periods and total purification of pollutants • No need to return sludge
Membrane-coupled sequencing batch reactor	• Reduce SBR cycle length, smaller footprint, less sludge production and higher volumetric loading rates • Avoiding the formation of byproducts • Compactness and superior water reuse potential • Shorter HRT and longer SRT • Ease and economical in operation
Ultrasound-induced sequencing batch reactor	• Technological flexibility and superior economic efficiency • Suitable for wastewater co-treatment with a significantly larger percentage of leachate • Increases biodegradability of mature landfill leachate and decomposition of recalcitrant organic pollutants • No chemical reagents are required
Photo-sequencing batch reactors (PSBRs)	• Reduced carbon dioxide generation • Energy saving due to low aeration requirement • Easy cultivation of algal-bacterial granules
Photo-fermentative sequencing batch reactor (PFSBR)	• High theoretical hydrogen yield, none oxygen evolution and utilization of metabolites from dark fermentation. • Ability to convert a wide spectrum of light into hydrogen gas

in a pilot-scale SBBR plant. As microbes are retained on the media, a short HRT is sufficient in a small reactor or more treatment capacity in a given size of the reactor as compared to the basic SBR unit. In conclusion, biofilm-based SBR reactors are highly resilient, well suited for different types of wastewater and generate less sludge wasting (Ozturk et al., 2019). The choice of support media is important to help avoid loading shocks, for instance, activated carbon media can be used for high organic carbon influent or zeolite for wastewater with high ammonia content. They adsorb shock load temporarily and later slowly perform desorption of shock load constituents, together with their biodegradation. Use of powdered activated carbon (PAC) in the treatment of landfill leachate resulted in better color, COD and NH_3-N removal compared to the conventional SBR. It has been demonstrated that the use of advanced control tools rather than simple time-based control tools has improved the removal efficiencies of total phosphorus (TP), total nitrogen (TN) and COD with significantly fewer energy demands (Jin et al., 2012).

The bio-floc technology (BFT) is a modification to the SBR system. It has demands in the aquaculture industry where wastewater treatment and protein-rich food for fish are expensive. Bio-floc is a macro-aggregate of microbes that are able to convert nitrogenous compounds present in wastewater to organic proteins. Thus, bio-floc microbes can be fed to fish. A lab-scale study of the SBR plant was used for external growth of bio-floc with up to 98% nitrogen removal at the C/N ratio in a range of 10–15 (Luo et al., 2013). Additionally, the bio-floc also converts nitrogen present in suspended solids in aquaculture water into microbial protein, which is later fed to fish thereby enabling the efficiency of nitrogen removal to reach nearly 100% within six hours. SBBR and SBR processes have found applications for treating industrial wastewater contaminated with phenol and its derivatives like para-nitrophenol (PNP), which is chemical hazard commonly used in pharmaceutical, agricultural and in dye factories as a process intermediate. Complete removal of PNP was demonstrated using SBBR and SBR processes for the influent water containing 350 mg/L of PNP. The flow rate was $0.368 \, kg/m^3$ per day, and polyethylene rings were used as support media. Although the average efficiency of NH_3-N elimination was slightly reduced in SBBR and SBR, it eliminated up to 96% and 86% of NH_3-N concentration, respectively (De Schryver and Verstraete, 2009). Successful implementation of SBR for treating raw wastewater containing high concentrations of nitrogen and low organic carbon has been demonstrated. For instance, anaerobic SBR (ASBR) was used to treat wastewater from the opto-electronic industry having the C/N ratio of 0.2 (Daverey et al., 2012). It was found that simultaneous ammonia oxidation, partial nitrification and denitrification occurred during the treatment process. Similarly, in another study, wastewater from the thin-film transistor liquid crystal display production unit containing monoethanolamine (MEA), tetra-methyl ammonium hydroxide (TMAH) and dimethyl sulfoxide (DMSO) was treated using two SBR systems. The MEA degradation was achieved readily under aerobic, anoxic and anaerobic conditions, while TMAH and DMSO degradation occurred effectively under aerobic and anaerobic conditions, respectively (Lei et al., 2010). The porous biomass carrier SBR (PBCSBR) is a hybrid system having high biomass and employing time-sequential oxic/anoxic phases. The PBCSBR has been investigating the efficient removal of nutrients. In a study, dairy manure was treated in PBCSBR that utilized natural fiber carriers for biofilm as a novel biomass retention method. Similarly, fiber biofilm carriers were used in another study to treat flushed dairy manure in ASBR. In this study, high methane yield was obtained at psychrophilic conditions and short HRT (6 days). ASBRs are able to uncouple SRT with HRT for enhanced retention in biomass. Further, ASBR depends on a particular sequence of process phases for exerting selection pressure on microorganisms for immobilization (Ma et al., 2013). Aerobic SBRs are able to couple the photo-fenton process with reverse osmosis (RO) for reclamation of textile industry wastewater, thereby facilitating complete water recycling (Blanco et al., 2014). Various SBR operations like famine regime, cyclic feast, less settling time and large shear stress enable promotion of floc granule formation, which is dense consortia made of diverse species of microorganisms that work together to perform biodegradation of complex wastes. Using granular SBR, it was proved that step-feeding mode coupled with anoxic/oxic phases in an alternate fashion is an effective method to remove nitrogen. Unlike activated sludge, granular sludge has the advantages of excellent settlability, biosorption, high biomass retention, indifference to changes in organic load and exhibits diverse metabolism for simultaneous removal of N, P and COD (Jagaba et al., 2021). The application of granular sludge for toxic compounds like organofluorine specifically 2-fluorophenol (2-FP) degradation in SBR has been shown. The biodegradation of fluorine-containing aromatic compounds occurs via halo-catechol formation. The sufficient number of halo-aromatic compound-degrading microorganisms in an SBR is desirable owing to the sporadic nature and low concentration of halo-aromatic compounds. Granular SBR offers the advantages of high microbial retention and therefore is a promising technology for the treatment of influent-containing harmful compounds. The conventional SBR process utilizing sludge/flocculation can opt for granular sludge instead, as it offers a reduction in the settling time (Duque et al., 2011). Azo dye-containing wastewater was treated in SBR for COD removal and decolorization. The SBR consists of anoxic-aerobic reaction stages. The study concluded that a long anoxic phase stimulates bio decolorization and the short anoxic phase was better suited for COD removal. The granular-activated carbon-SBR (GAC-SBR) is a potential treatment option for textile wastewater contaminated with various dyes and was removed by adsorption into granular activated carbon (GAC). The addition of external carbon supplement is correlated with an increase in the dye removal efficiency (Al-Amrani et al., 2014). Endocrine disruptor compounds (EDCs) are a class of chemical compounds that cause disruption of normal functions of the endocrine system of humans and animals even at minute concentrations (μg/L). In recent times, EDCs have been found in water supplies, surface water and wastewater, thereby raising public concerns. The SBR process is a promising treatment technique for wastewater containing EDCs because of the simultaneous occurrence of anoxic/aerobic/anaerobic phases in a single reactor tank. This dynamic condition within a single basin promotes EDC biodegradation via the stimulation of various microbial activities. The longer duration of HRT and SRT is related to a surge in the removal rate of EDCs largely because it provides avenues for the growth of slow-growing microbes with scope for utilizing EDCs. Post-treatment steps like ozonation may be required in the case of critical final products and where STP effluent dilution does not take place (Siegrist et al., 2012).

3.4.1 ALGAE-BASED SEQUENCING BATCH SUSPENDED BIOFILM REACTOR (A-SBSBR)

This system consists of a biofilm that can float to the surface when aeration is switched off. The sun's illumination on the biofilm assists in algae growth and enrichment. During aeration, the microbial population on the biofilm can trade substances between sludge, wastewater and algae. Under light illumination, the gaseous or soluble CO_2 is captured during algal photosynthesis to generate O_2 molecules, which are taken up by bacteria for degradation of pollutants (Tang et al., 2018).

3.4.2 AN AIRLIFT LOOP SEQUENCING BATCH BIOFILM REACTOR

The SBBR with an airlift loop is a fixed reactor partitioned into two zones: aeration and reverse flow. The reactor design facilitates the coupling of nitrification, denitrification and phosphorus removal. Like SBR, the three operational phases of airlift loop SBBR are anaerobic, aerobic and anoxic phase. The carrier materials are packed in the two zones that promote the growth of denitrifying phosphorus accumulating organisms in a predominant manner. The packing density can be changed to regulate the amount of sludge decanted, thereby facilitating the optimal rate of phosphorus removal (Zhan et al., 2006).

3.4.3 PRESSURIZED SEQUENCING BATCH REACTOR

The pressurized aeration process is used to increase the oxygen transfer rate. The activated sludge is pressurized to increase oxygen solubility resulting in enhanced transfer momentum of oxygen. Compared to conventional aeration of the biofilm and activated sludge, the pressurized aeration method is much more effective, resulting in a dramatic increase in the degradation of organic materials. Further, the aeration under pressure causes a decrease in the volume of the aeration tank and hydraulic loading time, thereby allowing the process to run under high concentrations of organic carbon. The pressurized aeration process can give considerable savings, particularly for large-scale wastewater treatment. However, under high pressure (~100 bars), microbial growth is inhibited. This tendency of microbial cells can be used to eliminate and inactivate microbes, subsequently providing long shelf life for many packaged food and materials. Conversely, in case bioreactors, the consequences of high pressure are not relevant and the pressure in the reactor is controlled at a moderate level (<1 MPa). The moderate level of pressure does not cause damage to microbial processes (Zhang et al., 2017). The study by Elkaramany et al. (2018) demonstrated that the DO level in wastewater increased in pressurized SBR compared to the traditional SBR process. The recirculation of pressurized air in the reactor leads to a threefold increase in the time period for the contact between liquid phase and air bubbles.

3.4.4 MICRO-ELECTROLYSIS IN SEQUENCING BATCH REACTOR

The micro-electrolysis process involves iron reduction or internal electrolysis. The chemistry of the process is based on metal corrosion. It is a coupled reaction of electro-coagulation and electro-aggregation by making use of microbial battery electrode reaction formed in the electrolyte solution. The reactor has suspended filler made from non-reactive carbon particles and iron scrapers. It is a potential treatment technology for landfill leachate. It is particularly effective for the degradation of color, metal ion sequestration and humic acid decomposition. The internal micro-electrolysis (IME)-based SBR has the capacity to integrate both oxidative and reductive IME in a single set of reactor. The reconstruction of traditional SBR can be done to incorporate an iron-carbon unit that is suited for small to medium-scale treatment plants. The micro-electrolytic configuration of SBR needs frequent cleaning with acid leading to the consumption of large amount of iron and thus having an inadequate capacity for treatment (Duque et al., 2011).

3.4.5 GRANULAR SEQUENCING BATCH REACTOR

The aerobic granular sludge is referred to an aggregate of microbial mass formed without a carrier material in aerobic conditions (as shown in Figure 3.3). The granular sludge has a dense and uniform structure, homogenous morphology, shock bearing physical structure with the ability to act as a buffer for toxic compounds. Numerous studies have reported that granules may get fragmented over long periods of an operation mainly due to substrate unbalance, overloading, unsuitable process configurations and loss of key microbial groups, stress-causing compounds, over growth of filamentous microbes, EPS secretion and anaerobic core hydrolysis. However, stable and strengthened granules can be formed by application of the following strategies: the selection of process configuration suitable for the enrichment of slow-growing microbes while inhibiting the growth of anaerobic microbes. The aerobic and anoxic conditions influence the penetration of DO in the granules, thereby affecting nitrogen removal due to fluctuations in nitrification and denitrification functions. The diffusion of DO in a single granule is dependent on the spatial distribution, density, size and microbial diversity of the granules. Conversely, low temperature and high salt

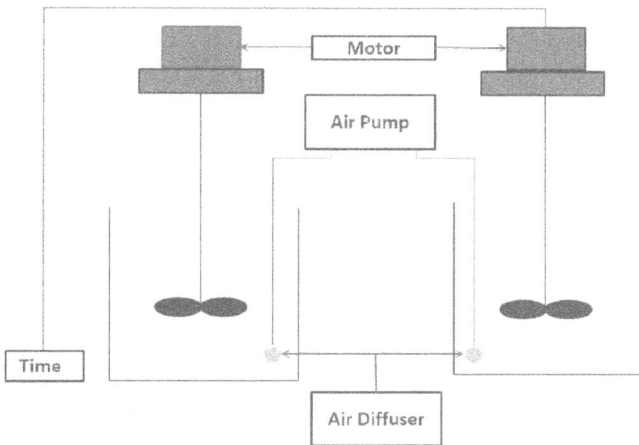

FIGURE 3.3 Schematic diagram showing an aerobic granular sequencing batch reactor.

concentrations negatively influence the strength, stability and performance of aerobic granules (He et al., 2020).

3.4.6 Fixed Bed Sequencing Batch Reactor (FBSBR)

The FBSBR can operate effectively at high organic loading rates (OLR). The biofilm promotes buffering for dealing with fluctuations in hydraulic loadings. Also, the biofilm system influences microbial growth and up keeping of microbial activity. Further, the gradient of DO in biofilm layers can facilitate higher efficiency of phosphorus removal. The FBSBR has problems of clogging, backwashing requirements, stabilization ratio and high nutrient concentration (Hosseini Koupaie et al., 2011).

3.4.7 Moving Bed Sequencing Batch Reactor (MBSBR)

The MBSBR consists of an attached microbial growth system developed on the basis of traditional activated sludge and biofiltration process for wastewater treatment. In MBSBR, the mixer of biomass and activated sludge is added to biofilm carriers. The carrier media have less density than water and therefore float on the surface and circulate by aeration in the liquid of the reactor basin. The choice of biofilm support media is critical in enhancing the efficiency of nitrogen removal. This is because the process efficiency is limited by media clogging, hydraulic loading instability and head loss (Tan et al., 2016). The study by Malakootian et al. (2020) concluded that suspended biomass supports a high rate of degradation, and SND efficiency is dependent on the thickness of the biofilm, DO level and organic carbon concentration in the influent water. As such, the more the biofilm thickness, the more the SND efficiency. The MBSBR system is suitable for the treatment of wastewater concentrated with non-biodegradable pollutants.

3.4.8 Integrated Fixed-Film-Activated Sludge Sequencing Batch Reactor (IFAS-SBR)

The IFAS-SBR is a system consisting of biofilm carriers as part of an activated sludge reactor. The configuration of the IFAS-SBR process provides ample surface area for the fixation of bacterial cells and their proliferation. The investigation into the structure and function of fixed microbes on the biocarrier is essential for understanding the functions of individual and combined microbial communities. The main application of the IFAS-SBR system is for the treatment of wastewater with low concentrations of COD. Biological degradation and absorption are chief mechanisms for the removal of contaminants in IFAS-SBR. Thereby making the conditions conducive for the growth of denitrifying bacteria. The modification in the reaction and sludge-settling duration is suggested to facilitate nitrate removal. Unlike fixed media, moving ones are better able to promote the transfer of oxygen and nutrients in the reactor. It is demonstrated that the biofilm is conducive to the flourishing of nitrifying bacteria as the EPS present in the suspended flocs or attached to biofilm is affected by variations in OLRs (Shao et al., 2018).

3.4.9 MEMBRANE-COUPLED SEQUENCING BATCH REACTOR

The membrane-coupled sequencing batch reactor is a versatile engineered system for the removal of diverse compounds including organic matter, nutrients and toxic contaminants from the wastewater. It is a promising alternative to settling and decanting phases in SBR treatment. It also offers accommodation for an increase in particle size of sludge and concentration of soluble microbial exudes. Additionally, it can be regarded as a tertiary treatment for eliminating coliforms (Frank et al., 2017). The membrane-based reactors have problems of foul smell and, thus, need frequent cleaning. It is reported that operational conditions, MLSS, feed, F/M ratio, microbial exudes, floc and membrane type and characteristics are probable reasons for fouling. The fouling can be kept in check by either operating plant less than critical flux levels or maintaining the turbulent environment. In addition, fouling can be alleviated by imparting feast famine conditions that will promote sludge granulation in the SBR. The membrane can be maintained clean for some time with the aid of air backwashing. Afterwards, mechanical cleaning becomes critical for reducing the fouling of the membrane (Shariati et al., 2011).

3.4.10 ULTRASOUND-INDUCED SEQUENCING BATCH REACTOR

The ultrasound-induced SBR is a novel technology for high throughput and cost-effective biological treatment of wastewater. Ultrasonic waves are sound or acoustic waves having >20-kHz frequency which is more than the hearing limit of human. The critical parameters of ultrasonic sound are irradiation intensity, frequency, cycle, time and proportion (Jin et al., 2013). In the ultrasonic-induced SBR process, radical generation in the cavitation bubble aids in the removal of pollutants. The ultrasonic waves cause cyclic rarefaction and compression of the liquid medium. Moreover, the bubble diameter is determined by the frequency of the ultrasound waves. In case of the high concentration of organic mass, a high-frequency ultrasound generates more stable and smaller bubbles. This has a remarkable positive impact on sludge gravitating speed with no adverse effect on microbial metabolism, thus making the biological wastewater treatment highly stable. Conversely, low-frequency ultrasound (20–110 kHz) generates strong hydrokinetic shearing forces causing sludge instability. However, low ultrasound waves help to improve the development and functional profile of microbial cells. This is due to improvement in enzyme activity, enhancement in oxygen and nutrient transfer rates, increase in membrane permeability, stimulation of biosynthesis and cellular growth and enhancement rate of cell excretion. In several studies, the excess sludge formation was decreased by applying low-frequency (<100 kHz) and low-intensity (<2 W/cm^2) ultrasound. It was found that at a short duration of ultrasound irradiation, the floc agglomerate distribution in the liquid phase ensued without any cellular damage. However, longer irradiation time causes the weakening of bacterial cell wall and materials inside the cell extruded out in the liquid phase (Jin et al., 2013).

3.4.11 PHOTO-SEQUENCING BATCH REACTORS (PSBRS)

The PSBR is a closed and compact system with uniform distribution and direct transmission of high and low light in the bioreactor. The uncoupling of SRT from HRT is done to control nutritional dynamics and microbial composition that aids in the cell harvesting process. Similar to other hybrid SBR configurations, the increase in biomass flocculation and thereby large flocs formation is achieved in PSBR after the addition of an additional sedimentation stage in the process (Wang et al., 2015). In the PSBR process, various parameters of irradiated light including light intensity (LI), light quality and photoperiod have a significant effect on the removal of nutrients, algal growth, bioactivity and production of lipids in the case of algae culture. Especially, important is LI influencing the amount of photosynthesis and algal growth. This is because the LI affects the formation of oxygen, organic mass and biomass settlability. Moreover, low DO, poor LI, high ammonia and nitrite concentrations when persist at the same time are capable of significantly inhibiting the NOB. Photo inhibition of algae is stimulated by rich LI, consequently resulting in a decrease in organic mass and effluent quality. The variation in LI in the PSBR process affects the microbial community of granules resulting in the enrichment of different classes of bacteria and algae with different functional activities (Meng et al., 2019). Further, nitrogen metabolism is directly related to the concentration of DO. The PSBR system is deployed for production of algal-bacterial granules. Natural sunlight irradiation in aerobic conditions will facilitate a rapid growth of algae-bacteria granular consortia in the SBR reactor system. The findings from the study of He et al. (2018) concluded that the expansion of water-borne algal biomass negatively affects the sludge-settling characteristics and the average size of granules but significantly promotes the functional activity of microbes. During light irradiation, oxygen production during photosynthesis induces AOB. The dark period promotes denitritation due to the rapid consumption rate of DO during algal respiration and microbial metabolism. The study by Arun et al. (2019) has shown that by altering the dark and light periods, a complete BNR can be done without the need for aeration.

3.4.12 PHOTO-FERMENTATIVE SEQUENCING BATCH REACTOR (PFSBR)

The PFSBR has the potential for effective hydrogen production by photo-fermentation. The low yield and slow rate of hydrogen production are some challenges of the photo-fermentation process. This is because of poor microbial flocculation resulting in low capacity for biomass retention. The support materials such as activated carbon or solar optical fibers can be employed for immobilization of microbes that will aid in continuous production of hydrogen gas (Xie et al., 2012).

3.4.13 PHOTOCATALYTIC HYBRID SEQUENCING BATCH REACTOR (PHSBR)

The PHSBR is an integration of SBR and photocatalysis in a single reactor. Therefore, biodegradation and photodegradation take place simultaneously. The photocatalytic reactions convert persistent pollutants into readily degradable intermediates through the partial oxidation process. The lab-scale studies suggested that the simultaneous process has advantages of stable biodegradation performance and enhanced mineralization rates. With time in the reactor, the process efficiency gradually increased as the microbial population has adapted to the contaminant toxicity (Yusoff et al., 2018).

There are several other hybrid systems in place such as the expended bed reactor, fluidized bed reactor, porous media system, immersed media system, micro-aeration system, acidogenic co-fermentation system, iron–flocculation SBR, sludge tank halved SBR, fix-growth SBR, double sludge switching SBR, double-layer-pack biofilm SBR, internal-circulate SBR, smart SBR, alternative pumping SBR, and membrane-aerated biofilm SBR.

3.5 CONCLUSION

The rise in global demand for water has led to an increase in the recycling of wastewater and the implementation of stringent discharge standards. Conventional treatment processes are unable to meet the required criteria for effluent, whereas SBR-based treatment plants achieve the required effluent with or without retrofitting in the existing treatment plants. Due to automation, operational flexibility and online systems, SBR appears to be a promising technology for retrofitting existing activated sludge process-based treatment plants. SBR can be controlled by simply altering the control parameters of the phases of SBR cycle. In SBR, the operating conditions of each phase, along with the time of each phase, can be altered as per the objectives of treatment. The only drawback for the application of SBR at a large scale is due to sophistication in operation. Additionally, there exists a problem with aeration devices at specific operating conditions. Currently, various approaches have been tried for optimization of the existing system to improve and evaluate the reactor design and operational approach. SBR has been found reliable for the transformation of nitrogen in wastewater treatments. Recently, assessments of various combinations of wastewater treatment technologies at the lab scale have also been done that further extend the scope of SBR technology. The biofilm-based SBR is a prospective modification that has found applications for treating industrial wastewater contaminated with phenol and other pharmaceutical and in dye factories wastewater treatment.

REFERENCES

Al-Amrani, W.A., Lim, P.E., Seng, C.E., Wan Ngah, W.S., 2014. Factors affecting bio-decolorization of azo dyes and COD removal in anoxic-aerobic REACT operated sequencing batch reactor. *J. Taiwan Inst. Chem. Eng.* 45, 609–616. doi: 10.1016/j.jtice.2013.06.032.

Arun, S., Manikandan, N.A., Pakshirajan, K., Pugazhenthi, G., 2019. Novel shortcut biological nitrogen removal method using an algae-bacterial consortium in a photo-sequencing batch reactor: Process optimization and kinetic modelling. *J. Environ. Manage.* 250, 109401. doi: 10.1016/j.jenvman.2019.109401.

Blackburne, R., Yuan, Z., Keller, J., 2008. Demonstration of nitrogen removal via nitrite in a sequencing batch reactor treating domestic wastewater. *Water Res.* 42, 2166–2176. doi: 10.1016/j.watres.2007.11.029.

Blanco, J., Torrades, F., Morón, M., Brouta-Agnésa, M., García-Montaño, J., 2014. Photo-Fenton and sequencing batch reactor coupled to photo-Fenton processes for textile wastewater reclamation: Feasibility of reuse in dyeing processes. *Chem. Eng. J.* 240, 469–475. doi: 10.1016/j.cej.2013.10.101.

Bonassa, G., Bolsan, A.C., Hollas, C.E., Venturin, B., Candido, D., Chini, A., De Prá, M.C., Antes, F.G., Campos, J.L., Kunz, A., 2021. Organic carbon bioavailability: Is it a good driver to choose the best biological nitrogen removal process? *Sci. Total Environ.* doi: 10.1016/j.scitotenv.2021.147390.

Choi, D., Cho, K., Jung, J., 2019. Optimization of nitrogen removal performance in a single-stage SBR based on partial nitritation and ANAMMOX. *Water Res.* 162, 105–114. doi: 10.1016/j.watres.2019.06.044.

Daverey, A., Su, S.H., Huang, Y.T., Lin, J.G., 2012. Nitrogen removal from opto-electronic wastewater using the simultaneous partial nitrification, anaerobic ammonium oxidation and denitrification (SNAD) process in sequencing batch reactor. *Bioresour. Technol.* 113, 225–231. doi: 10.1016/j.biortech.2011.12.004.

De Schryver, P., Verstraete, W., 2009. Nitrogen removal from aquaculture pond water by heterotrophic nitrogen assimilation in lab-scale sequencing batch reactors. *Bioresour. Technol.* 100, 1162–1167. doi: 10.1016/j.biortech.2008.08.043.

Du, R., Cao, S., Niu, M., Li, B., Wang, S., Peng, Y., 2017. Performance of partial-denitrification process providing nitrite for anammox in sequencing batch reactor (SBR) and upflow sludge blanket (USB) reactor. *Int. Biodeterior. Biodegrad.* 122, 38–46. doi: 10.1016/j.ibiod.2017.04.018.

Duque, A.F., Bessa, V.S., Carvalho, M.F., de Kreuk, M.K., van Loosdrecht, M.C.M., Castro, P.M.L., 2011. 2-Fluorophenol degradation by aerobic granular sludge in a sequencing batch reactor. *Water Res.* 45, 6745–6752. doi: 10.1016/j.watres.2011.10.033.

Dutta, A., Sarkar, S., 2015. Sequencing batch reactor for wastewater treatment: Recent advances. *Curr. Pollut. Rep.* 1, 177–190. doi: 10.1007/s40726-015-0016-y.

Elkaramany, H.M., Elbaz, A.A., Mohamed, A.N., Sakr, A.H., 2018. Improving the biological treatment of waste water using pressurised sequencing batch reactor. *J. Environ. Eng. Sci.* 13, 37–43. doi: 10.1680/jenes.17.00029.

Frank, V.B., Regnery, J., Chan, K.E., Ramey, D.F., Spear, J.R., Cath, T.Y., 2017. Co-treatment of residential and oil and gas production wastewater with a hybrid sequencing batch reactor-membrane bioreactor process. *J. Water Process Eng.* 17, 82–94. doi: 10.1016/j.jwpe.2017.03.003.

Frison, N., Katsou, E., Malamis, S., Fatone, F., 2016. A novel scheme for denitrifying biological phosphorus removal via nitrite from nutrient-rich anaerobic effluents in a short-cut sequencing batch reactor. *J. Chem. Technol. Biotechnol.* 91, 190–197. doi: 10.1002/jctb.4561.

Ge, H., Batstone, D.J., Keller, J., 2015. Biological phosphorus removal from abattoir wastewater at very short sludge ages mediated bynovel PAO clade Comamonadaceae. *Water Res.* 69, 173–182. doi: 10.1016/j.watres.2014.11.026.

He, Q., Chen, L., Zhang, S., Chen, R., Wang, H., Zhang, W., Song, J., 2018. Natural sunlight induced rapid formation of water-born algal-bacterial granules in an aerobic bacterial granular photo-sequencing batch reactor. *J. Hazard. Mater.* 359, 222–230. doi: 10.1016/j.jhazmat.2018.07.051.

He, Q., Wang, H., Chen, L., Gao, S., Zhang, W., Song, J., Yu, J., 2020. Robustness of an aerobic granular sludge sequencing batch reactor for low strength and salinity wastewater treatment at ambient to winter temperatures. *J. Hazard. Mater.* 384, 121454. doi: 10.1016/j.jhazmat.2019.121454.

Hosseini Koupaie, E., Alavi Moghaddam, M.R., Hashemi, H., 2011. Comparison of overall performance between moving-bed and conventional sequencing batch reactor. *Iran. J. Environ. Heal. Sci. Eng.* 8, 235–244.

Hu, X., Sobotka, D., Czerwionka, K., Zhou, Q., Xie, L., Makinia, J., 2018. Effects of different external carbon sources and electron acceptors on interactions between denitrification and phosphorus removal in biological nutrient removal processes. *J. Zhejiang Univ. Sci. B* 19, 305–316. doi: 10.1631/jzus.B1700064.

Jagaba, A.H., Kutty, S.R.M., Lawal, I.M., Abubakar, S., Hassan, I., Zubairu, I., Umaru, I., Abdurrasheed, A.S., Adam, A.A., Ghaleb, A.A.S., Almahbashi, N.M.Y., Al-dhawi, B.N.S., Noor, A., 2021. Sequencing batch reactor technology for landfill leachate treatment: A state-of-the-art review. *J. Environ. Manage.* 282, 111946. doi: 10.1016/j.jenvman.2021.111946.

Jaramillo, F., Orchard, M., Muñoz, C., Zamorano, M., Antileo, C., 2018. Advanced strategies to improve nitrification process in sequencing batch reactors: A review. *J. Environ. Manage.* 218, 154–164. doi: 10.1016/j.jenvman.2018.04.019.

Jin, R., Liu, G., Li, C., Xu, R., Li, H., Zhang, L., Zhou, J., 2013. Effects of carbon-nitrogen ratio on nitrogen removal in a sequencing batch reactor enhanced with low-intensity ultrasound. *Bioresour. Technol.* 148, 128–134. doi: 10.1016/j.biortech.2013.08.141.

Jin, Y., Ding, D., Feng, C., Tong, S., Suemura, T., Zhang, F., 2012. Performance of sequencing batch biofilm reactors with different control systems in treating synthetic municipal wastewater. *Bioresour. Technol.* 104, 12–18. doi: 10.1016/j.biortech.2011.08.086.

Lei, C.N., Whang, L.M., Chen, P.C., 2010. Biological treatment of thin-film transistor liquid crystal display (TFT-LCD) wastewater using aerobic and anoxic/oxic sequencing batch reactors. *Chemosphere* 81, 57–64. doi: 10.1016/j.chemosphere.2010.07.001.

Li, Q., Wang, S., Zhang, P., Yu, J., Qiu, C., Zheng, J., 2018. Influence of temperature on an Anammox sequencing batch reactor (SBR) system under lower nitrogen load. *Bioresour. Technol.* 269, 50–56. doi: 10.1016/j.biortech.2018.08.057.

Luo, G., Avnimelech, Y., Pan, Y. feng, Tan, H., 2013. Inorganic nitrogen dynamics in sequencing batch reactors using biofloc technology to treat aquaculture sludge. *Aquac. Eng.* 52, 73–79. doi: 10.1016/j.aquaeng.2012.09.003.

Lv, X.M., Shao, M.F., Li, C.L., Li, J., Gao, X.L., Sun, F.Y., 2014. A comparative study of the bacterial community in denitrifying and traditional enhanced biological phosphorus removal processes. *Microbes Environ.* 29, 261–268. doi: 10.1264/jsme2.ME13132.

Ma, J., Yu, L., Frear, C., Zhao, Q., Li, X., Chen, S., 2013. Kinetics of psychrophilic anaerobic sequencing batch reactor treating flushed dairy manure. *Bioresour. Technol.* 131, 6–12. doi: 10.1016/j.biortech.2012.11.147.

Malakootian, M., Shahamat, Y.D., Mahdizadeh, H., 2020. Purification of diazinon pesticide by sequencing batch moving-bed biofilm reactor after ozonation/Mg-Al layered double hydroxides pre-treated effluent. *Sep. Purif. Technol.* 242, 116754. doi: 10.1016/j.seppur.2020.116754.

Meng, F., Xi, L., Liu, D., Huang, Weiwei, Lei, Z., Zhang, Z., Huang, Wenli, 2019. Effects of light intensity on oxygen distribution, lipid production and biological community of algal-bacterial granules in photo-sequencing batch reactors. *Bioresour. Technol.* 272, 473–481. doi: 10.1016/j.biortech.2018.10.059.

Meng, Q., Yang, F., Liu, L., Meng, F., 2008. Effects of COD/N ratio and DO concentration on simultaneous nitrifcation and denitrifcation in an airlift internal circulation membrane bioreactor. *J. Environ. Sci.* 20, 933–939. doi: 10.1016/S1001-0742(08)62189-0.

Ni, M., Pan, Y., Chen, Y., Zhang, X., Huang, Y., Song, Z., 2021. Effects of seasonal temperature variations on phosphorus removal, recovery, and key metabolic pathways in the suspended biofilm. *Biochem. Eng. J.* 176, 108187. doi: 10.1016/j.bej.2021.108187.

Nielsen, P.H., McIlroy, S.J., Albertsen, M., Nierychlo, M., 2019. Re-evaluating the microbiology of the enhanced biological phosphorus removal process. *Curr. Opin. Biotechnol.* 57, 111–118. doi: 10.1016/j.copbio.2019.03.008.

Ozturk, A., Aygun, A., Nas, B., 2019. Application of sequencing batch biofilm reactor (SBBR) in dairy wastewater treatment. *Korean J. Chem. Eng.* 36, 248–254. doi: 10.1007/s11814-018-0198-2.

Qi, R., Qin, D., Yu, T., Chen, M., Wei, Y., 2020. Start-up control for nitrogen removal via nitrite under low temperature conditions for swine wastewater treatment in sequencing batch reactors. *N. Biotechnol.* 59, 80–87. doi: 10.1016/j.nbt.2020.05.005.

Rajab, A.R., Salim, M.R., Sohaili, J., Anuar, A.N., 2022. Feasibility of nutrients removal and its pathways using integrated anaerobic-aerobic sequencing batch reactor. *Bioresour. Technol. Rep.* 17, 100912. doi: 10.1016/j.biteb.2021.100912.

Shao, Y., Yang, S., Mohammed, A., Liu, Y., 2018. Impacts of ammonium loading on nitritation stability and microbial community dynamics in the integrated fixed-film activated sludge sequencing batch reactor (IFAS-SBR). *Int. Biodeterior. Biodegrad.* 133, 63–69. doi: 10.1016/j.ibiod.2018.06.002.

Shariati, S.R.P., Bonakdarpour, B., Zare, N., Ashtiani, F.Z., 2011. The effect of hydraulic retention time on the performance and fouling characteristics of membrane sequencing batch reactors used for the treatment of synthetic petroleum refinery wastewater. *Bioresour. Technol.* 102, 7692–7699. doi: 10.1016/j.biortech.2011.05.065.

Siegrist, H., Joss, A., Ternes, T., Oehlmann, J., 2012. Fate of edcs in wastewater treatment and Eu perspective on Edc regulation. *Proc. Water Environ. Fed.* 2005, 3142–3165. doi: 10.2175/193864705783865640.

Soejima, K., Matsumoto, S., Ohgushi, S., Naraki, K., Terada, A., Tsuneda, S., Hirata, A., 2008. Modeling and experimental study on the anaerobic/aerobic/anoxic process for simultaneous nitrogen and phosphorus removal: The effect of acetate addition. *Process Biochem.* 43, 605–614. doi: 10.1016/j.procbio.2008.01.022.

Tan, K.C., Seng, C.E., Lim, P.E., Oo, C.W., Lim, J.W., Kew, S.L., 2016. Alteration of moving bed sequencing batch reactor operational strategies for the enhancement of nitrogen removal from stabilized landfill leachate. *Desalin. Water Treat.* 57, 15979–15988. doi: 10.1080/19443994.2015.1075427.

Tang, C.C., Tian, Y., He, Z.W., Zuo, W., Zhang, J., 2018. Performance and mechanism of a novel algal-bacterial symbiosis system based on sequencing batch suspended biofilm reactor treating domestic wastewater. *Bioresour. Technol.* 265, 422–431. doi: 10.1016/j.biortech.2018.06.033.

Wang, D., Xu, Q., Yang, W., Chen, H., Li, X., Liao, D., Yang, G., Yang, Q., Zeng, G., 2014. A new configuration of sequencing batch reactor operated as a modified aerobic/extended-idle regime for simultaneously saving reactor volume and enhancing biological phosphorus removal. *Biochem. Eng. J.* 87, 15–24. doi: 10.1016/j.bej.2014.03.009.

Wang, M., Yang, H., Ergas, S.J., van der Steen, P., 2015. A novel shortcut nitrogen removal process using an algal-bacterial consortium in a photo-sequencing batch reactor (PSBR). *Water Res.* 87, 38–48. doi: 10.1016/j.watres.2015.09.016.

Xie, G.J., Liu, B.F., Guo, W.Q., Ding, J., Xing, D.F., Nan, J., Ren, H.Y., Ren, N.Q., 2012. Feasibility studies on continuous hydrogen production using photo-fermentative sequencing batch reactor. *Int. J. Hydrogen Energy* 37, 13689–13695. doi: 10.1016/j.ijhydene.2012.02.107.

Xu, G., Zhou, Y., Yang, Q., Lee, Z.M.P., Gu, J., Lay, W., Cao, Y., Liu, Y., 2015. The challenges of mainstream deammonification process for municipal used water treatment. *Appl. Microbiol. Biotechnol.* 99, 2485–2490. doi: 10.1007/s00253-015-6423-6.

Yusoff, N.A., Ong, S.A., Ho, L.N., Rashid, N.A., Wong, Y.S., Saad, F.N.M., Khalik, W.F., Lee, S.L., 2018. Development of simultaneous photo-biodegradation in the photocatalytic hybrid sequencing batch reactor (PHSBR) for mineralization of phenol. *Biochem. Eng. J.* 138, 131–140. doi: 10.1016/j.bej.2018.07.015.

Zhan, X.M., Rodgers, M., O'Reilly, E., 2006. Biofilm growth and characteristics in an alternating pumped sequencing batch biofilm reactor (APSBBR). *Water Res.* 40, 817–825. doi: 10.1016/j.watres.2005.12.003.

Zhang, Y., Jiang, W.L., Qin, Y., Wang, G.X., Xu, R.X., Xie, B., 2017. Dynamic changes of bacterial community in activated sludge with pressurized aeration in a sequencing batch reactor. *Water Sci. Technol.* 75, 2639–2648. doi: 10.2166/wst.2017.147.

4 Zero Liquid Discharge in Industries

Aiman Malik, Mohd Salim Mahtab,
and Izharul Haq Farooqi
Aligarh Muslim University

CONTENTS

4.1   Introduction 75
4.2   Existing ZLD Systems 77
    4.2.1   Thermal ZLD 78
    4.2.2   Thermal ZLD Incorporated with an RO System 79
    4.2.3   Different ZLD Systems in Combination with Membrane-Based
            Techniques 79
4.3   Importance of ZLD Techniques 80
4.4   Challenges and Environmental Aspects of ZLD Technology 82
4.5   Conclusion 83
Acknowledgment 83
References 83

4.1   INTRODUCTION

Due to rapid urbanization and industrialization, pollution has increased many folds and water resources are becoming threatened. Among the environmental issues, this is one of the biggest challenges for environmentalists and water researchers (Grant et al., 2012; Vörösmarty et al., 2010). Water reclamation and reuse options are some of the growing solutions which could be helpful to meet the high water demand for industries, irrigation, agricultural fields, and other recreational purposes. Freshwater is mostly used in industries and produces lots of wastewater that need to be efficiently treated for further reuse and discharge (Yaqub and Lee, 2019). Usually, satisfying stringent discharge standards is challenging and requires advanced treatment technologies. Otherwise, improper treatment and disposal of industrial effluents would heavily contaminate ground and surface water bodies, which affects aquatic lives as well as humans (Bhargava and Bhargava, 2020; Khan et al., 2015). Among the domain of wastewater treatment, various methods have been adopted viz. conventional biological treatments, chemical oxidation (Cheremisinoff, 2019), coagulation-flocculation, carbon adsorption, advanced oxidation processes (Mo et al., 2018; Lu et al., 2019), and membrane processes (Wang et al., 2008). Nowadays, advanced and sustainable treatment strategies for industrial effluents are in demand to reduce the

DOI: 10.1201/9781003202431-4

consumption of freshwater and, henceforth, minimize contamination of receiving water bodies (Xiong and Wei, 2017; Abdelhamid, 2015).

In this scenario, zero liquid discharge (ZLD) technology is one of the appealing and versatile solutions for treating wastewater from industries by maximizing water recycling and minimizing the volume of wastewater produced (Mays, 2007). It is a closed water cycle where no water is discharged from a system if there is the possibility of it being reused after appropriate treatment. This improves efficiency and reduces risks associated with contamination (Byers, 1995). ZLD is a strategic wastewater management system that ensures that there will be no discharge of industrial wastewater into the environment. It is achieved by treating wastewater through recycling and then recovery and reuse for industrial purposes (www.saltworkstech.com). In this method, maximum recovery was obtained from the treatment of wastewater and leaving behind only solid waste as a residue for disposal. There were many treatments that were designed to follow this condition, but only ZLD is able to achieve a reduction in waste. However, the cost and challenges of recovery increase in this technique because wastewater is highly concentrated so it needs more treatments (Al-Obaidani et al., 2008). Since the conventional way of ZLD consumes a large amount of energy as several thermal processes are used, therefore, new development of membrane-based processes like reverse osmosis (RO) is being done to lower the energy requirements (Elimelech and Phillip, 2011). However, incorporating RO technology improves energy efficiency only in relation to thermal-driven ZLD techniques, but also to a certain salinity range limit. More recently, salt-concentrating techniques have emerged as alternative ZLD technologies, including membrane distillation (MD), forward osmosis (FO), and electrodialysis (ED) (Tong and Elimelech, 2016).

The reasons for its use across the globe are due to the increasing value of freshwater, severe water shortages, contamination of aquatic environments, new environmental regulation policies, and the increase in the cost of disposing of wastewater. So, the multi-purpose nature of this technology is an efficient solution for industrial effluent treatment and management. Figure 4.1 shows the basic process flow diagram in

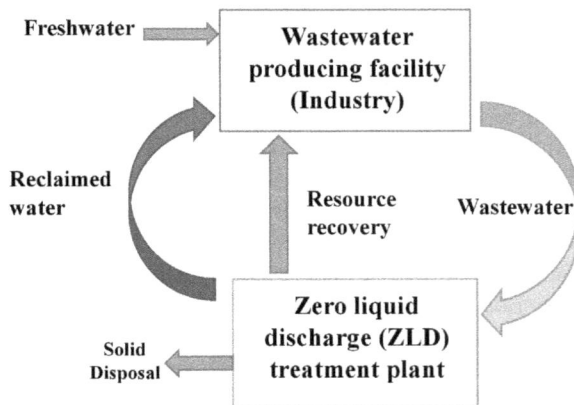

FIGURE 4.1 The basic process flow diagram in the treatment of wastewater using ZLD technologies.

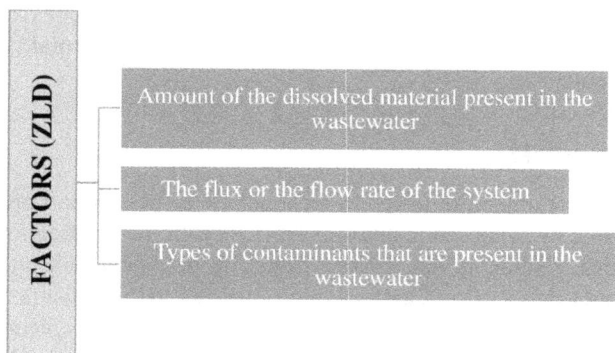

FIGURE 4.2 Important factors influencing the components of the ZLD treatment plant (Bhargava and Bhargava, 2020).

TABLE 4.1

Different Components of ZLD Treatment System (Bhargava and Bhargava, 2020)

S.No.	Components/ Treatment Techniques	Roles/Characteristics/ Parameters Removal	Remarks
1	Clarifier or reactor	Metals and hardness of water	Helps in primary treatment
2	Chemical feed	Suspended solids and metals	Helps in the process of precipitation, flocculation, and coagulation
3	Filter press	Secondary solid waste	Works in combination with an evaporator
4	Ultrafiltration (UF)	Suspended solids	Prevent scaling, fouling, and corrosion.
5	Reverse osmosis (RO)	Dissolved solids	Use semipermeable membrane
6	Brine concentrator	Waste volume reduction	Concentrate on the rejected RO stream
7	Evaporator	Vapors out excess water	It works before the use of a crystallizer
8	Crystallizer	Remaining liquid	Produces dry and solid cake as a residue

the treatment of wastewater using ZLD technologies. Different components of the ZLD treatment plant largely depend on the following factors, as shown in Figure 4.2 (Bhargava and Bhargava, 2020). In addition, Table 4.1 shows different components and their roles in ZLD treatment systems.

So, the general objective of the present chapter is to provide an overview of ZLD treatment techniques. This includes ZLD types, their importance, and challenges in the domain of wastewater treatment and in particular industrial effluent. Lastly, the concluding remarks and some future research directions are also emphasized with the help of available literature.

4.2 EXISTING ZLD SYSTEMS

Globally, conventional thermal ZLD process schemes are employed that are expensive as well as not much efficient, hence urging researchers to find new alternatives with better output (Yaqub and Lee, 2019). Three different types of existing technologies are listed here which focus on the systems combining thermal, RO, and other emerging technologies.

4.2.1 Thermal ZLD

It consists of pre-treatment, followed by treatment in a concentrator and crystallizer, which is followed by evaporation and solid recovery. In the pre-treatment, wastewater goes through treatments like normal filtration, pH adjustment in the optimum range, de-aeration process, and anti-scaling if needed. After these processes, water is further concentrated using a brine concentrator followed by a brine crystallizer, as shown in Figure 4.3. This results in the production of distillates that are further recycled and reused as clean water, whereas solids are often recovered as valuable by-products or in some cases sent to evaporation ponds for disposal after processing (Tong and Elimelech, 2016). Brine concentrators function by mechanical vapor compression (MVC), and in this, water from the feed is mixed with brine slurry, which travels within the concentrator from the heat exchanger, as seen in Tong and Elimelech (2016). After this, the slurry is passed into the heat exchanger and evaporation occurs. This latent heat is produced by superheated water vapor, which vaporizes the brine slurry. It was observed that evaporation ponds consume solar energy, which is inexpensive as compared to brine crystallizers, exclusively where less water is being treated and land is low in cost.

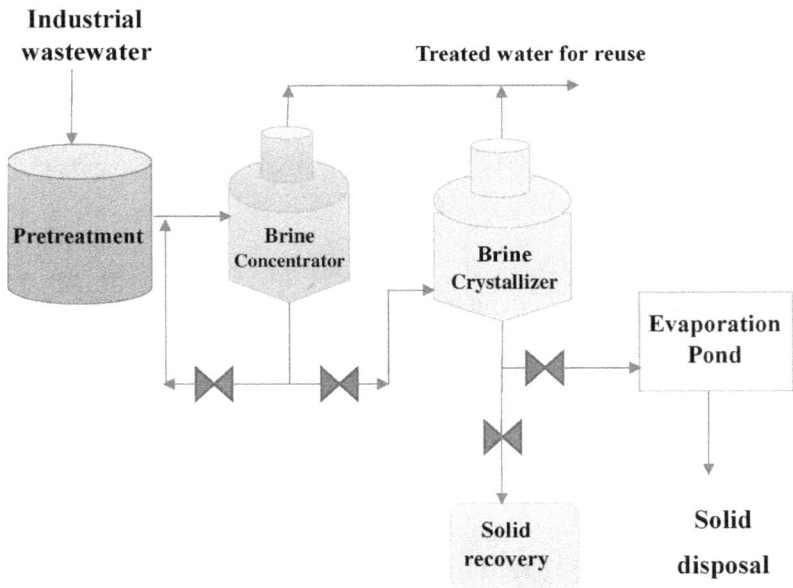

FIGURE 4.3   Process flow diagram of thermal ZLD (Yaqub and Lee, 2019).

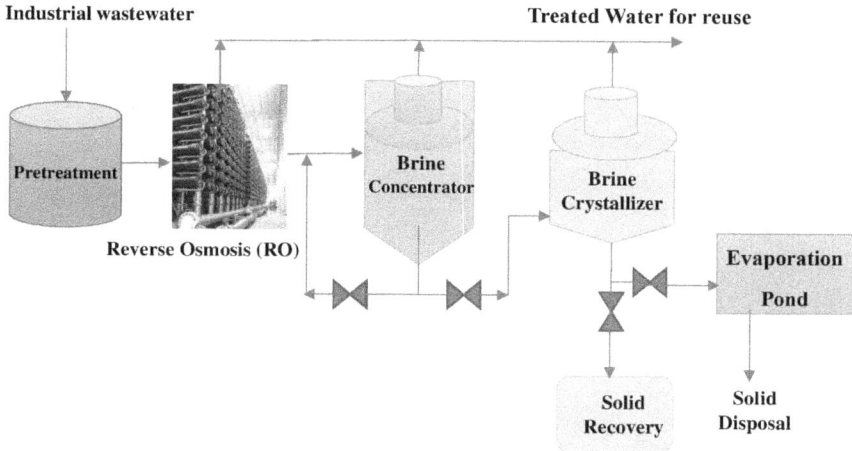

FIGURE 4.4 Process flow of ZLD system with RO (Yaqub and Lee, 2019).

4.2.2 THERMAL ZLD INCORPORATED WITH AN RO SYSTEM

In this system, RO, a well-established, pressure-driven desalination technology, is combined with thermal ZLD systems to reduce the volume of brine slurry present in the crystallizer or brine concentrator; hence, energy consumption is reduced, as shown in Figure 4.4. The energy consumption by RO is around 2 kWh_e/m^3 of product water which is an ~50% recovery in desalination of seawater (Yaqub and Lee, 2019). As compared to brine concentrators and crystallizers, this is quite a low value. From previous studies, it was seen that the cost of operation is reduced for the RO by around 48%–67% as compared to the brine concentrator-evaporation pond setup (Bond and Veerapaneni, 2007; Bond and Veerapaneni, 2008). Also, the operational cost and input of brine concentrators are high even after adding a secondary RO plant (Bond and Veerapaneni, 2008). Here, wastewater is concentrated by RO before entering thermal processes, minimizing capital and operational costs. The application of RO has certain problems like salinity limit and fouling problems that lead to decreased water flux and the lifespan of membranes in ZLD systems. Therefore, various techniques can be employed during pre-treatment including chemical softening, pH adjustment, and ion exchange. During pre-treatment, intensive use of chemicals results in additional solid waste production and ultimately increases operational costs. Alternatively, ultrafiltration (UF) may be an efficient pre-treatment technology.

4.2.3 DIFFERENT ZLD SYSTEMS IN COMBINATION
WITH MEMBRANE-BASED TECHNIQUES

Earlier it was seen that ZLD systems worked on thermal processes; however, RO was combined with ZLD at a later stage, but its problem was the narrow range of salinity on which it works. So, new techniques were researched, which could give better results like membrane-based techniques. Some of these are FO, MD, and ED, which

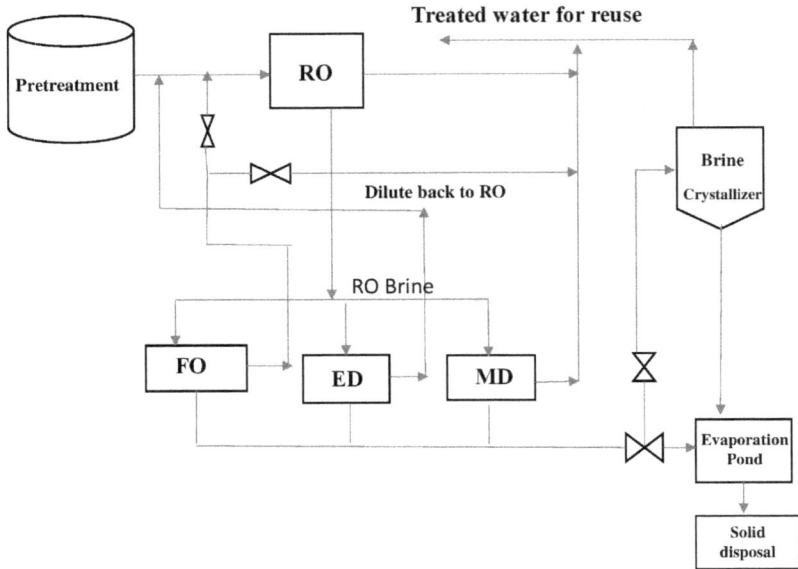

FIGURE 4.5 Process flow showing different ZLD membrane-based systems like FO, RO, MD, and ED (Yaqub and Lee, 2019).

are now used in new ZLD systems. They have shown better results as compared to thermal and RO-based techniques. Figure 4.5 shows the different approaches through which the RO brine concentrates can be treated further. After that, they are fed into the evaporation pond or brine crystallizer. A detailed comparison of these technologies is listed in Table 4.2. MD and FO work on the principle of both thermal and membrane processes. Hence, in this case, membranes are the important elements and thermal energy is supplied from the source to the system (Yaqub and Lee, 2019).

4.3 IMPORTANCE OF ZLD TECHNIQUES

ZLD is a technique where no liquid discharge occurs and contaminations are reduced to solid waste. This technique basically uses a water cycle in a closed manner so that no water is discharged from a system; instead, it is reused after appropriate treatment. This is helpful in improving water usage efficiency and also eliminates the risk of water contamination via wastewater discharge. Industrial process and facility ZLD have shown many benefits, some of which are listed below (Al-Obaidani et al., 2008):

- Waste volume is reduced in this technique which reduces the expenses related to wastewater management
- Water acquisition costs and risks are reduced due to the recycling of water on site
- Greenhouse gas impact is reduced as the number of vehicles associated with wastewater disposal off-site is also reduced

TABLE 4.2

Technologies Used Incorporated with ZLD System (Yaqub and Lee, 2019)

Technologies	Process Description	Benefits	Limitations	References
Mechanical vapor compression (MVC)	Feedwater mixed with brine slurry travels in the brine concentrator through the tubes of the heat exchanger	High salinity limit >200,000 mg/L	High capital and operational costs	Charisiadis (2018) and Ghaffour et al. (2013)
Reverse osmosis (RO)	RO is a process of desalinating seawater using a semipermeable membrane, which is permeable only to water and impermeable to the salt present in the seawater	Energy-efficient	Limited salinity limit and scaling problem	Ibrahim et al. (2020), Elimelech and Phillip (2011), and Charisiadis (2018)
Membrane distillation (MD)	In MD, separation is thermally driven through a hydrophobic, highly porous membrane that only permits the vapor produced due to the temperature difference on both sides of the membrane	Higher salinity limit >200,000 mg/L	Limited area of application; low flux and recovery	Wang et al. (2008) and Charisiadis (2018)
Forward osmosis (FO)	Water molecules pass through a semipermeable membrane due to difference in osmotic pressure	High salinity limit >200,000 mg/L; requires less fouling and low-grade heat	Low flux at high salinity, reverse solute flux, limited use	Suwaileh et al. (2020) and McGinnis et al. (2013)
Electrodialysis (ED)	Dissolve ions are eliminated using ion-exchange membranes, where the electric potential is the driving force. These membranes allow transport of selective counter-ions but prevent the carrying of co-ions	Salinity limit ~100,000 mg/L; also, low fouling	High energy consumption	Loganathan et al. (2015)

- Showing sustainable use by improving environmental performance and risks associated with future permitting
- Many of the processes recover valuable resources, like salt in the form of sodium chloride and ammonium sulfate fertilizer
- ZLD also has the potential to recover resources that are present in wastewater. The solids produced from waste can be reused or sold as a part of their industrial process and facilities
- Some examples shown are the following: gypsum can be recovered from mine water and flue gas desalinization (FGD) wastewater, which is then sold to use in drywall manufacturing, and in another example, in USA oil field brines, lithium is found at almost the same level as South American solars.

Regardless of various new organizations that are quite motivated to target ZLD, however, its achievement demonstrates sustainable development, good economics, environmental stewardship, and corporate responsibility. Due to certain advantages like reduced disposal cost, more amount of water is reused and fewer greenhouse gases, the operation of an in-house ZLD plant is used commercially. By operating this, the overall impact on local ecosystems and climate is minimized.

4.4 CHALLENGES AND ENVIRONMENTAL ASPECTS OF ZLD TECHNOLOGY

All techniques have pros and cons, and the same is the case with ZLD. It eliminates water contaminants and also reduces pollution, but some challenges still need to be addressed.

1. The recalcitrant organics are often present in the wastewater, which are highly toxic as well as chemically stable and hence not easily decomposed even through ZLD techniques. This is a major challenge that needs to be addressed. Therefore, they lead to problems like membrane fouling and compromise the system's stability (Yaqub and Lee, 2019; Xiong and Wei, 2017).
2. The second challenging task is the high operational cost which needs to be reduced for large-scale implementation (Yaqub and Lee, 2019; Xiong and Wei, 2017). This is due to the pre-treatment systems that are required in ZLD for a high water recovery rate. Ion exchange and chemical and biological reactions are used in pre-treatments to completely remove the hardness present in water. The design becomes complicated due to this high cost which often leads to operation. Also, the quantity of sludge generated is increased, followed by an increment in the salinity of wastewater. Therefore, nowadays, researchers are trying to make hybrid membranes at room temperature for high efficiency (Xiong and Wei, 2017).
3. Thermal systems are required for high water recovery in the pre-concentration step, which further adds to the cost and energy consumption. This challenge is not easy to remove as they are needed. The thermal system with a big capacity like 100–300 m^3/hour is used mostly. Henceforth, such a large thermal system would require heavy investment and will also consume more energy (Xiong and Wei, 2017).
4. From an environmental perspective too, this technique is quite challenging as it releases CO_2 into the atmosphere, which leads to air pollution (Grant et al., 2012). This is due to pre-treatments such as acidification and degasification which are used for the conventional removal of waste. Further, CO_2 is also released during the concentration of RO brine in ED techniques. To control scaling, de-carbonation is also used, which releases this gas (Zhang et al., 2012). Also, greenhouse emission increases due to the large consumption of energy. The use of RO with ZLD uses renewable energy, which can help in the reduction of greenhouse emissions and hence improve the air quality index (AQI) (Xiong and Wei, 2017).

4.5 CONCLUSION

All around the world, ZLD systems are implemented at a rapid rate for industrial waste water treatment and management that shows a reduction in water contamination. It also helps in improving the efficiency of wastewater reclamation and reuse. However, it possesses certain problems when used on a large-scale like high capital cost and energy consumption. When compared to conventional wastewater treatment, this technique's energy requirements are quite high along with other expenses. Its solid waste disposal is also a big issue that hinders its use. All the effluents produced need to be treated in such a manner that they meet the standard discharge limits. Treated water is used afterward in different processes, and the waste produced is classified into organic and inorganic waste which is further treated. Also, overexploitation of water resources and freshwater scarcity due to climatic changes lead to the implementation of the ZLD system in the future. For example, continuous drought conditions are being experienced in the Southwestern United States (Cook et al., 2015), and water deficiency is a problem in areas with high-water-consuming industries (e.g., coal-fired power plants) in China (Zhang et al., 2016). Basically, for sustainable water supply, techniques like RO have great potential for treating high salinity ranges in wastewater. In addition, hybrid membranes can be designed, which are advantageous in terms of higher resistance, due to which fouling/scaling problems will be reduced. Thus, ZLD becomes more efficient and feasible because energy consumption is reduced. Furthermore, using membrane-based technologies also helps in the reduction of costs by reducing pre-treatment costs and also by improving output water quality which can be reused. Currently, for the concentration of RO brine, new technologies can be used like MD, FO, and ED/EDR. But these techniques still need more research for large-scale applications.

ACKNOWLEDGMENT

The authors would like to express their gratitude to the Department of Civil Engineering (Aligarh Muslim University) for providing technical assistance with this chapter.

REFERENCES

Abdelhamid, A. (2015). India uses zero liquid discharge (ZLD) to clean the ganges river.
Al-Obaidani, S., Curcio, E., Macedonio, F., Di Profio, G., Al-Hinai, H., & Drioli, E. (2008). Potential of membrane distillation in seawater desalination: Thermal efficiency, sensitivity study and cost estimation. *Journal of Membrane Science*, 323(1), 85–98.
Bhargava, R., & Bhargava, P. (2020). A zero liquid discharge in chemical industry: A review. *The SAGE Journal of Innovative Research in Computing*, 1(1), 1–4.
Bond, R., & Veerapaneni, S. (2007). *Zero Liquid Discharge for Inland Desalination*. AWWA Research Foundation: Denver, CO.
Bond, R., & Veerapaneni, S. (2008). Zeroing in on ZLD technologies for inland desalination. *Journal-American Water Works Association*, 100(9), 76–89.
Byers, B. (1995). Zero discharge: A systematic approach to water reuse. *Chemical Engineering*, 102(7), 96.
Charisiadis, C. (2018). Brine Zero Liquid Discharge (ZLD) Fundamentals and Design. South Miami, FL.

Cheremisinoff, P. N. (2019). *Handbook of Water and Wastewater Treatment Technology.* Routledge: London.

Cook, B. I., Ault, T. R., & Smerdon, J. E. (2015). Unprecedented 21st-century drought risk in the American Southwest and Central Plains. *Science Advances,* 1, e1400082.

Elimelech, M., & Phillip, W. A. (2011). The future of seawater desalination: Energy, technology, and the environment. *Science,* 333(6043), 712–717.

Ghaffour, N., Missimer, T. M., & Amy, G. L. (2013). Technical review and evaluation of the economics of water desalination: Current and future challenges for better water supply sustainability. *Desalination,* 309, 197–207.

Grant, S. B., Saphores, J. D., Feldman, D. L., Hamilton, A. J., Fletcher, T. D., Cook, P. L., ... & Marusic, I. (2012). Taking the "waste" out of "wastewater" for human water security and ecosystem sustainability. *Science,* 337(6095), 681–686.

Ibrahim, G. S., Isloor, A. M., & Farnood, R. (2020). Fundamentals and basics of reverse osmosis. In: Basile, A., & Ghasemzadeh, K. (Eds.), Current Trends and Future Developments on (Bio-) Membranes (pp. 141–163). Elsevier: Amsterdam, Netherlands.

Khan, S. U., Noor, A., & Farooqi, I. H. (2015). GIS application for groundwater management and quality mapping in rural areas of District Agra, India. *International Journal of Water Resources & Arid Environments (IJWRAE),* 4(1), 89–96.

Loganathan, K., Chelme-Ayala, P., & El-Din, M. G. (2015). Treatment of basal water using a hybrid electrodialysis reversal–reverse osmosis system combined with a low-temperature crystallizer for near-zero liquid discharge. *Desalination,* 363, 92–98.

Lu, K. J., Cheng, Z. L., Chang, J., Luo, L., & Chung, T. S. (2019). Design of zero liquid discharge desalination (ZLDD) systems consisting of freeze desalination, membrane distillation, and crystallization powered by green energies. *Desalination,* 458, 66–75.

Mays, L. W. (2007). *Water Resources Sustainability.* McGraw-Hill Education: New York.

McGinnis, R. L., Hancock, N. T., Nowosielski-Slepowron, M. S., & McGurgan, G. D. (2013). Pilot demonstration of the NH3/CO2 forward osmosis desalination process on high salinity brines. *Desalination,* 312, 67–74.

Mo, J., Yang, Q., Zhang, N., Zhang, W., Zheng, Y., & Zhang, Z. (2018). A review on agro-industrial waste (AIW) derived adsorbents for water and wastewater treatment. *Journal of Environmental Management,* 227, 395–405.

Suwaileh, W., Pathak, N., Shon, H., & Hilal, N. (2020). Forward osmosis membranes and processes: A comprehensive review of research trends and future outlook. *Desalination,* 485, 114455.

Tong, T., & Elimelech, M. (2016). The global rise of zero liquid discharge for wastewater management: Drivers, technologies, and future directions. *Environmental Science & Technology,* 50(13), 6846–6855.

Vörösmarty, C. J., McIntyre, P. B., Gessner, M. O., Dudgeon, D., Prusevich, A., Green, P., ... & Davies, P. M. (2010). Global threats to human water security and river biodiversity. *Nature,* 467(7315), 555–561.

Wang, L. K., Chen, J. P., Hung, Y. T., & Shammas, N. K. (Eds.). (2008). *Membrane and Desalination Technologies* (Vol. 13). Springer Science + Business Media, LLC: Berlin/ Heidelberg, Germany.

Xiong, R., & Wei, C. (2017). Current status and technology trends of zero liquid discharge at coal chemical industry in China. *Journal of Water Process Engineering,* 19, 346–351.

Yaqub, M., & Lee, W. (2019). Zero-liquid discharge (ZLD) technology for resource recovery from wastewater: A review. *Science of the Total Environment,* 681, 551–563.

Zhang, C., Zhong, L., Fu, X., Wang, J., & Wu, J. (2016). Revealing water stress by the thermal power industry in China based on a high spatial resolution water withdrawal and consumption inventory. *Environmental Science & Technology,* 50, 1642–1652.

Zhang, Y., Ghyselbrecht, K., Vanherpe, R., Meesschaert, B., Pinoy, L., & Van der Bruggen, B. (2012). RO concentrate minimization by electrodialysis: Techno-economic analysis and environmental concerns. *Journal of Environmental Management,* 107, 28–36.

5 Advanced Oxidation Processes and Their Applications

Mohd Salim Mahtab and Izharul Haq Farooqi
Aligarh Muslim University

Anwar Khursheed
King Saud University

Mohd Azfar Shaida
Aligarh Muslim University

CONTENTS

5.1 Introduction 85
5.2 Types of Different Advanced Oxidation Processes....................................88
    5.2.1 Fenton-based AOPs...88
        5.2.1.1 Classical Fenton Process (CFP) 88
        5.2.1.2 Fenton-like Process..89
        5.2.1.3 Photo-Fenton Process 89
        5.2.1.4 Electro-Fenton Process 90
        5.2.1.5 Heterogeneous Fenton Catalysis 91
    5.2.2 Ozon-based AOPs..92
        5.2.2.1 Peroxone Process (O_3/H_2O_2) 93
        5.2.2.2 Ozonation at Elevated pH 94
5.3 Applications of Advanced Oxidation Processes...94
5.4 Conclusion 96
Acknowledgment 96
References...96

5.1 INTRODUCTION

As a result of rising industrialization, wastewater production has increased dramatically, necessitating effective treatment to safeguard ground and surface water pollution (Khan et al., 2015). On the other hand, the heightened demand for water reclamation and reuse requires innovative treatment techniques. Besides, the stringent discharge standards further enhance the requirement for advanced treatment

DOI: 10.1201/9781003202431-5

options. The ineffective treatment creates many problems for the environment and pollutes the water bodies, which is problematic for aquatic life and public health (Mahtab et al., 2021). In recent years, trace organic compounds (TOrCs) have been reported in the aquatic environment, including pharmaceuticals, consumer items, and industrial chemicals (Khan et al., 2021). Aside from urban and agricultural run-offs, wastewater treatment plant effluents are thought to be the largest source of TOrC emissions (Gros et al., 2010; Luo et al., 2014). Because conventional physical and biological wastewater treatment can only partially remove TOrCs, they remain in wastewater treatment plant effluents released into surface waters (Luo et al., 2014). As a result, environmentalists and researchers are concerned about their suitable treatment. Advanced oxidation processes (AOPs) are considered highly effective and viable choices for the degradation and removal of a wide range of contaminants and TOrCs in these situations (Comninellis et al., 2008; Klavarioti et al., 2009; Yang et al., 2014; Stefan, 2018; Hussain et al., 2020; Hussain et al., 2021).

AOPs form powerful oxidants in situ to oxidize organic compounds (Huang et al., 1993; Hussain et al., 2020). These include processes that use OHradicals ($^\bullet$OH), which account for most AOPs, and those that use other oxidizing species, such as sulfate or chlorine radicals. Figure 5.1 depicts various oxidizing agents. Hydroxyl radical-based treatments have multiple advantages, as widely documented in earlier research. The hydroxyl radicals ($^\bullet$OH) are the principal reactive species in AOPs, and Figure 5.2 depicts several favorable characteristics of the hydroxyl radicals ($^\bullet$OH).

A comprehensive analysis of AOPs regarding running costs, sustainability, and general viability make selecting the most appropriate treatment techniques among AOPs challenging. Several AOPs are well-established and in full-scale operation in drinking water treatment and water reuse facilities, notably those involving ozonation and UV irradiation. Various researchers are constantly presenting innovative investigations of a range of evolving AOPs for water treatment (for example, electrochemical AOPs, plasma, electron beam, ultrasound, or microwave-based AOPs)

Oxidation Potential (E^0) (V)

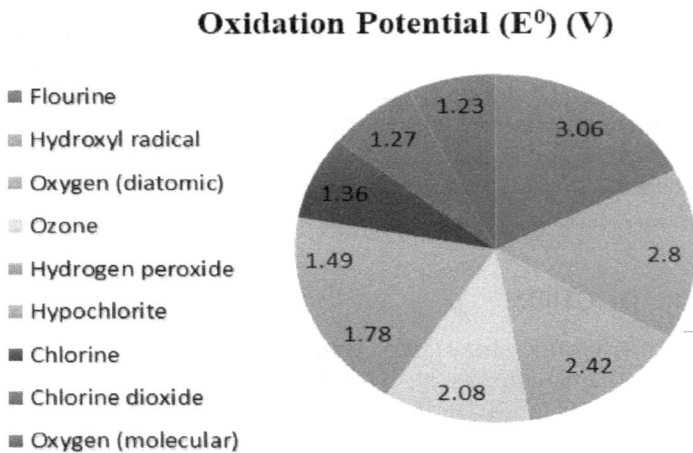

FIGURE 5.1   Various oxidizing agents' oxidation potential.

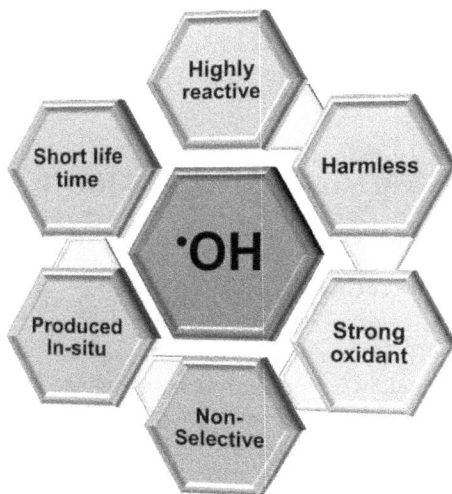

FIGURE 5.2   Characteristics of hydroxyl radicals.

TABLE 5.1

List of Various Advanced Oxidation Processes Reported in the Literature

Type of AOPs	Classification
Fenton-based	Fe^{2+}/H_2O_2, Fenton like
Ozone-based	O_3/H_2O_2, O_3/UV
Photochemical	UV/H_2O_2, UV/O_3, Photo-Fenton, Photocatalysis
Sonochemical	US/H_2O_2, US/O_3, Sono-Fenton
Electrochemical	Electro-Fenton
Sono-Photo Chemical	Sono-Photo-Fenton
Photo-Electro Chemical	Photo-Electro-Fenton
Sono-Electro Chemical	Sono-Electro-Fenton

(Stefan, 2018). When it comes to choosing the optimal treatment options, the characteristics of wastewater samples are crucial.

For samples with high biodegradable contents, such as high biological oxygen demand (BOD) and low toxicity, traditional biological treatment approaches (aerobic or anaerobic) are preferable. However, samples with low biodegradability will very often require chemical treatment. AOPs are used to treat complex wastewater containing refractory chemicals in general. AOPs' applicability is further enhanced by their high treatment efficacies and fast treatment times. These AOPs take less time to complete than conventional treatment techniques. Table 5.1 lists the many forms of AOPs that have been reported in the literature. The generation of reactive oxidative species in situ and the interaction of oxidants with target pollutants are two stages in all AOPs. The mechanisms of radical production are influenced by the proposed system and water quality and are dependent on process-specific character- istics. Other factors influence the effectiveness of contaminant removal in addition to radical scavenging.

5.2   TYPES OF DIFFERENT ADVANCED OXIDATION PROCESSES

5.2.1   Fenton-based AOPs

5.2.1.1   Classical Fenton Process (CFP)

The Fenton process has been considered the oldest method among AOPs given by the British chemist H.J.H. Fenton in 1894. The process was well-utilized for several wastewater treatments containing recalcitrant compounds. The process only utilizes two chemicals, namely ferrous ion (as the catalyst) and hydrogen peroxide (as an oxidizing agent). The combined chemicals (Fe^{2+} and H_2O_2) of the CFP are called Fenton's reagent. The advantages of the CFP are well-reported in the literature, like the ease in application, fewer chemicals requirement, quick degradation of a variety of pollutants, readily available and non-toxic chemicals requirement, etc.

On the other hand, the reported drawbacks of the CFP are restricting its wide- spread applications, especially for full-scale. The reported disadvantages are a large amount of iron sludge production after the treatment, a narrow working pH range requirement, and the high dosage of chemicals required for high treatment effica- cies. The dosage of the reagents varies depending on the sample type and required treatment efficacies. Reaction pH and reagent dosage are the main influencing fac- tors in the CFP. The pH of around 2.5–4 is effective, as per the reported studies. The reactions involved in the CFP in the absence of the organic compounds have been summarized below:

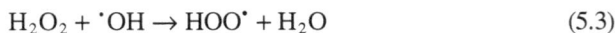

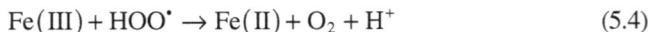

$$Fe(II) + H_2O_2 \rightarrow Fe(III) + {}^{\bullet}OH + OH^- \tag{5.1}$$

$$Fe(III) + H_2O_2 \rightarrow Fe(II) + HOO^{\bullet} + H^+ \tag{5.2}$$

$$H_2O_2 + {}^{\bullet}OH \rightarrow HOO^{\bullet} + H_2O \tag{5.3}$$

$$Fe(III) + HOO^{\bullet} \rightarrow Fe(II) + O_2 + H^+ \tag{5.4}$$

$$Fe(II) + {}^{\bullet}OH \rightarrow Fe(III) + OH^- \tag{5.5}$$

$$Fe(II) + HOO^{\bullet} \rightarrow Fe(III) + HO_2 \tag{5.6}$$

$$HOO^{\bullet} + HOO^{\bullet} \rightarrow H_2O_2 + O_2 \tag{5.7}$$

$${}^{\bullet}OH + HOO^{\bullet} \rightarrow H_2O + O_2 \tag{5.8}$$

$$^\cdot OH + {}^\cdot OH \rightarrow H_2O_2 \qquad (5.9)$$

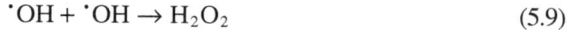

The hydroxyl radicals degrade the organic compounds by the following three mechanisms, i.e. hydrogen abstraction, hydroxyl addition, and electron transfer, as shown in reactions (11–13) (Huang et al., 1993; Bello et al., 2019).

$$^\cdot OH + Organic\, compounds \rightarrow Products \qquad (5.10)$$

$$^\cdot OH + RH \rightarrow R^\cdot + H_2O \qquad (5.11)$$

$$^\cdot OH + R \rightarrow R(OH) \qquad (5.12)$$

$$^\cdot OH + Fe^{2+} \rightarrow Fe^{3+} + OH^- \qquad (5.13)$$

5.2.1.2   Fenton-like Process

Several wastewaters have been treated using the Fenton reaction and the Fenton-like reaction initiated by Fe^{3+} and H_2O_2. Differentiating these processes is pointless from a mechanistic perspective because Fe^{2+} and Fe^{3+} are present in the chain of Fenton reactions depicted in the initial reactions. Once Fenton oxidation begins, all initially added Fe^{2+} is quickly oxidized to Fe^{3+}, resulting in a system that acts independently of iron oxidation states (Pignatello et al., 2006). However, a significant distinction in actuality is that at the start of Fenton oxidation, the rapid development of hydroxyl radicals may occur. In contrast, Fenton-like oxidation has a moderate generation rate of hydroxyl radicals. Because the rate constant in reaction (1) is substantially greater than in reaction (2), the latter reaction becomes a rate-limiting step, slowing the release of hydroxyl radicals. Fenton and Fenton-like reactions have similar organics removal efficiency, according to Rivas et al. (2003). According to Kim et al. (2001), the Fenton reaction removed more COD and had a higher BOD5/COD ratio than the Fenton-like reaction. Furthermore, the optimal pH of 3.0 for Fenton oxidation was lower than the optimal pH of 4.5 for the Fenton-like reaction.

5.2.1.3   Photo-Fenton Process

This process involves UV light radiations to attain the higher production of •OH and to regenerate Fe^{2+} ions (Kim et al., 1997). The UV or visible light radiation of wavelength below 450 nm is preferably used in the process (Zepp et al., 1992; Mahtab and Farooqi, 2020). In this process, the photoreduction of Fe^{3+} by UV irradiation causes the photochemical regeneration of Fe^{2+}, which reacts with H_2O_2 and produces •OH and Fe^{3+} ions, as shown in Eq. (5.14). The regeneration of Fe^{3+} continues the cycle and leads to higher •OH production, which enhances the Fenton's oxidation performance (Faust and Hoigné, 1990). The process also accompanies the direct photolysis of H_2O_2 to generate the •OH, as shown in Eq. (5.15). However, the presence of iron complexes in a solution absorbs a large part of radiation and affects the photolysis of H_2O_2 (Safarzadeh-Amiri et al., 1997). The role of pH is vital in the photo-Fenton process (PFP), which determines the formation of different iron complexes. At a pH value of 3, Fe^{3+} ions effectively converted into the most photoreactive ferric ion water complex, i.e. $[Fe(OH)]^{2+}$ species. The metal charge transfer excitation of $[Fe(OH)]^{2+}$ by UV radiation regenerates Fe^{2+} and produces $^{\cdot}OH$, as shown in Eq. (5.16) (Faust and Hoigné, 1990; Avetta et al., 2015). Acidic conditions (pH = 3) also favor the conversion of carbonates and bicarbonates into carbonic acid, which comparatively exhibits low susceptibility toward $^{\cdot}OH$ radicals (Legrini et al., 1993).

$$Fe^{3+} + hv + H_2O \rightarrow Fe^{2+} + {}^{\cdot}OH + H^+ \qquad (5.14)$$

$$H_2O_2 + hv \rightarrow 2\,{}^{\cdot}OH \qquad (5.15)$$

$$\left[Fe(OH)\right]^{2+} + hv \rightarrow Fe^{2+} + {}^{\cdot}OH \qquad (5.16)$$

The addition of ligands may further enhance the regeneration of Fe^{2+}. These complexes under UV irradiation follow the ligand to metal charge transfer step and regenerate Fe^{2+} ions, as shown in Eq. (5.17). In general, the combination of photochemistry and the Fenton process is a very compelling technology.

$$Fe^{3+} - L + hv \rightarrow Fe^{2+} + L^{\cdot+} \qquad (5.17)$$

5.2.1.4 Electro-Fenton Process

This process involves using electrons to complement the CFP. The Electro-Fenton process (EFP) works on the principle of cathodic reduction of Fe^{3+} and O_2 to generate Fenton's reagents, i.e., Fe^{2+} and H_2O_2 (He and Zhou, 2017). The EFP can be classified into four types based on Fenton's reagent formation. Type 1 involves using oxygen sparging cathode and sacrificial anode for the generation of H_2O_2 and Fe^{2+}, respectively, with no external addition of reagents (Ting et al., 2008). In Type 2, Fe^{2+} is generated from the sacrificial anode while H_2O_2 is externally added. In Type 3, oxygen sparging cathodes are used for the electro-generation of H_2O_2, and Fe^{2+} is externally added (Bello et al., 2019). Type 4 involves the electrolytic regeneration of Fe^{2+} by the cathodic reduction of Fe^{3+} ions (Zhang et al., 2006). However, type 3 is the most popular EFP that is used for the continuous electro-generation of H_2O_2. In a typical process, a constant oxygen gas supply at the cathode in an acidic medium causes its two-electron reduction. It leads to the formation of H_2O_2 as shown in Eq. (5.18) (Pliego et al., 2015). Initially, a small quantity of ferrous salts is added to the cell to react with H_2O_2 and generate Fe^{3+}, which continues the cathodic electro-regeneration of Fe^{2+}, as shown in Eq. (5.19) (Brillas et al., 2009). The sacrificial anode oxidation of iron is also significant in terms of the production of Fe^{2+}, as shown in Eq. (5.20) (Varank et al., 2020).

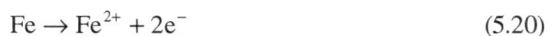

$$O_2 + 2H^+ + 2e^- \rightarrow H_2O_2 \qquad (5.18)$$

$$Fe^{3+} + e^- \rightarrow H_2O_2 \qquad (5.19)$$

$$Fe \rightarrow Fe^{2+} + 2e^- \qquad (5.20)$$

The process is very advantageous over the CFP. The electro-generation of H_2O_2 could lead to an 80% cost reduction and save the associated transport and handling cost. The effective utilization of Fe^{3+} and continuous regeneration of Fe^{2+} minimizes the problem of sludge production and enhances the production of •OH (Huang and Chu, 2012; Pliego et al., 2015). However, several factors like pH, current density, dissolved oxygen level, catalyst concentration, electrolytes, electrode nature, and temperature affect the efficiency of the process. EFP showed the same trend of results for solution pH, temperature, and initial concentration of pollutants as exhibited by the CFP. Applied current is an essential factor determining the electron generation and regeneration of H_2O_2 and Fe^{2+}, respectively. The higher applied current leads to higher efficiency but up to a specific limit. A value higher than certain pre-determined levels causes parasitic reactions and adversely affects the performance of the process. Lin and Chang (2000) have reported results in 69% of COD removal and 15.82% of NH_3-N reduction for treating landfill leachate by the EFP. The process further increased the biodegradability of leachate from a value of 0.1 to 0.29. In general, the EFP with the high in-situ generation of H_2O_2, electro-regeneration of Fe^{2+}, and low sludge production is very advantageous but requires high energy. The operational costs involved in the EFP include labor, material, cost of energy consumption, fixed, and disposal costs, but the major part of these costs in the EFP comes from the consumption of electric energy (Tirado et al., 2018). The high treatment cost due to the high electricity consumption is considered the main drawback of the EFP. The higher duration of treatment for the adequate mineralization of the resistant intermediates formed in the process leads to higher associated treatment costs (Monteil et al., 2019). The higher currents lead to the adequate mineralization of contaminants and add up to the higher electric energy consumption. Hence, it is essentially required to correctly set the applied current density that marks the balance between the energy-related costs and the efficiency of the process (He and Zhou, 2017). The electricity consumption in the electro-Fenton treatment process is analyzed by Eq. (5.21) (Tirado et al., 2018).

$$\text{Energy consumption} = \left\{ U \cdot I \cdot T \cdot 1{,}000 \, / \left(\text{COD}_o - \text{COD} \right) V \right\} \qquad (5.21)$$

where U = consumed electric energy (kWh/kg COD), I = current intensity (Amp), T = time (h), V = volume of water (L), COD_o = initial COD (mg/L), and COD = final COD (mg/L).

5.2.1.5 Heterogeneous Fenton Catalysis

On the other hand, heterogeneous Fenton catalysis, one of the sophisticated oxidation technologies, is of great interest for pollutant removal due to its intrinsic procedure and extensive application (Xia et al., 2011). Heterogeneous Fenton-like reactions on solid catalysts may efficiently catalyze the oxidation of organic pollutants over a wide pH range, which is beneficial for in-situ treatment of polluted groundwater and soil and can be reused for consecutive cycles. A surface-controlled reaction, a heterogeneous Fenton-like response, is governed by the catalyst surface area, H_2O_2 concentration, reaction temperature, pH, and ionic strength of the solution (Matta et al., 2007). When only Fe^{3+} is present initially, Fe^{2+} slowly regenerates and commences oxidation processes.

5.2.2 Ozon-based AOPs

Ozone-based AOPs are also widely used to treat a variety of wastewaters. The oxidative power of ozone is high ($E° = 2.08$ V) (Figure 5.1). The molecular structure of organic substances is altered by ozone, which oxidizes them into more biodegradable compounds that may be eliminated by biological treatment. Ozone-based AOPs significantly reduce COD and BOD levels in leachate and other wastewaters. Rivas et al. (2003) used an ozone dosage of 1.3–1.5 g O_3/g of COD for one hour to produce a 30% COD reduction, but Hagman et al. (2008) used 4 g/L O_3 to obtain a 22% COD reduction at pH 8–9. Wang et al. (2004) also found a drop in leachate alkalinity from 4,030 to 2,900 mg/L by 12.5 g O_3/L. The ozonation process is pH-dependent and can take place either through molecular ozone reactions (direct electrophilic attack on refractory contaminants) or by the formation of •OH radicals (indirect attack due to ozone breakdown) (Kurniawan et al., 2006). The following are the reactions that occur during the ozonation process:

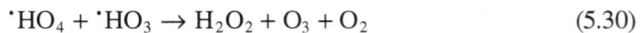

$$O_3 + OH^- \rightarrow {}^{\bullet}HO_2 + {}^{\bullet}O_2^- \tag{5.22}$$

$$HO_2 \leftrightarrow {}^{\bullet}O_2^- + H^+ \tag{5.23}$$

$$O_3 + {}^{\bullet}O_2^- \rightarrow {}^{\bullet}O_3^- + O_2 \tag{5.24}$$

$${}^{\bullet}O_3^- + H^+ \leftrightarrow {}^{\bullet}HO_3 \tag{5.25}$$

$${}^{\bullet}HO_3 \rightarrow {}^{\bullet}OH + O_2 \tag{5.26}$$

$${}^{\bullet}OH + O_3 \rightarrow {}^{\bullet}HO_4 \tag{5.27}$$

$${}^{\bullet}HO_4 \rightarrow {}^{\bullet}HO_2 + O_2 \tag{5.28}$$

$${}^{\bullet}HO_4 + {}^{\bullet}HO_4 \rightarrow H_2O_2 + 2O_3 \tag{5.29}$$

$${}^{\bullet}HO_4 + {}^{\bullet}HO_3 \rightarrow H_2O_2 + O_3 + O_2 \tag{5.30}$$

When the pH of a reaction increases over 9.0, ozone-resistant compounds known as hydroxyl radical scavengers form, which prevent oxidation; as illustrated in reactions (31) and (32), carbonate ions generated from bicarbonate ions act as scavengers, slowing down the rate of oxidation (Kurniawan et al., 2006).

$$CO_3^{-2} + {}^{\bullet}OH \rightarrow OH^- + {}^{\bullet}CO_3^- \tag{5.31}$$

$$HCO_3^- + {}^{\bullet}OH \rightarrow OH^- + HCO_3^{\bullet} \tag{5.32}$$

5.2.2.1 Peroxone Process (O_3/H_2O_2)

The use of hydrogen peroxide in conjunction with ozonation resulted in a more significant reduction in COD. The introduction of the radical system can be used to oxidize refractory compounds because it allows for selective molecular ozone reactions before the process is changed to free radical attack (non-selective). Ozone and hydrogen peroxide combine in a complicated series of reactions to produce •OH radicals, as demonstrated in reactions (33–39) (Langlais et al., 1991). Two hydroxyl radicals are produced as a result of these reactions, which comprise one H_2O_2 and two O_3 molecules (Schulte et al., 1995).

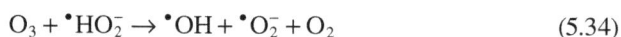

$$H_2O_2 + H_2O \leftrightarrow {}^\bullet HO_2^- + H_3O^+ \qquad (5.33)$$

$$O_3 + {}^\bullet HO_2^- \rightarrow {}^\bullet OH + {}^\bullet O_2^- + O_2 \qquad (5.34)$$

$$^\bullet O_2^- + H^+ \leftrightarrow {}^\bullet HO_2 \qquad (5.35)$$

$$^\bullet O_2^- + O_3 \rightarrow {}^\bullet O_3^- + O_2 \qquad (5.36)$$

$$^\bullet O_3^- + H^+ \leftrightarrow {}^\bullet HO_3 \qquad (5.37)$$

$$^\bullet HO_3 \rightarrow {}^\bullet OH + O_2 \qquad (5.38)$$

$$H_2O_2 + 2O_3 \rightarrow 2{}^\bullet OH + 3O_2 \qquad (5.39)$$

At an initial pH of 7, 60 minutes of 5.6 gm O_3/hour ozone injection followed by 400 mg/L H_2O_2 resulted in a 72% COD reduction and an increase in the BOD_5/COD ratio from 0.01 to 0.24 (Cortez et al., 2011). Ozone has long been utilized in water treatment as an oxidant and disinfectant. Ozone is an electron-rich oxidant that mostly affects double bonds, amines, and activated aromatic rings (e.g., phenol). Because its reactions in actual aqueous solutions frequently contain the formation of •OH, ozonation is commonly classified as an AOP or AOP-like process. The reaction's initiation, on the other hand, is rather sluggish, with a second-order rate constant of 70 $M^{-1}s^{-1}$.

Although peroxide is produced due to ozone interactions with the aqueous matrix, its contribution to the overall •OH formation during wastewater ozonation is insignificant (Nöthe et al., 2009). The O_3/H_2O_2 technique is well-established in water treatment and reuse applications. However, studies have shown that its benefits for wastewater applications are restricted due to severe competitive reactions (Hübner et al., 2015). It may, however, be a viable treatment option for reducing bromate generation during ozonation.

5.2.2.2 Ozonation at Elevated pH

Ozonation at high pH is a useful AOP (Buffle et al., 2006). The pH of treated water influences the effectiveness of direct ozonation (Calderara et al., 2002). If calcium carbonate precipitation is not a problem, ozonation at pH > 8 may be feasible. Because leachate is a complex matrix with high organic content, ozone treatment cannot satisfy the discharge standards alone. A high ozone dose is required to reduce COD, which makes this approach energy-intensive. Because some ozone is lost in the off-gas entering the ozone reactor, all ozone-based AOPs have a lower ozone mass transfer from gas to liquid. Although efforts are being made to increase ozone mass transfer efficiency, ozonation remains a viable treatment option as a pre- or post-treatment of leachate (Miklos et al., 2018).

5.3 APPLICATIONS OF ADVANCED OXIDATION PROCESSES

The diverse uses of different AOPs for several wastewater treatments, including landfill leachate, pharmaceutical wastewater, municipal wastewater, textile wastewater, and other industrial effluents, have been described in the literature. Table 5.2 summarizes previous studies on AOPs.

TABLE 5.2
The Various Applications of AOPs Reported in the Literature

Sample/Wastewater Type	Type of AOPs	References
Municipal landfill leachate	Classical Fenton process	Pieczykolan et al. (2013)
Sanitary landfill leachate	Coagulation-Fenton process	Moradi and Ghanbari (2014)
Urban landfill leachate	Fenton-like process	Martins et al. (2012)
Landfill leachate	Photo-electro-Fenton process	Altin (2008)
Municipal landfill leachate	Classical Fenton process	Gau and Chang (1996)
Municipal landfill leachate	Classical Fenton process	Kim and Huh (1997)
Biologically pre-treated leachate	Classical Fenton process	Welander and Henrysson (1998)
Pre-treated leachate	Electro-Fenton process	Lin and Chang (2000)
Biologically pre-treated leachate	Photo-Fenton process	Lau et al. (2002)
Mature leachate	Photo-Fenton process	De Morais and Zamora (2005)
Landfill leachate	Electro-Fenton process	Zhang et al. (2006)
Mature landfill leachate	Classical Fenton process	Deng (2007)
Mature leachate	Integrated Fenton-ultrafiltration process	Primo et al. (2008)
Fenton-ultrafiltration process	Classical Fenton process	Cortez et al. (2010)
Municipal landfill leachate	Classical Fenton process	Yilmaz et al. (2010)
Stabilized landfill leachate	Classical Fenton process	Mohajeri et al. (2011)
Mature landfill leachate	Photo-Fenton process	Jain et al. (2018)
Raw leachate	Sono-photo-Fenton process	Zha et al. (2016)
Pre-coagulated leachate membrane concentrates	Classical Fenton process	Xu et al. (2009)

(Continued)

TABLE 5.2 (*Continued*)
The Various Applications of AOPs Reported in the Literature

Sample/Wastewater Type	Type of AOPs	References
Stabilized landfill leachate	Heterogeneous electro-Fenton process	Sruthi et al. (2018)
Stabilized landfill leachate	Heterogeneous Fenton process	Niveditha and Gandhimathi (2020)
Municipal wastewater	Ozonation process	Tofani and Richard (1995)
Synthetic wastewater	Ozonation process	Beltran et al. (2000)
Landfill leachate	Classical Fenton process	Mahtab et al. (2021)
Pharmaceutical wastewater	Ozonation and peroxonation process	Alaton et al. (2004)
Industrial wastewater (textile)	Ozonation + ferrous sulfate (coagulation) process	Selcuk (2005)
Industrial wastewater (pharmaceuticals)	Ozonation process	Hernando et al. (2007)
Landfill leachate	Ozone/hydrogen peroxide process	Tizaoui et al. (2007)
Industrial wastewater (steel)	Ozonation process	Chang et al. (2008)
Effluent (dyes and textile industries)	Ozonation process	Pachhade et al. (2009)
Wastewater	Ozonation process	Turhan and Ozturkcan (2013)
Landfill leachate	Ozonation process	Amr et al. (2014)
Winery wastewater	Photo-electro-Fenton process	Díez et al. (2016)
Winery wastewater	Adsorption and photo-Fenton process	Guimarães et al. (2019)
Winery wastewater	Electro-Fenton-photolytic reactor	Díez et al. (2017)
Olive oil mill wastewater origin	Photo-Fenton	García and Hodaifa (2017)
Olive oil mill wastewater origin	Coagulation/flocculation followed by solar photo-Fenton oxidation	Ioannou-Ttofa et al. (2017)
Olive oil mill wastewater origin	Coagulation/flocculation followed by Fenton oxidation and biological treatment (only industry)	Amaral-Silva et al. (2017)
Olive oil mill wastewater Origin	Combined electrocoagulation (ECR)-photocatalytic (PCR) degradation system	Ates et al. (2017)
Olive oil mill wastewater origin	Combined ozone/Fenton process	Kirmaci et al. (2018)
Meat processing plants wastewater origin	Classical Fenton process	Masoumi et al. (2015)
Dairy wastewater	Electro-Fenton process	Davarnejad and Nikseresht (2016)
Dairy wastewater	Electro-Fenton process	Akkaya et al. (2019)
Food industry	Ultrasonic irradiation	Yılmaz and Fındık (2017)
Food industry	Ozone-based processes	Guzmán et al. (2016)
Food dye	Heterogeneous E-Fenton process	Barros et al. (2016)
Food dye	Photocatalytic process	Júnior et al. (2019)

5.4   CONCLUSION

This chapter overviews various AOPs' basic details and applications. The general reaction mechanisms and specific information regarding AOPs are also highlighted. It can be concluded that the versatile applications of AOPs are still in demand and need to be further explored to reduce the associated drawbacks. The studies on the disadvantages of specific AOPs and their sustainable solutions could be the likely domain for further research. The extensive full-scale applications of AOPs are limited and need to be explored further. It was observed that integrated treatment technologies are much more efficient and environmentally friendly. Hence, the combined treatment technologies should be implemented for complex wastewater treatments. The single treatment options are difficult to achieve stringent discharge standards, which further favors the requirement of combined treatment technologies. The generated sludge after the treatment should also be efficiently handled or managed to avoid the secondary pollution of soil, ground, and surface water. Several AOPs are efficient only as a pre- or post-treatment option; hence, a suitable selection of AOPs is essential for overall processes' performance.

ACKNOWLEDGMENT

The authors would like to express their gratitude to the Department of Civil Engineering (Aligarh Muslim University) for providing technical assistance with this chapter.

REFERENCES

Akkaya, G. K., Erkan, H. S., Sekman, E., et al. (2019). Modeling and optimizing Fenton and electro-Fenton processes for dairy wastewater treatment using response surface methodology. *Int. J. Environ. Sci. Technol.* 16, 2343–2358.

Alaton, I. A., Dogruel, S., Baykal, E., and Gerone, G. (2004). Combined chemical and biological oxidation of penicillin formulation effluent. *J. Environ. Manage.* 73(2), 155–163. doi: 10.1016/j.jenvman.2004.06.007.

Altin, A. (2008). An alternative type of photoelectro-Fenton process for the treatment of landfill leachate. *Sep. Purif. Technol.* 61(3), 391–397. doi: 10.1016/j.seppur.2007.12.004.

Amaral-Silva, N., Martins, R. C., Nunes, P., et al. (2017). From a lab test to industrial 18 advanced oxidation processes – applications, trends, and prospects application: Scale-up of Fenton process for real olive mill wastewater treatment. *J. Chem. Technol. Biotechnol.* 92, 1336–1344.

Amr, S. S. A., Aziz, H. A., and Bashir, M. J. (2014). Application of response surface methodology (RSM) for optimization of semi-aerobic landfill leachate treatment using ozone. *Appl. Water Sci.* 4(3), 231–239. doi: 10.1007/s13201-014-0156-z.

Ates, H., Dizge, N., and Cengiz, Y. H. (2017). Combined process of electrocoagulation and photocatalytic degradation for the treatment of olive washing wastewater. *Water Sci. Technol.* 75(1), 141–154.

Avetta, P., Pensato, A., Minella, M., Malandrino, M., Maurino, V., Minero, C., and Vione, D. (2015). Activation of persulfate by irradiated magnetite: Implications for the degradation of phenol under heterogeneous photo-Fenton-like conditions. *Environ. Sci. Technol.* 49(2), 1043–1050. doi: 10.1021/es503741d.

Barros, W. R. P., Steter, J. R., Lanza, M. R. V., et al. (2016). Catalytic activity of Fe$_{3-x}$Cu$_x$O$_4$ 20 advanced oxidation processes-applications, trends, and prospects ($0 \leq x \leq 0.25$) nanoparticles for the degradation of Amaranth food dye by heterogeneous electro-Fenton process. *Appl. Catal. B Environ.* 180, 434–441.

Bello, M. M., Raman, A. A. A., and Asghar, A. (2019). A review on approaches for addressing the limitations of Fenton oxidation for recalcitrant wastewater treatment. *Process Saf. Environ. Protect.* 126, 119–140. doi: 10.1016/j.psep.2019.03.028.

Beltran, F. J., García-Araya, J. F., and Álvarez, P. M. (2000). Sodium dodecylbenzene sulfonate removal from water and wastewater. 1. Kinetics of decomposition by ozonation. *Ind. Eng. Chem. Res.* 39(7), 2214–2220. doi: 10.1021/ie990721a.

Brillas, E., Sirés, I., and Oturan, M. A. (2009). Electro-Fenton process and related electrochemical technologies based on Fenton's reaction chemistry. *Chem. Rev.* 109(12), 6570–6631. doi: 10.1021/cr900136g.

Buffle, M.-O., Schumacher, J., Meylan, S., Jekel, M., and von Gunten, U. (2006). Ozonation and advanced oxidation of wastewater: Effect of O3 dose, pH, DOM and HO -scavengers on ozone decomposition and HO• generation. *Ozone Sci. Eng.* 28(4), 247–259. doi: 10.1080/01919510600718825.

Calderara, V., Jekel, M., and Zaror, C. (2002). Ozonation of 1-naphthalene, 1,5-naphthalene, and 3- nitrobenzene sulphonic acids in aqueous solutions. *Environ. Technol.* 23(4), 373–380. doi: 10.1080/09593332508618403.

Chang, E. E., Hsing, H. J., Chiang, P. C., Chen, M. Y., and Shyng, J. Y. (2008). The chemical and biological characteristics of coke-oven wastewater by ozonation. *J. Hazard. Mater.* 156(1–3), 560–567. doi: 10.1016/j.jhazmat.2007.12.106.

Comninellis, C., Kapalka, A., Malato, S., Parsons, S. A., Poulios, I., and Mantzavinos, D. (2008). Advanced oxidation processes for water treatment: Advances and trends for R&D. *J. Chem. Technol. Biotechnol.* 83(6), 769–776. doi: 10.1002/jctb.1873.

Cortez, S., Teixeira, P., Oliveira, R., and Mota, M. (2010). Fenton's oxidation as post-treatment of a mature municipal landfill leachate. *Int. J. Environ. Sci. Eng.* 2(1), 40–43.

Cortez, S., Teixeira, P., Oliveira, R., and Mota, M. (2011). Evaluation of fenton and ozone-based advance oxidation processes as mature landfill leachate pre-treatments. *J. Environ. Manage.* 92, 749–755. doi: 10.1016/j.jenvman.2010.10.035.

Davarnejad, R. and Nikseresht, M. (2016). Dairy wastewater treatment using an electrochemical method: Experimental and statistical study. *J. Electroanal. Chem.* 775, 364–373.

De Morais, J. L. and Zamora, P. P. (2005). Use of advanced oxidation processes to improve the biodegradability of mature landfill leachates. *J. Hazard. Mater.* 123(1–3), 181–186. doi: 10.1016/j.jhazmat.2005.03.041.

Deng, Y. (2007). Physical and oxidative removal of organics during Fenton treatment of mature municipal landfill leachate. *J. Hazard. Mater.* 146(1–2), 334–340. doi: 10.1016/j.jhazmat.2006.12.026.

Díez, A. M., Iglesias, O., Rosales, E., et al. (2016). Optimization of two-chamber photo electro Fenton reactor for the treatment of winery wastewater. *Process Saf. Environ. Prot.* 101, 72–79.

Díez, A. M., Rosales, E., Sanromán, M. A., et al. (2017). Assessment of LED-assisted electro-Fenton reactor for the treatment of winery wastewater. *Chem. Eng. J.* 310, 399–406.

Faust, B. C. and Hoigné, J. (1990). Photolysis of Fe (III)-hydroxy complexes as sources of OH radicals in clouds, fog and rain. *Atmos. Environ. Part A. General Topics* 24(1), 79–89. doi: 10.1016/0960-1686(90)90443-Q.

García, C. A. and Hodaifa, G. (2017). Real olive oil mill wastewater treatment by photo Fenton system using artificial ultraviolet light lamps. *J. Cleaner Prod.* 162, 743–753.

Gau, S. H. and Chang, F. S. (1996). Improved Fenton method to remove recalcitrant organics in landfill leachate. *Water Sci. Technol.* 34(7–8), 455–462. doi: 10.1016/S0273-1223(97)8 1411-4.

Gros, M., Petrović, M., Ginebreda, A., and Barceló, D. (2010). Removal of pharmaceuticals during wastewater treatment and environmental risk assessment using hazard indexes. *Environ. Int.* 36(1), 15–26. doi: 10.1016/j.envint.2009.09.002.

Guimarães, V., Lucas, M..S., and Peres, J. A. (2019). Combination of adsorption and heterogeneous photo-Fenton processes for the treatment of winery wastewater. *Environ. Sci. Pollut. Res.* 26, 31000–31013.

Guzmán, J., Mosteo, R., Sarasa, J., et al. (2016). Evaluation of solar photo-Fenton and ozone based processes as citrus wastewater pre-treatments. *Sep. Purif. Technol.* 164, 155–162.

Hagman, M., Heander, E., and Jansen, J. L. C. (2008). Advanced oxidation of refractory organics in leachate–potential methods and evaluation of biodegradability of the remaining substrate. *Environ. Technol.* 29, 941–946.

He, H. and Zhou, Z. (2017). Electro-Fenton process for water and wastewater treatment. *Crit. Rev. Env. Sci. Tec.* 47(21), 2100–2131. doi: 10.1016/j.jes.2015.12.003.

Hernando, M. D., Petrovic, M., Radjenovic, J., Fernandez-Alba, A. R., and Barceló, D. (2007). Removal of pharmaceuticals by advanced treatment technologies. *Comprehens. Anal. Chem.* 50, 451–474. doi: 10.1016/S0166-526X(07)50014-0.

Huang, C., Dong, C., and Tang, Z. (1993). Advanced chemical oxidation: Its present role and potential future in hazardous waste treatment. *Waste Manage.* 13(5), 361–377. doi: 0.1016/0956-053X(93)90070-D.

Huang, C. P. and Chu, C. S. (2012). Indirect electrochemical oxidation of chlorophenols in dilute aqueous solutions. *J. Environ. Eng.* 138(3), 375–385. doi: 10.1061/(ASCE) EE.1943-7870.0000518.

Hübner, U., Zucker, I., and Jekel, M. (2015). Options and limitations of hydrogen peroxide addition to enhance radical formation during ozonation of secondary effluents. *J. Water Reuse Desalin.* 5(1), 8. doi: 10.2166/wrd.2014.036.

Hussain, M., Mahtab, M. S., and Farooqi, I. H. (2020). The applications of ozone-based advanced oxidation processes for wastewater treatment: A review. *Adv. Environ. Res.* 9(3), 191–214.

Hussain, M., Mahtab, M. S., and Farooqi, I. H. (2021). A comprehensive review of the Fenton-based approaches focusing on landfill leachate treatment. *Adv. Environ. Res.* 10(1), 59–86.

Ioannou-Ttofa, L., Michael-Kordatou, I., Fattas, S. C., et al. (2017). Treatment efficiency 17 application of advanced oxidation process in the food industry and economic feasibility of biological oxidation, membrane filtration and separation processes, and advanced oxidation for the purification and valorization of olive mill wastewater. *Water Res.* 114, 1–13.

Jain, B., Singh, A. K., Kim, H., Lichtfouse, E., and Sharma, V. K. (2018). Treatment of organic pollutants by homogeneous and heterogeneous Fenton reaction processes. *Environ. Chem. Lett.* 16(3), 947–967. doi: 10.1007/s10311-018-0738-3.

Júnior, W. J., Júnior, N., Aquino, R. V. S, et al. (2019). Development of a new PET flow reactor applied to food dyes removal with advanced oxidative processes. *J Water Process Eng.* 31, 100823.

Khan, S. U., Noor, A., and Farooqi, I. H. (2015). GIS application for groundwater management and quality mapping in rural areas of District Agra, India. *Int. J. Water Res. Arid. Environ.* 4(1), 89–96.

Khan, S. U., Rameez, H., Basheer, F., and Farooqi, I. H. (2021). Eco-toxicity and health issues associated with the pharmaceuticals in aqueous environments: A global scenario. In: Khan, N. A., Ahmed, S., Vambol, V., and Vambol, S. (Eds.), Pharmaceutical Wastewater Treatment Technologies: Concepts and Implementation Strategies (pp. 145–179). IWA Publishing, London.

Kim, J. S., Kim, H. Y., Won, C. H., and Kim, J. G. (2001). Treatment of leachate produced in stabilized landfills by coagulation and Fenton oxidation process. *J. Chin. Inst. Chem. Eng.* 32(5), 425–429.

Kim, S. M., Geissen, S. U., and Vogelpohl, A. (1997). Landfill leachate treatment by a photoassisted Fenton reaction. *Water Sci. Technol.* 35(4), 239–248.

Kim, Y. K. and Huh, I. R. (1997). Enhancing biological treatability of landfill leachate by chemical oxidation. *Environ. Eng. Sci.* 14(1), 73–79. doi: 10.1089/ees.1997.14.73.

Kirmaci, A., Duyar, A., Akgul, V., et al. (2018). Optimization of combined ozone/Fenton process on olive mill wastewater treatment. *Aksaray Univ. J. Sci. Eng.* 2, 52–62.

Klavarioti, M., Mantzavinos, D., and Kassinos, D. (2009). Removal of residual pharmaceuticals from aqueous systems by advanced oxidation processes. *Environ. Int.* 35(2), 402–417. doi: 10.1016/j.envint.2008.07.009.

Kurniawan, T. A., Lo, W. H., and Chan, G. Y. S. (2006). Radicals catalyzed oxidation reactions for degradation of recalcitrant compounds from landfill leachate. *Chem. Eng. J.* 125(1), 35–57.

Langlais, B., Reckhow, D. A., and Brink, D. R. (1991). *Ozone in Water Treatment: Application and Engineering*. Lewis Publishers, Inc., Chelsea, MI.

Lau, I. W., Wang, P., Chiu, S. S., and Fang, H. H. (2002). Photoassisted Fenton oxidation of refractory organics in UASB-pretreated leachate. *J. Environ. Sci.* 14(3), 388–392.

Legrini, O., Oliveros, E., and Braun, A. M. (1993). Photochemical processes for water treatment. *Chem. Rev.* 93(2), 671–698. doi: 10.1021/cr00018a003.

Lin, S. H. and Chang, C. C. (2000). Treatment of landfill leachate by combined electro-Fenton oxidation and sequencing batch reactor method. *Water Res.* 34(17), 4243–4249. doi: 10.1016/S0043-1354(00)00185-8.

Luo, Y., Guo, W., Ngo, H. H., Nghiem, L. D., Hai, F. I., Zhang, J., Liang, S., and Wang, X. C. (2014). A review on the occurrence of micropollutants in the aquatic environment and their fate and removal during wastewater treatment. *Sci. Total Environ.* 473–474, 619–641. doi: 10.1016/j.scitotenv.2013.12.065.

Mahtab, M. S. and Farooqi, I. H. (2020). UV-TiO2 process for landfill leachate treatment: Optimization by response surface methodology. *Int. J. Res. Eng. Appl. Manage.* 5(12), 14–18.

Mahtab, M. S., Islam, D. T., and Farooqi, I. H. (2021). Optimization of the process variables for landfill leachate treatment using Fenton based advanced oxidation technique. *Eng. Sci. Technol. Int. J.* 24(2), 428–435. doi: 10.1016/j.jestch.2020.08.013.

Martins, R. C., Lopes, D. V., Quina, M. J., and Quinta-Ferreira, R. M. (2012). Treatment improvement of urban landfill leachates by Fenton-like process using ZVI. *Chem. Eng. J.* 192, 219–225. doi: 10.1016/j.cej.2012.03.053.

Masoumi, Z., Shokohi, R., Atashzaban, Z., et al. (2015). Stabilization of excess sludge from poultry slaughterhouse wastewater treatment plant by the Fenton process. *Avicenna J. Environ. Health Eng.* 2, 3239–3239.

Matta, R., Hanna, K., and Chiron, S. (2007). Fenton-like oxidation of 2,4,6-trinitrotoluene using different iron minerals. *Sci. Total Environ.* 385, 242–251.

Miklos, D. B., Remy, C., Jekel, M., Linden, K. G., Drewes, Jö. E., and Hübner, U. (2018). Evaluation of advanced oxidation processes for water and wastewater treatment: A critical review. *Water Res.* doi: 10.1016/j.watres.2018.03.042.

Mohajeri, S., Aziz, H. A., Zahed, M. A., Mohajeri, L., Bashir, M. J., Aziz, S. Q., and Isa, M. H. (2011). Multiple responses analysis and modeling of Fenton process for treatment of high strength landfill leachate. *Water Sci. Technol.* 64(8), 1652–1660. doi: 10.2166/wst.2011.489.

Monteil, H., Péchaud, Y., Oturan, N., and Oturan, M. A. (2019). A review on efficiency and cost effectiveness of electro-and bio-electro-Fenton processes: Application to the treatment of pharmaceutical pollutants in water. *Chem. Eng. J.* 376, 119577. doi: 10.1016/j.cej.2018.07.179.

Moradi, M. and Ghanbari, F. (2014). Application of response surface method for coagulation process in leachate treatment as pretreatment for Fenton process: Biodegradability improvement. *J. Water Process Eng.* 4, 67–73. doi: 10.1016/j.jwpe.2014.09.002.

Niveditha, S. V. and Gandhimathi, R. (2020). Mineralization of stabilized landfill leachate by heterogeneous Fenton process with RSM optimization. *Sep. Sci. Technol.* 1–10. doi: 10.1080/01496395.2020.1725573.

Nöthe, T., Fahlenkamp, H., and von Sonntag, C. (2009). Ozonation of wastewater: Rate of ozone consumption and hydroxyl radical yield. *Environ. Sci. Technol.* 43(15), 5990–5995. doi: 10.1021/es900825f.

Pachhade, K., Sandhya, S., and Swaminathan, K. (2009). Ozonation of reactive dye, Procion red MX-5B catalyzed by metal ions. *J. Hazard. Mater.* 167(1–3), 313–318. doi: 10.1016/j.jhazmat.2008.12.126.

Pieczykolan, B., Płonka, I., Barbusiński, K., and Amalio-Kosel, M. (2013). Comparison of landfill leachate treatment efficiency using the advanced oxidation processes. *Arch. Environ. Protect.* 39(2), 107–115. doi: 10.2478/aep-2013-0016.

Pignatello, J. J., Oliveros, E., and MacKay, A. (2006). Advanced oxidation processes for organic contaminant destruction based on the Fenton reaction and related chemistry. *Crit. Rev. Env. Sci. Technol.* 36(1), 1–84. doi: 10.1080/10643380500326564.

Pliego, G., Zazo, J. A., Garcia-Muñoz, P., Munoz, M., Casas, J. A., and Rodriguez, J. J. (2015). Trends in the intensification of the Fenton process for wastewater treatment: An overview. *Crit. Rev. Environ. Sci. Technol.* 45(24), 2611–2692. doi: 10.1080/10643389.2015.1025646.

Primo, O., Rueda, A., Rivero, M. J., and Ortiz, I. (2008). An integrated process, Fenton reaction− ultrafiltration, for the treatment of landfill leachate: Pilot plant operation and analysis. *Ind. Eng. Chem. Res.* 47(3), 946–952. doi: 10.1021/ie071111a.

Rivas, F. J., Beltran, F., Gimeno, O., and Carvalho, F. (2003). Fenton-like oxidation of landfill leachate. *J. Environ. Sci. Health Part A: Environ. Sci. Eng.* 38(2), 371–379.

Safarzadeh-Amiri, A., Bolton, J. R., and Cater, S. R. (1997). Ferrioxalate-mediated photodegradation of organic pollutants in contaminated water. *Water Res.* 31(4), 787–798. doi: 10.1016/S0043-1354(96)00373-9.

Schulte, P., Bayer, A., Kuhn, F., Luy, T., and Volkmer, M. (1995). H_2O_2/O_3, H_2O_2/UV, and H_2O_2/Fe^{2+} processes for the oxidation of hazardous wastes. *Ozone Sci. Eng.* 17(2), 119–134.

Selcuk, H. (2005). Decolorization and detoxification of textile wastewater by ozonation and coagulation processes. *Dyes Pigments* 64(3), 217–222. doi: 10.1016/j.dyepig.2004.03.020.

Sruthi, T., Gandhimathi, R., Ramesh, S. T., and Nidheesh, P. V. (2018). Stabilized landfill leachate treatment using heterogeneous Fenton and electro-Fenton processes. *Chemosphere* 210, 38–43. doi: 10.1016/j.chemosphere.2018.06.172.

Stefan, M. I. (Ed.), 2018. *Advanced Oxidation Processes for Water Treatment: Fundamentals and Applications*. IWA Publishing, London. doi: 10.2166/9781780407197.

Ting, W. P., Lu, M. C., and Huang, Y. H. (2008). The reactor design and comparison of Fenton, electro-Fenton and photoelectro-Fenton processes for mineralization of benzene sulfonic acid (BSA). *J. Hazard. Mater.* 156(1–3), 421–427. doi: 10.1016/j.jhazmat.2007.12.031.

Tirado, L., Gökkuş, Ö., Brillas, E., and Sirés, I. (2018). Treatment of cheese whey wastewater by combined electrochemical processes. *J. Appl. Electrochem.* 48(12), 1307–1319. doi: 10.1007/s10800-018-1218-y.

Tizaoui, C., Bouselmi, L., Mansouri, L., and Ghrabi, A. (2007). Landfill leachate treatment with ozone and ozone/hydrogen peroxide systems. *J. Hazard. Mater.* 140(1–2), 316–324. doi: 10.1016/j.jhazmat.2006.09.023.

Tofani, G. and Richard, Y. (1995). Use of ozone for the treatment of a combined urban and industrial effluent: A case history. *Ozone Sci. Eng.* 17(3), 345–354. doi: 10.1080/01919519508547540.

Turhan, K. and Ozturkcan, S. A. (2013). Decolorization and degradation of reactive dye in aqueous solution by ozonation in a semi-batch bubble column reactor. *Water Air Soil Pollut.* 224(1), 1353. doi: 10.1007/s11270-012-1353-8.

Varank, G., Guvenc, S. Y., Dincer, K., and Demir, A. (2020). Concentrated leachate treatment by Electro-Fenton and Electro-Persulfate processes using central composite design. *Int. J. Environ. Res.* 1–23. doi: 10.1007/s41742-020-00269-y.

Wang, F., Gamal El-Din, M., and Smith, D. W. (2004). Oxidation of aged raw landfill leachate with O3 Only and O3/H2O2: Treatment efficiency and molecular size distribution analysis. *Ozone Sci. Eng.* 26, 287–298. doi: 10.1080/01919510490455971.

Welander, U. and Henrysson, T. (1998). Physical and chemical treatment of a nitrified leachate from a municipal landfill. *Environ. Technol.* 19(6), 591–599. doi: 10.1080/09593331908616715.

Xia, M., Chen, C., Long, M., Chen, C., Cai, W., and Zhou, B. (2011). Magnetically separable mesoporous silica nanocomposite and its application in Fenton catalysis. *Microporous Mesoporous Mater.* 145, 217–223.

Xu, X. R., Li, X. Y., Li, X. Z., and Li, H. B. (2009). Degradation of melatonin by UV, UV/ H_2O_2, Fe^{2+}/H_2O_2 and $UV/Fe^{2+}/H_2O_2$ processes. *Sep. Purif. Technol.* 68(2), 261–266. doi: 10.1016/j.seppur.2009.05.013.

Yang, Y., Pignatello, J. J., Ma, J., and Mitch, W. A. (2014). Comparison of halide impacts on the efficiency of contaminant degradation by sulfate and hydroxyl radical-based advanced oxidation processes (AOPs). *Environ. Sci. Technol.* 48(4), 2344–2351. doi: 10.1021/es404118q.

Yılmaz, E., Fındık, S. (2017). Sonocatalytic treatment of baker's yeast effluent. *J. Water Reuse Desalinat.* 7, 88–96.

Yilmaz, T., Aygün, A., Berktay, A., and Nas, B. (2010). Removal of COD and colour from young municipal landfill leachate by Fenton process. *Environ. Technol.* 31(14), 1635–1640. doi: 10.1080/09593330.2010.494692.

Zepp, R. G., Faust, B. C., and Hoigne, J. (1992). Hydroxyl radical formation in aqueous reactions (pH 3–8) of iron (II) with hydrogen peroxide: The photo-Fenton reaction. *Environ. Sci. Technol.* 26(2), 313–319. doi: 10.1021/es00026a011.

Zha, F. G., Yao, D. X., Hu, Y. B., Gao, L. M., and Wang, X. M. (2016). Integration of U.S./Fe^{2+} and photo Fenton in sequencing for degradation of landfill leachate. *Water Sci. Technol.* 73(2), 260–266. doi: 10.2166/wst.2015.487.

Zhang, H., Zhang, D., and Zhou, J. (2006). Removal of COD from landfill leachate by electro-Fenton method. *J. Hazard. Mater.* 135(1–3), 106–111. doi: 10.1016/j.jhazmat.2005.11.025.

6 An Overview Exploring Electrochemical Technologies for Wastewater Treatment

Saif Ullah Khan, Rahat Alam, and Amir Aslam
Aligarh Muslim University

CONTENTS

6.1 Introduction 103
6.2 Electrochemical-Based Approaches 104
    6.2.1 Electro-Oxidation 104
           6.2.1.1 Mechanism 105
    6.2.2 Electrodeposition 107
    6.2.3 Electrodialysis 110
    6.2.4 Electrocoagulation 112
6.3 Conclusion 114
References 114

6.1 INTRODUCTION

Population explosion, rapid industrialization, and urbanization have resulted in excessive consumption of water which has laid down the need for wastewater treatment for its reuse (Usmani et al. 2021). Thus, treating wastewater has been the most critical challenge of the 21st century. Most of the industries responsible for wastewater generation include food processing industries, pulp and paper industries, tanneries, distilleries, sugar industries, textile industries, petrochemical industries, etc. (Khan et al. 2015). The wastewater flushed out of these industries has characteristics with high biochemical oxygen demand, chemical oxygen demand, suspended solids, salinity, color, turbidity, and pH imbalances, along with toxic compounds such as heavy metals, pharmaceutical compounds, phenols, and surfactants (Faheem et al. 2021).

For the efficient treatment of these effluents, several technologies like adsorption, membrane filtration, ion exchange, and advanced oxidation have been used conventionally (Khan et al. 2021b). However, these technologies are successful in removing most of the pollutants, but some concentrations of these still remain in the treated effluents due to incomplete degradation caused by operational limitations and technical issues; few other drawbacks of these include high operational costs, complex

DOI: 10.1201/9781003202431-6

procedures, excessive sludge generation, etc. (Zaidi et al. 2021). To overcome these issues, electrochemical treatment procedures have been explored (Alam et al. 2021a). These electrochemical-based treatment methods are quite effective in the treatment of several types of wastewater. The obtained effluent is colorless, odorless, and free from target pollutants to meet the desired quality parameters.

Electrochemical techniques have been implemented with several enhancements in the form of electrochemical oxidation, electrodialysis, electrocoagulation, etc. Electrochemical treatment is carried out in reactors by applying electric current with oxidation taking place at the anode while reduction at the cathode. Reactor designs along with operational process variables such as voltage, current, electrode material, and electrode spacing are chosen and optimized according to the target wastewater type, and its composition as well as considering the degree of treatment required. This review explores the following electrochemical-based techniques along with the mechanism involved in the effective treatment of wastewater and sludge as described in the section below.

6.2   ELECTROCHEMICAL-BASED APPROACHES

6.2.1   ELECTRO-OXIDATION

The concept of using electrochemical oxidation or anodic oxidation for wastewater treatment came into light in the 19th century after the investigation of the electrochemical decomposition of cyanide (Rajeshwar, Ibanez, and Swain 1994). Electrochemical oxidation is a type of advanced oxidation process, which is nowadays being extensively used in the treatment of wastewater obtained from different industries. In this, an oxidation reaction occurs at the anode, whereas the reduction reaction takes place at the cathode, and the redox reaction involved behaves as the fundamental that is used in the elimination of the pollutants. There are different researchers around the world, who have emphasized the importance of electrochemical oxidation, as a process for the disinfection of water, removal of organic or inorganic compounds (such as cyanide) from water, etc. During the 1970s, this method was investigated extensively for the removal of phenolic compounds (Papouchado et al. 1975; Koile and Johnson 1979). By the 1980s, electro-oxidation has been explored for various metal anodes and for different pollutants (Kolesnikov et al. 2015).

Electrochemical oxidation consisting of the conductive diamond anode was utilized for the disinfection of wastewater, without using any additional chemical reagent (Cano et al. 2012). They removed *Escherichia coli* (or *E. coli*, bacteria present in the intestines of both humans and animals for assisting in the proper digestion of food but can be harmful in certain cases). The technology was used to treat three different effluents obtained from a municipal wastewater treatment plant (WWTP), and in each instance, full eradication of *E. coli* was accomplished. Furthermore, working within the present densities described in this work, which is in the range of 1.3–13 A/m², it has been discovered that the method used does not produce any harmful byproducts (not even perchlorates or halo compounds).

In another study, disinfection of water was carried out by applying electrochemical oxidation (EO), along with the elimination of antibiotic-resistant bacteria (ARB) with the ultraviolet (UV) disinfection method combined with EO, thus overcoming the shortcomings of both processes when operated individually while removing such contaminants (Herraiz-Carboné et al. 2021). These processes were studied separately as well as in a combined form. The anodes used for disinfection were boron-doped diamond (BDD) and mixed metal oxides (MMOs). The results revealed that the maximum disinfection of the ARB was achieved when both processes were combined. In another study carried out using photo-assisted-electrochemical-oxidation (PAEO), the removal of sulfides from the tannery lime wastewater was analyzed (Selvaraj et al. 2020). Individually, electro-oxidation was inefficient for the proper elimination of chemical oxygen demand (COD) as well as trace organic compound (TOC), but the combination of EO with UV light investigated at various current densities (15, 20, and 25 mA/cm²), along with MMOs and Ti sheet as electrodes, and the results obtained indicated that PAEO at a current density of 25 mA/cm², effectively eliminated 100% sulfides, 92% COD, and 70% of TOC. Similarly, another group of researchers investigated the elimination of Procion Red MX-5B dye from the wastewater with the help of a BDD anode under various experimental conditions (Cotillas et al. 2018). They found that with the use of low current densities, natural pH, and high flow rates, the dye concentration, as well as the COD, were totally eradicated, and the procedure's overall efficiency got enhanced.

Since the past three decades, electrocatalytic improvement, its efficiency as well as electrochemical activity of various electrode materials have been extensively investigated for different pollutant sources, e.g., wastewater, landfill leachate, and sludge—mainly of industries (Chen 2004) (Figure 6.1).

The novelty in this field has led to invention of various new materials and methods with different electro-oxidation mechanisms which can be subdivided into two categories: direct and indirect electro-oxidation techniques (Jüttner, Galla, and Schmieder 2000).

6.2.1.1   Mechanism

6.2.1.1.1   Direct/Indirect Electro-Oxidation

The direct electro-oxidation of pollutants is based on the evolution of intermediate oxygen near the anode, mainly hydroxyl radicals ($^\cdot$OH) when an electric potential

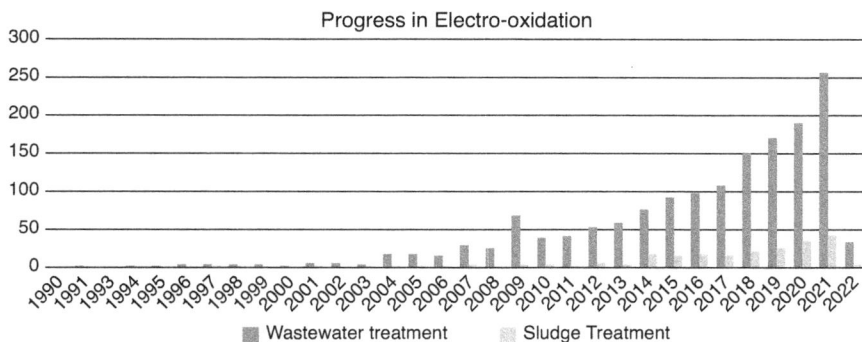

FIGURE 6.1   A graphical representation showing the increasing application of electro-oxidation in the field of wastewater treatment during the past two decades.

is applied between electrodes. These radicals degrade several stubborn pollutants present in wastewater/sludge because of their high redox potential. Direct electro-oxidation does not need an additional oxidizing agent; hence, it is simple and occurs near the anode where operating conditions are generally critical. On the other hand, an oxidizing agent is added in indirect electro-oxidation to promote the formation of oxidants such as Fenton's reagent, peroxide, Cl_2, hypochlorite, and ozone. This is the reason that the indirect electro-oxidation process occurs on the anode as well as in the bulk (Jüttner, Galla, and Schmieder 2000). A schematic diagram for direct/indirect electro-oxidation is illustrated in Figure 6.2.

The wastewater treatment setup by electro-oxidation consists of an electrochemical cell containing wastewater/sludge and a submerged electrode. Usually, cathode materials are the same, but anodes can vary depending on the type and degree of pollutant removal. For a considerable amount of radical generation, the current density is taken at 10–100 mA/cm². Also, the conductivity should be optimized to maintain higher efficiency and energy reduction. However, it should not exceed the recommended limit (1,000 mS/cm). By applying a potential difference between electrodes, reactive species formed mainly a hydroxyl radical, which further oxidizes the pollutants present in wastewater/sludge.

On the basis of adsorbing capacity/bond strength of anodes toward hydroxyl radicals, anodes can be divided into "active" (strongly attached to the hydroxyl radical) and "non-active" (loosely attached to the hydroxyl radical) anodes (Marselli et al. 2003). Active anodes make strong bonds with hydroxyl radicals by chemisorption,

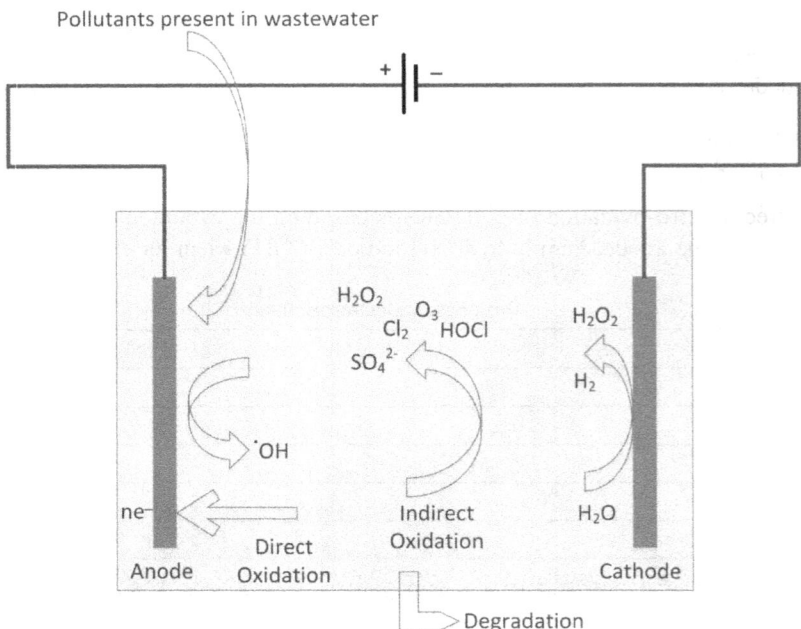

FIGURE 6.2   A schematic diagram for direct/indirect electro-oxidation is illustrated.

and the MO/M redox couple acts as a mediator in the oxidation of pollutants according to Eqs. (6.1)–(6.5) (Martínez-Huitle and Ferro 2006).

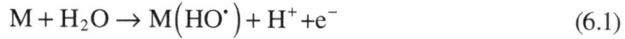

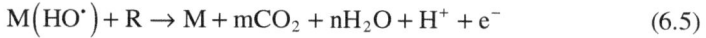

$$M + H_2O \rightarrow M(HO^{\cdot}) + H^+ + e^- \tag{6.1}$$

$$M(HO^{\cdot}) \rightarrow MO + H^+ + e^- \tag{6.2}$$

$$MO + R \rightarrow M + RO \tag{6.3}$$

$$MO \rightarrow M + 1/2O_2 \tag{6.4}$$

$$M(HO^{\cdot}) + R \rightarrow M + mCO_2 + nH_2O + H^+ + e^- \tag{6.5}$$

where M denotes anodes and R denotes pollutants.

In the case of non-active anodes, the oxidation of pollutants is mediated by loosely bonded (by physio-sorption with the anode) hydroxyl radicals. Here, anodes work just as a sink for electrons Eq. (6.5) (Marselli et al. 2003). Several anode materials, such as steel (Al-Malack and Siddiqui 2013), Pt (Mousset et al. 2014), IrO_2, and SnO_2 (Sun et al. 2020), have been investigated in the past. Still, new anode materials are explored to improve their activity and stability. Few recent studies using novel anodic materials for electro-oxidation are summarized in Table 6.1.

As the studies enlisted in Table 6.1 suggest, the removal efficiency achieved is above 90% in the majority of the cases. Another advantage of using this method is the mitigation of the sludge problem, as, unlike biological methods, this method generates sludge in negligible amounts, and hence, it is easy to handle.

6.2.2 Electrodeposition

Electrodeposition is based on the system consisting of an electrochemical cell that reduces dissolved metal cations and deposits on the cathode when electricity is passed. This is an old process, invented back in 1805, also known as electroplating, cathodic reduction, and electrolytic recovery (Maarof, Daud, and Aroua 2017). However, the application of this technology for wastewater/sludge treatment is new (Delgado, Fernández-Morales, and Llanos 2021). In this field, electrodeposition serves several purposes such as sludge handling (Trinh et al. 2021; Isabel et al. 2014), metal removal from wastewater (Stando et al. 2021; Gu et al. 2020; Ning, Yang, and Wu 2019), and precious metal recovery (Gouyon et al. 2020; Yliniemi et al. 2018; Lekka et al. 2015). As there is no addition of chemicals in this process, the generated sludge here is considered to be cleaner (Natsui, Yamaguchi, and Einaga 2016). Along with this, the recovery of common metals from the sludge of the electroplating industry could be beneficial by this process, as metal concentration is higher (de Nepel et al. 2020; Li et al. 2019). However, to remove organic pollutants, a combined process is needed along with electrodeposition (Wang et al. 2021).

TABLE 6.1

Recent Works Carried Out Using Various Novel Anodic Materials for the Electro-Oxidation Treatment of Wastewater

Effluent Type	Anode Type	Method of EO	Operating Conditions	Removal (%)	References
Landfill leachate	Boron-doped diamond	Coagulation + EO	40 mA/cm²; 6 hours	COD: 93.5%	GilPavas, Dobrosz-Gómez, and Gómez-García (2018)
Landfill leachate	mZVI-RuO₂-IrO₂/Ti	EC + EO	10V; 3 hours	COD: 66.7%; NH₄-N: 95.8%	Sun et al. (2020)
Landfill leachate	dimensionally stable anodes (DSA)	SBBGR + EO	83 mA/cm²; 4 hours	COD: 87%	Del Moro et al. (2016)
Landfill leachate	Ti/SnO₂-Sb₂O₅	UASB-A/O-ANR + EO	10 mA/cm²; 45 minutes	COD: 95.4%; NH₄-N: 97.2%	Wu et al. (2016)
Landfill leachate	Ti/Pt	MBR + EO	40 mA/cm²; 60 minutes	COD: 63%; NH₄-N: 80%	Feki et al. (2009)
Synthetic wastewater	Pt/MnO₂	EO (non-active anode)	7 mA/cm²; 3 hours	COD: 70%; Methylene blue: 90%	Alaoui et al. (2015)
Synthetic wastewater	Pt	EO (non-active anode)	20 mA/cm²; 20 hours	TOC: 20%; Phenol: 100%	Li et al. (2005)
Synthetic wastewater	BDD	EO (non-active anode)	20 mA/cm²; 0.5 hours	TOC: 50%; dimethyl phthalate: 100%	Li et al. (2010)
Industrial wastewater	BDD	EC-EO	120 mA/cm²; 1.5 hours	COD: 95%; phosphorous: 97%	Sanni et al. (2022)
Industrial wastewater	IrO₂–RuO₂	Electrochemical oxidation-in-situ coagulation (ECO-IC)	20 mA/cm²; 1 hour	Methyl orange: 97%; Fe: 96%	Ganzoury et al. (2022)
Industrial wastewater	BDD	Electro-Fenton	45.4 mA/cm²; 7 hours	Acid black 210 dye: 95%	Cruz et al. (2021)
Industrial wastewater	Ti	EO	30 mA/cm²; 3 minutes	COD: 64%; color: 90%	Ramesh, Gnanamangai, and Mohanraj (2021)

The electrodeposition mechanism follows basic oxidation and reduction reactions on the anode and cathode, respectively. On applying electricity, heavy metal ions present in electrolytes are deposited on the cathode surface (Korolev et al. 2020). However, a sophisticated electrode material is needed to recover a considerable amount in case of lower metal concentration in the electrolyte (Li et al. 2019; Lou et al. 2018; Su et al. 2017). A general cathodic reaction is shown by Eq. (6.6), where M and n denote the metal and its valence, respectively.

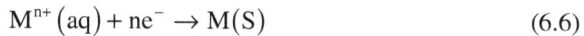

$$M^{n+}(aq) + ne^- \rightarrow M(S) \tag{6.6}$$

It gives a pure metal deposit on a cathode. However, other side reactions at the cathode and anode as well might occur during wastewater treatment as it consists of several ions. The gas evolution here might loosen the deposited metal and undermine the efficiency which should be prevented by appropriate design (Paul Chen and Lim 2005). A typical electrodeposition setup for wastewater treatment is shown in Figure 6.3.

Wastewater treatment via an electrodeposition system contains an electrochemical reactor including a cathode, an anode, and wastewater, laden with heavy metals, taken as electrolytes. Here, positive metal ions present in wastewater receive electrons from the cathode–electrolyte interface, supplied by an external source, and deposited there as a metal atom. Mainly, this process occurs in three steps: the transportation of metal ions from bulk to the cathode–electrolyte interface by convection and diffusion, stripping at the cathode–electrolyte interface, charge transfer, the crystal nuclei formation, and finally the formation of a stable metallic layer (Maarof, Daud, and Aroua 2017).

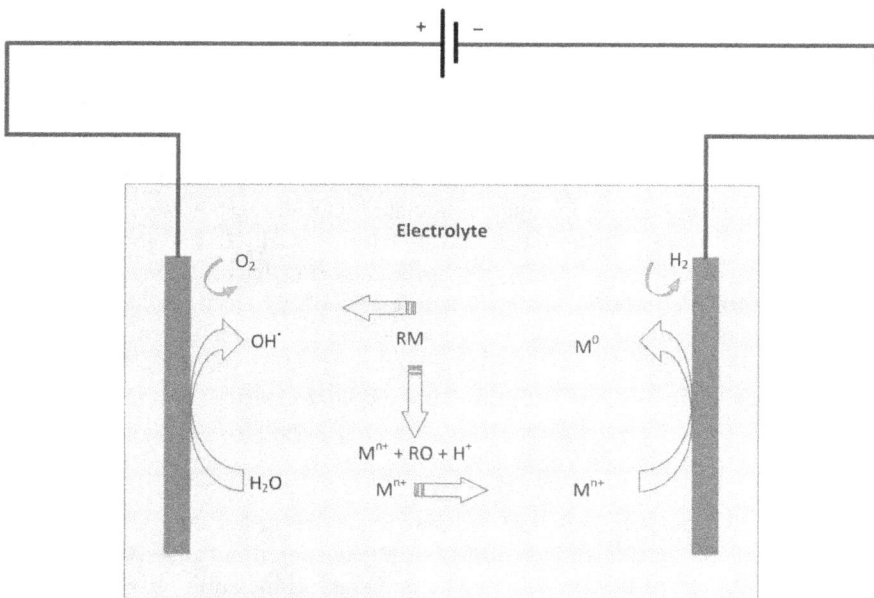

FIGURE 6.3   A typical electrodeposition setup for wastewater treatment is shown.

The design of the reactor could vary depending upon electrode configurations (two/three dimensional and static/moving electrodes), electrical connections (monopolar/bipolar), compartment division, and feeding mode (Bebelis et al. 2013). Each design has some advantages and shortcomings depending on the initial condition. However, some characteristics, such as large active surface area per unit reactor volume, easy mass transfer, high current density, and low cell voltage can make the design more efficient (Baghban, Mehrabani-Zeinabad, and Moheb 2014). In addition, the performance of the reactor could be quantified using electrochemical parameters such as current efficiency (CE), Y_{st} (space-time yield), and E_c (energy consumption) Eqs. (6.7)–(6.9) (Chen 2004).

$$CE = \frac{zF\Delta m}{MI\Delta t} \times 100 \tag{6.7}$$

$$Y_{st} = \frac{iaM}{1,000zF} CE \tag{6.8}$$

$$E_c = \frac{VIt}{m} \tag{6.9}$$

where z=charge number, F=Faraday constant, Δm=deposited mass in time interval Δt, M=molecular mass (g/mol), I=current passed (A), i=current density (A/m^2), a=specific electrode area (m^2/m^3), V=cell voltage (V), t=electrolysis time (hours), and m=product mass (kg).

In other words, CE is the ratio of consumed current while depositing the target metal to the total current. Y_{st} (kg/m^3/second) represents the metal production rate per unit reactor volume, while E_c (kWh/kg) is the energy consumption in this process.

Usually, cathodes used in the electrodeposition process are made up of metals (platinum, aluminum, and copper), metal oxides, stainless steel, and carbonaceous material. Recently, some modifications in electrodes are being made using nanomaterials to improve their performance (Liu et al. 2013). Researchers are trying to improve electrodeposition efficiency and metal recovery from wastewater by using a microporous electrode, for instance, reticulated vitreous carbon which has a high specific surface area (Ramalan et al. 2012; Dell'Era et al. 2014; Gao et al. 2021). Another group of researchers tried mitigating mass transfer limitations to improve the electrodeposition process using a rotating cylindrical electrode (RCE) as this system provides recirculation and homogeneity to electrolytes (Rosales and Nava 2017; Rivera et al. 2021). However, these materials are still expensive. Thus, further research is needed in this field to discover cheaper and superior novel electrodes.

6.2.3 ELECTRODIALYSIS

In recent times, electrodialysis (ED) is being widely regarded as an effective electrochemical membrane technique for nutrient concentration and recovery from wastewater. It is a type of separation technique that has the ability to demineralize a wide range of effluents obtained from the pharmaceutical, chemical, and food industries. The benefit of using the ED process is that it can be carried out at room temperature without causing any effect on the constituent's thermal breakdown.

Transport of ionizable species can also be done without impacting the concentration of non-ionizable species.

Besides its ability to efficiently eliminate pollutants such as emerging contaminants, heavy metals, and so on, electrodialysis is also being used for the desalination of water. ED is frequently used for the desalination of brackish water which has a salinity less than seawater, but more than that of freshwater. Approximately, 4% of global water desalination is performed by electrodialysis and 64% of it is done by another membrane-based separation process, namely, reverse osmosis (Almarzooqi et al. 2014). However, there has been an increase in the use of electrodialysis in recent times due to the various advantages it has over reverse osmosis such as prolonged lifespan of membrane, ease in implementation, better recovery rate for water, successful execution even at higher temperatures, and so on. Also, ED does not require any additional treatment before or after the purification of wastewater, which is needed in the case of RO.

In 2015, a group of researchers conducted review research on desalination procedures and discovered that electrodialysis had been used to treat water (brackish water), with salinities >1,5000 mg/L (Burn et al. 2015). In another study, electrodialysis was utilized to desalinate the seawater to get a high ratio of water recovery. The study was conducted in Morocco, which has fresh water in abundance, and lastly, the research also monitored the effects that the desalted water had when it was used for various agricultural purposes. The results obtained, indicated that under the given circumstances, such desalted water can be used for germination as well as irrigation (El Malki et al. 2007). Although the above studies along with several other studies on the desalination of water by ED look promising, still the problems associated with it such as membrane fouling need to be encountered more appropriately in order to use ED at a larger scale for water desalination, just like RO.

The process of electrodialysis can also be used for the elimination of different metal impurities from water. Few researchers explored how ED may be used for the treatment of drainage obtained from acid mines (Cardoso et al. 2013). The study concluded, cations of metals such as iron (Fe), aluminum (Al), manganese (Mn), lead (Pb), zinc (Zn), and copper (Cu) can be removed successfully using ED with an effectiveness higher than 97%.

Although different strategies have been devised, for decreasing ion exchange membranes (IEM) clogging, the typical electrodialysis process still requires cleansing on a regular basis, which leads to a raise in operational expenses. In order to get rid of these issues, researchers have established an electrodialysis reversal (EDR) method, which has provided better working conditions in terms of less fouling and reduction in pre-treatment and cleansing processes (Campione et al. 2018). The arrangement of the EDR system is quite similar to the electrodialysis, with the only difference being the existence of reversing valves in the case of EDR. These valves are responsible for the reversal of the electrical charges on the membrane after a certain period of time, which avoids the production of scales on the membranes.

There have been several studies in which the process of electrodialysis was used for the eradication of emerging pollutants (such as pharmaceutical compounds, personal care products, endocrine disruptors, etc.) that cannot be eliminated by simple conventional water treatment processes. Another group of researchers took into treatment consideration, 49 pharmaceutical compounds, endocrine disruptors, and other related compounds present in the wastewater supplied to the drinking water treatment plant (Gabarrón et al. 2016). Although the results displayed that most of the ECs were eliminated either by chlorine dioxide oxidation or by filtration (filters made up of granular activated carbon), EDR was still able to remove ionized ECs to some extent, and therefore, the study concluded that EDR can be further used as a secondary treatment process for emerging contaminants.

6.2.4 ELECTROCOAGULATION

Electrocoagulation has been one of the most prominent and successfully adopted electrochemical methods for the treatment of wastewater having several types of contaminants including heavy metals (Mateen et al. 2020), organic matter from landfill leachate (Faheem et al. 2021), saline wastewater (Alam et al. 2021b), fluoride (Khan et al. 2020), textile dyes (Xiong et al. 2001), foodstuff (Chen et al. 2000), etc. Overall EC has a quite complex mechanism, simultaneously including various physio-chemical reactions and using sacrificial electrodes so as to provide ions in the

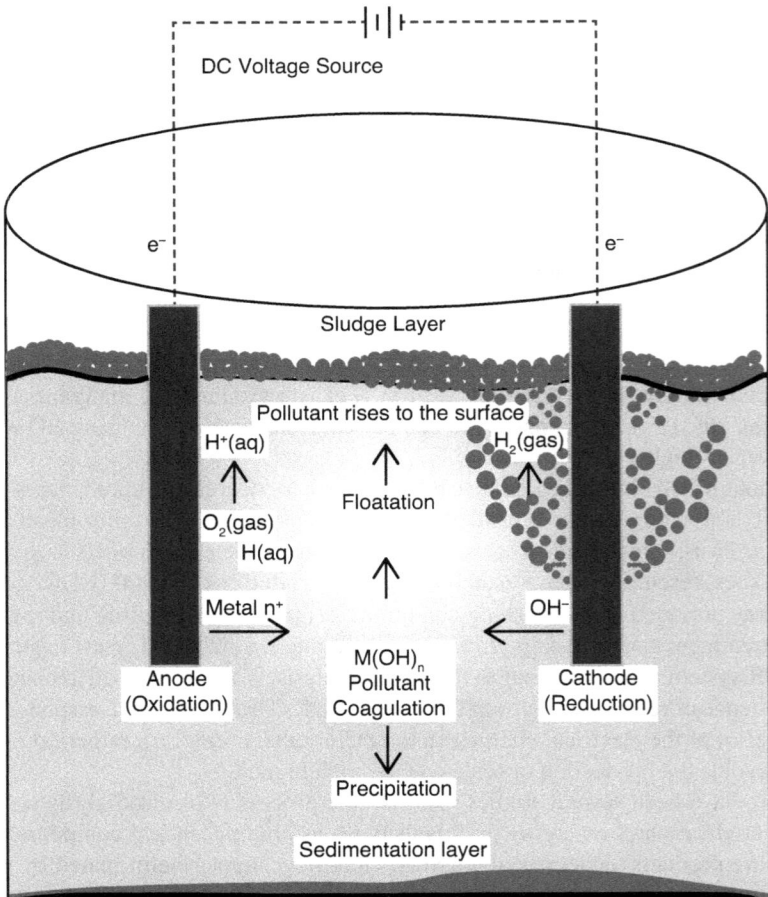

FIGURE 6.4 A typical electrocoagulation reactor along with a mechanism for wastewater treatment is shown.

form of coagulants as shown in Figure 6.4. The process involves "*in situ*" production of coagulating ions with three consecutive phases: (1) production of coagulant by electrolytic oxidation of dissipating electrode (generally Al or Fe), (2) destabilization of the target pollutant, particulate suspension, and (3) aggregation of destabilized particles together to form flocs. The general mechanism undergoing inside an EC cell with dissipative electrodes of metal M can be described in the form of chemical reactions as follows (Mousazadeh et al. 2021):

At the anode:

$$M_{(s)} \rightarrow M_{(aq)}^{n+} + ne^- \tag{6.10}$$

$$2H_2O_{(l)} \rightarrow 4H_{(aq)}^+ + O_{2(g)} + 4e^- \tag{6.11}$$

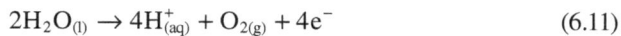

At the cathode:

$$M_{(aq)}^{n+} + ne^- \rightarrow M_{(s)} \tag{6.12}$$

$$2H_2O_{(l)} + 2e^- \rightarrow H_{2(g)} + 2OH^- \tag{6.13}$$

Considering iron-made rods as electrodes, $Fe_{(aq)}^{3+}$ ions are formed by EO which further reacts to form the resultant hydroxide [$Fe(OH)_3$] and poly-hydroxide species, namely, $Fe(H_2O)_6^{3+}$, $Fe(H_2O)_5(OH)^{2+}$, $Fe(H_2O)_4(OH)_2^+$, $Fe_2(H_2O)_8(OH)_2^{4+}$, and $Fe_2(H_2O)_6(OH)_4^{2+}$ based on the pH conditions (Khan et al. 2019). These hydroxides/poly-hydroxy-metallic complexes show strong empathy toward distributed ions and particles therefore start coagulation. The gasses evolved result in the up-flowing of the coagulated particles. The reactions may be shown as follows (Khan et al. 2021a):

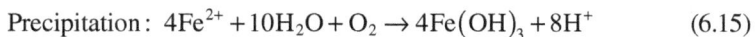

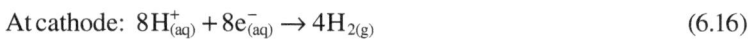

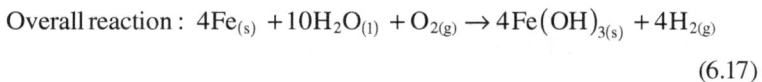

$$\text{At anode: } 4Fe_{(s)} \rightarrow 4Fe_{(aq)}^{2+} + 8e^- \tag{6.14}$$

$$\text{Precipitation: } 4Fe^{2+} + 10H_2O + O_2 \rightarrow 4Fe(OH)_3 + 8H^+ \tag{6.15}$$

$$\text{At cathode: } 8H_{(aq)}^+ + 8e_{(aq)}^- \rightarrow 4H_{2(g)} \tag{6.16}$$

$$\text{Overall reaction: } 4Fe_{(s)} + 10H_2O_{(l)} + O_{2(g)} \rightarrow 4Fe(OH)_{3(s)} + 4H_{2(g)} \tag{6.17}$$

Most of the studies undertaken in the past have undertaken only a few parameters to optimize the electrocoagulation process. However, for attaining the maximum removal efficiency with minimum energy consumption, various factors that influence the process, such as the initial pH, applied current, operational time, initial metal ion concentration, electrode material, electrode size and shape, spacing and arrangement, the conductivity of the solution, temperature, agitation rate, and nature of power supply (DC or AC). All these parameters have a particular role to play and are needed to be adjusted accordingly for the target removal of contaminants.

6.3 CONCLUSION

Electrochemical methods have gained wider acceptance as an alternative to conventional wastewater treatment methods, although their potential needs further exploration. EC techniques are widely adopted for removing heavy metals, besides other contaminants, such as organic pollutants, suspended and dissolved solids, colloidal materials, etc. However, their efficiency varies significantly, depending on the operating conditions. Although most of the investigations so far are limited at the laboratory level with artificially prepared solutions or industrial effluent lacking full- and field-scale studies, the success of the process depends a lot on optimizing the process variable. Considering, electrocoagulation, to attain maximum removal efficiency with minimum energy consumption, various factors that influence the process, include initial concentration, initial pH, applied current, operational time, electrode material, electrode size, spacing and arrangement, the conductivity of the solution, agitation rate, and nature of power supply. All these parameters have their particular role to play and need to be adjusted accordingly for the target removal of contaminants Although most of the electrochemical methods are self-sufficient in treating effluents with only a few exceptions, it is used as a hybrid process, along with some other treatment methods in a continuous phase. However, several studies have used the EC methods either as a pre-treatment or a polishing step, indicating its suitability to be applied as a preceding or polishing step.

REFERENCES

Alam, R., M. Sheob, B. Saeed, S. U. Khan, M. Shirinkar, Z. Frontistis, F. Basheer, and I. H. Farooqi, Use of electrocoagulation for treatment of pharmaceutical compounds in water/wastewater: A review exploring opportunities and challenges, *Water*, vol. 13, no. 15, p. 2105, 2021a. doi: 10.3390/w13152105.

Alam, R., S. U. Khan, M. Usman, M. Asif, and I. H. Farooqi, A critical review on treatment of saline wastewater with emphasis on Electrochemical based approaches. *Process Safety Environ. Prot.*, vol. 158, pp. 625–643, 2021b. doi: 10.1016/j.psep.2021.11.054.

Alaoui, A., K. El Kacemi, K. El Ass, S. Kitane, and S. El Bouzidi, Activity of Pt/MnO2 electrode in the electrochemical degradation of methylene blue in aqueous solution, *Sep. Purif. Technol.*, vol. 154, pp. 281–289, 2015. doi: 10.1016/j.seppur.2015.09.049.

Al-Malack, M. H. and M. Siddiqui, Treatment of synthetic petroleum refinery wastewater in a continuous electro-oxidation process, *Desalin. Water Treat.*, 2013. doi: 10.1080/19443994.2013.767215.

Almarzooqi, F. A., A. A. Al, I. Saadat, and N. Hilal, Application of capacitive deionisation in water desalination: A review, *DES*, vol. 342, pp. 3–15, 2014. doi: 10.1016/j.desal.2014.02.031.

Baghban, E., A. Mehrabani-Zeinabad, and A. Moheb, The effects of operational parameters on the electrochemical removal of cadmium ion from dilute aqueous solutions, *Hydrometallurgy*, 2014. doi: 10.1016/j.hydromet.2014.07.013.

Bebelis, S. et al., Highlights during the development of electrochemical engineering, *Chem. Eng. Res. Des.*, 2013. doi: 10.1016/j.cherd.2013.08.029.

Burn, S., M. Hoang, D. Zarzo, F. Olewniak, E. Campos, B. Bolto, and O. Barron, Desalination techniques: A review of the opportunities for desalination in agriculture, *DES*, vol. 364, pp. 2–16, 2015. doi: 10.1016/j.desal.2015.01.041.

Campione, A., L. Gurreri, M. Ciofalo, G. Micale, A. Tamburini, and A. Cipollina, Electrodialysis for water desalination: A critical assessment of recent developments on process fundamentals, models and applications, *Desalination*, vol. 434, pp. 121–160, 2018. doi: 10.1016/j.desal.2017.12.044.

Cano, A., P. Cañizares, C. Barrera-Díaz, C. Sáez, and M.A. Rodrigo, Use of conductive-diamond electrochemical-oxidation for the disinfection of several actual treated wastewaters, *Chem. Eng. J.,* vol. 211, pp. 463–469, 2012.

Cardoso, D., L. Stéphano, M. Antônio, S. Rodrigues, A. Moura, J. Alberto, and S. Tenório, Water recovery from acid mine drainage by electrodialysis, *Miner. Eng.*, vol. 40, pp. 82–89, 2013. doi:10.1016/j.mineng.2012.08.005.

Chen, G., Electrochemical technologies in wastewater treatment, *Sep. Purif. Technol.*, vol. 38, no. 1, pp. 11–41, 2004. doi: 10.1016/j.seppur.2003.10.006.

Chen, X., G. Chen, and P.L. Yue, Separation of pollutants from restaurant wastewater by electrocoagulation, *Sep. Purif. Technol.*, vol. 19, no. 1–2, pp. 65–76, 2000.

Cotillas, S., J. Llanos, P. Cañizares, D. Clematis, G. Cerisola, M. A. Rodrigo, and M. Panizza, Removal of Procion Red MX-5B dye from wastewater by conductive-diamond electrochemical oxidation, *Electrochim. Acta.*, vol. 263, pp. 1–7, 2018.

Cruz, D. R. S. et al., Magnetic nanostructured material as heterogeneous catalyst for degradation of AB210 dye in tannery wastewater by electro-Fenton process, *Chemosphere*, vol. 280, p. 130675, 2021. doi: 10.1016/j.chemosphere.2021.130675.

de Nepel, T. C., J. M. Costa, M. Gurgel Adeodato Vieira, and A. F. de Almeida Neto, Copper removal kinetic from electroplating industry wastewater using pulsed electrodeposition technique, *Environ. Technol. (United Kingdom)*, 2020. doi: 10.1080/09593330.2020.1793005.

Del Moro, G., L. Prieto-Rodríguez, M. De Sanctis, C. Di Iaconi, S. Malato, and G. Mascolo, Landfill leachate treatment: Comparison of standalone electrochemical degradation and combined with a novel biofilter, *Chem. Eng. J.*, vol. 288, pp. 87–98, 2016. doi: 10.1016/j.cej.2015.11.069.

Delgado, Y., F. J. Fernández-Morales, and J. Llanos, An old technique with a promising future: Recent advances in the use of electrodeposition for metal recovery, *Molecules,* 2021. doi: 10.3390/molecules26185525.

Dell'Era, A., M. Pasquali, C. Lupi, and F. Zaza, Purification of nickel or cobalt ion containing effluents by electrolysis on reticulated vitreous carbon cathode, *Hydrometallurgy*, 2014. doi: 10.1016/j.hydromet.2014.09.001.

El Malki, S., R. El Habbani, M. Tahaikt, M. Zeraouli, and A. Elmidaoui, The desalination of salt water destine to irrigation by electrodialysis and its effects on the germination, growth and seed yield of wheat, *African J. Agric. Res.*, vol. 2, pp. 41–46, 2007.

Faheem, K., S. U. Khan, and M. Washeem, Energy efficient removal of COD from landfill leachate wastewater using electrocoagulation: Parametric optimization using RSM. *Int. J. Environ. Sci. Technol.*, vol. 19, pp. 3625–3636, 2021. doi: 10.1007/s13762-021-03277-3.

Feki, F., F. Aloui, M. Feki, and S. Sayadi, Electrochemical oxidation post-treatment of landfill leachates treated with membrane bioreactor, *Chemosphere*, 2009. doi: 10.1016/j.chemosphere.2008.12.013.

Gabarrón, S., W. Gernjak, F. Valero, A. Barceló, M. Petrovic, and I. Rodríguez-Roda, Evaluation of emerging contaminants in a drinking water treatment plant using electrodialysis reversal technology. *J. Hazard. Mater.*, vol. 309, pp. 192–201, 2016.

Ganzoury, M. A., S. Ghasemian, N. Zhang, M. Yagar, and C.-F. deLannoy, Mixed metal oxide anodes used for the electrochemical degradation of a real mixed industrial wastewater, *Chemosphere*, vol. 286, p. 131600, 2022. doi: 10.1016/j.chemosphere.2021.131600.

Gao, Y., W. Zhu, Y. Li, J. Li, S. Yun, and T. Huang, Novel porous carbon felt cathode modified by cyclic voltammetric electrodeposited polypyrrole and anthraquinone 2-sulfonate for an efficient electro-Fenton process, *Int. J. Hydrogen Energy*, 2021. doi: 10.1016/j. ijhydene.2020.04.197.

GilPavas, E., I. Dobrosz-Gómez, and M. Á. Gómez-García, Optimization of sequential chemical coagulation: Electro-oxidation process for the treatment of an industrial textile wastewater, *J. Water Process Eng.*, vol. 22, pp. 73–79, 2018. doi: 10.1016/j. jwpe.2018.01.005.

Gouyon, J. et al., Reversible microfluidics device for precious metal electrodeposition and depletion yield studies, *Electrochim. Acta*, 2020. doi: 10.1016/j.electacta.2020.136474.

Herraiz-Carboné, M., S. Cotillas, E. Lacasa, P. Cañizares, M. A. Rodrigo, and C. Sáez, Enhancement of UV disinfection of urine matrixes by electrochemical oxidation. *J. Hazard. Mater.*, vol. 410, p. 124548, 2021.

Isabel, D. P., S. Brahima, M. Guy, and B. Jean-François, Simultaneous electrochemical leaching and electrodeposition of heavy metals in a single-cell process for wastewater sludge treatment, *J. Environ. Eng.*, vol. 140, no. 8, p. 4014030, 2014. doi: 10.1061/(ASCE) EE.1943-7870.0000856.

Jüttner, K., U. Galla, and H. Schmieder, Electrochemical approaches to environmental problems in the process industry, *Electrochim. Acta*, 2000, doi: 10.1016/S0013-4686(00)00339-X.

Khan, S. U., A. Noor, and I. H. Farooqi, GIS application for groundwater management and quality mapping in rural areas of District Agra, India, *Int. J. Water Res. Arid. Environ.*, vol. 4, no. 1, pp. 89–96, 2015.

Khan, S. U., D. T. Islam, I. H. Farooqi, S. Ayub, and F. Basheer, Hexavalent chromium removal in an electrocoagulation column reactor: Process optimization using CCD, adsorption kinetics and pH modulated sludge formation, *Process Safety Environ. Prot.*, vol. 122, pp. 118–130, 2019. doi: 10.1016/j.psep.2018.11.024.

Khan, S. U., M. Asif, F. Alam, N. A. Khan, and I. H. Farooqi, Optimizing fluoride removal and energy consumption in a batch reactor using electrocoagulation: A smart treatment technology. In: S. Ahmed, S. M. Abbas, and H. Zia (Eds.) *Smart Cities: Opportunities and Challenges* (pp. 767–778). Springer, Singapore, 2020.

Khan, S. U., M. Khalid, R. Zaidi, I. H. Farooqi, A. Azam, and S. Ayub, Applicability of Mn-Mg binary oxide nanoparticles for the adsorptive removal of copper and zinc from aqueous solution, *Mater. Today Proc.*, vol. 47, pp. 1500–1506, 2021a. doi: 10.1016/j. matpr.2021.05.430.

Khan, S. U., M. S. Mahtab, and I. H. Farooqi, Enhanced lead (II) removal with low energy consumption in an electrocoagulation column employing concentric electrodes: Process optimisation by RSM using CCD. *Int. J. Environ. Anal. Chem.*, pp. 1–18, 2021b. doi: 10.1080/03067319.2021.1873304

Koile, R. C. and D. C. Johnson, Electrochemical removal of phenolic films from a platinum anode, *Anal. Chem.*, vol. 51, pp. 741–744, 1979.

Kolesnikov, V. A., V. I. Il'in, V. A. Brodskii, T. V Guseva, and M. A. Vartanyan, Improvement of electroflotation treatment of waste waters from ceramic enterprises, *Glas. Ceram.*, vol. 71, p. 421, 2015 [Online]. Available: https://link.gale.com/apps/doc/A425765256/ AONE?u=anon~fbed9ed1&sid=googleScholar&xid=76cfa2fc.

Korolev, I., S. Spathariotis, K. Yliniemi, B. P. Wilson, A. P. Abbott, and M. Lundström, Mechanism of selective gold extraction from multi-metal chloride solutions by electrodeposition-redox replacement, *Green Chem.*, 2020. doi: 10.1039/d0gc00985g.

Lekka, M., I. Masavetas, A. V. Benedetti, A. Moutsatsou, and L. Fedrizzi, Gold recovery from waste electrical and electronic equipment by electrodeposition: A feasibility study, *Hydrometallurgy*, 2015. doi: 10.1016/j.hydromet.2015.07.017.

Li, H., X. Zhu, Y. Jiang, and J. Ni, Comparative electrochemical degradation of phthalic acid esters using boron-doped diamond and Pt anodes, *Chemosphere*, vol. 80, no. 8, pp. 845–851, 2010. doi: 10.1016/j.chemosphere.2010.06.006.

Li, T. et al., Recovery of Ni(II) from real electroplating wastewater using fixed-bed resin adsorption and subsequent electrodeposition, *Front. Environ. Sci. Eng.*, 2019. doi: 10.1007/s11783-019-1175-7.

Li, X.-Y., Y.-H. Cui, Y.-J. Feng, Z.-M. Xie, and J.-D. Gu, Reaction pathways and mechanisms of the electrochemical degradation of phenol on different electrodes, *Water Res.*, vol. 39, no. 10, pp. 1972–1981, 2005. doi: 10.1016/j.watres.2005.02.021.

Liu, Y., X. Wu, D. Yuan, and J. Yan, Removal of nickel from aqueous solution using cathodic deposition of nickel hydroxide at a modified electrode, *J. Chem. Technol. Biotechnol.*, vol. 88, no. 12, pp. 2193–2200, 2013. doi: 10.1002/jctb.4085.

Lou, W. et al., Additives-assisted electrodeposition of fine spherical copper powder from sulfuric acid solution, *Powder Technol.*, 2018. doi: 10.1016/j.powtec.2017.12.060.

Maarof, H. I., W. M. A. W. Daud, and M. K. D. Aroua, Recent trends in removal and recovery of heavy metals from wastewater by electrochemical technologies, *Rev. Chem. Eng.*, 2017. doi: 10.1515/revce-2016-0021.

Marselli, B., J. Garcia-Gomez, P.-A. Michaud, M. A. Rodrigo, and C. Comninellis, Electrogeneration of hydroxyl radicals on boron-doped diamond electrodes, *J. Electrochem. Soc.*, 2003. doi: 10.1149/1.1553790.

Martínez-Huitle, C. A. and S. Ferro, Electrochemical oxidation of organic pollutants for the wastewater treatment: Direct and indirect processes, *Chem. Soc. Rev.*, vol. 35, no. 12, pp. 1324–1340, 2006. doi: 10.1039/b517632h.

Mateen, Q. S., S. U. Khan, D. T. Islam, N. A. Khan, and I. H. Farooqi, Copper (II) removal in a column reactor using electrocoagulation: Parametric optimization by response surface methodology using central composite design. *Water Environ. Res.*, vol. 92, no. 9, pp. 1350–1362, 2020.

Mousazadeh, M., E. K. Niaragh, M. Usman, S. U. Khan, M. A. Sandoval, Z. Al-Qodah, Z. B. Khalid, V. Gilhotra, and M. M. Emamjomeh, A critical review of state-of-the-art electrocoagulation technique applied to COD-rich industrial wastewaters. *Environ. Sci. Pollut. Res.*, vol. 28, no. 32, pp. 43143–43172.

Mousset, E., N. Oturan, E. D. van Hullebusch, G. Guibaud, G. Esposito, and M. A. Oturan, Treatment of synthetic soil washing solutions containing phenanthrene and cyclodextrin by electro-oxidation. Influence of anode materials on toxicity removal and biodegradability enhancement, *Appl. Catal. B Environ.*, 2014. doi: 10.1016/j.apcatb.2014.06.018.

Gu, J. et al., Treatment of real deplating wastewater through an environmental friendly precipitation-electrodeposition-oxidation process: Recovery of silver and copper and reuse of wastewater, *Sep. Purif. Technol.*, 2020. doi: 10.1016/j.seppur.2020.117082.

Natsui, K., C. Yamaguchi, and Y. Einaga, Recovery of copper from dilute cupric sulfate solution by electrodeposition method using boron-doped diamond electrodes, *Phys. Status Solidi*, vol. 213, no. 8, pp. 2081–2086, 2016. doi: 10.1002/pssa.201600159.

Ning, D., C. Yang, and H. Wu, Ultrafast Cu^{2+} recovery from waste water by jet electrodeposition, *Sep. Purif. Technol.*, 2019. doi: 10.1016/j.seppur.2019.03.059.

Papouchado, L., R. W. Sandford, G. Petrie, and R. N. Adams, Anodic oxidation pathways of phenolic compounds Part 2. Stepwise electron transfers and coupled hydroxylations, *J. Electroanal. Chem.*, 1975. doi: 10.1016/0368-1874(75)85123-9.

Paul Chen, J. and L. L. Lim, Recovery of precious metals by an electrochemical deposition method, *Chemosphere*, 2005. doi: 10.1016/j.chemosphere.2005.02.001.

Rajeshwar, K., J. G. Ibanez, and G. M. Swain, Electrochemistry and the environment, *J. Appl. Electrochem.*, 1994. doi: 10.1007/BF00241305.

Ramalan, N. H. M., F. S. Karoonian, M. Etesami, S. Wen-Min, M. A. Hasnat, and N. Mohamed, Impulsive removal of Pb(II) at a 3-D reticulated vitreous carbon cathode, *Chem. Eng. J.*, 2012. doi: 10.1016/j.cej.2012.07.006.

Ramesh, K., B. M. Gnanamangai, and R. Mohanraj, Investigating techno-economic feasibility of biologically pretreated textile wastewater treatment by electrochemical oxidation process towards zero sludge concept, *J. Environ. Chem. Eng.*, vol. 9, no. 5, p. 106289, 2021. doi: 10.1016/j.jece.2021.106289.

Rivera, F. F., T. Pérez, L. F. Castañeda, and J. L. Nava, Mathematical modeling and simulation of electrochemical reactors: A critical review, *Chem. Eng. Sci.*, 2021. doi: 10.1016/j.ces.2021.116622.

Rosales, M. and J. L. Nava, Simulations of turbulent flow, mass transport, and tertiary current distribution on the cathode of a rotating cylinder electrode reactor in continuous operation mode during silver deposition, *J. Electrochem. Soc.*, vol. 164, no. 11, pp. E3345–E3353, 2017. doi: 10.1149/2.0351711jes.

Sanni, I., M. R. KarimiEstahbanati, A. Carabin, and P. Drogui, Coupling electrocoagulation with electro-oxidation for COD and phosphorus removal from industrial container wash water, *Sep. Purif. Technol.*, vol. 282, p. 119992, 2022. doi: 10.1016/j.seppur.2021.119992.

Selvaraj, H., P. Aravind, H. S. George, and M. Sundaram, Removal of sulfide and recycling of recovered product from tannery lime wastewater using photoassisted-electrochemical oxidation process, *J. Ind. Eng. Chem.*, vol. 83, pp. 164–172, 2020.

Stando, G., P. M. Hannula, B. Kumanek, M. Lundström, and D. Janas, Copper recovery from industrial wastewater: Synergistic electrodeposition onto nanocarbon materials, *Water Resour. Ind.*, 2021. doi: 10.1016/j.wri.2021.100156.

Su, J., X. Lin, S. Zheng, R. Ning, W. Lou, and W. Jin, Mass transport-enhanced electrodeposition for the efficient recovery of copper and selenium from sulfuric acid solution, *Sep. Purif. Technol.*, 2017. doi: 10.1016/j.seppur.2017.03.056.

Sun, D., X. Hong, K. Wu, K. S. Hui, Y. Du, and K. N. Hui, Simultaneous removal of ammonia and phosphate by electro-oxidation and electrocoagulation using RuO_2–IrO_2/Ti and microscale zero-valent iron composite electrode, *Water Res.*, 2020. doi: 10.1016/j.watres.2019.115239.

Trinh, H. B., J. Lee, S. Kim, J. C. Lee, J. C. F. Aceituno, and S. Oh, Selective recovery of copper from industrial sludge by integrated sulfuric leaching and electrodeposition, *Metals (Basel)*, 2021. doi: 10.3390/met11010022.

Usmani, S., M. Washeem, B. Rajput, M. S. Mahtab, S. U. Khan, and I. H. Farooqi, The fenton-based approaches focusing industrial saline wastewater treatment. In: S. Feroz, and D. W. Bahnemann (Eds.), *Removal of Pollutants from Saline Water* (pp. 375–398). CRC Press, Boca Raton, FL, 2021.

Wang, C., T. Li, G. Yu, and S. Deng, Removal of low concentrations of nickel ions in electroplating wastewater by combination of electrodialysis and electrodeposition, *Chemosphere*, 2021. doi: 10.1016/j.chemosphere.2020.128208.

Wu, L., D. Wei Liang, Y. Ying Xu, T. Liu, Y. Zhen Peng, and J. Zhang, A robust and cost-effective integrated process for nitrogen and bio-refractory organics removal from landfill leachate via short-cut nitrification, anaerobic ammonium oxidation in tandem with electrochemical oxidation, *Bioresour. Technol.*, 2016. doi: 10.1016/j.biortech.2016.04.041.

Xiong, Y.A., P.J. Strunk, H. Xia, X. Zhu, and H.T. Karlsson, Treatment of dye wastewater containing acid orange II using a cell with three-phase three-dimensional electrode, *Water Res.*, vol. 35, no. 17, pp. 4226–4230, 2001.

Yliniemi, K., Z. Wang, I. Korolev, P. Hannula, P. Halli, and M. Lundström, Effect of impurities in precious metal recovery by electrodeposition-redox replacement method from industrial side-streams and process streams, *ECS Trans.*, vol. 85, no. 4, pp. 59–67, 2018. doi: 10.1149/08504.0059ecst.

Zaidi, R., S. U. Khan, I. H. Farooqi, and A. Azam, Rapid adsorption of Pb (II) and Cr (VI) from aqueous solution by aluminum hydroxide nanoparticles: Equilibrium and kinetic evaluation, *Mater. Today Proc.*, vol. 47, pp. 1430–1437, 2021. doi: 10.1016/j.matpr.2021.03.224.

7 Constructed Wetlands for Wastewater Treatment

Rahat Alam and Saif Ullah Khan
Aligarh Muslim University

CONTENTS

7.1 Introduction 120
7.2 Constructed Wetland (CWL) Types.. 121
    7.2.1 Free Water Surface-Flow CWL.. 121
    7.2.2 Subsurface-Flow CWL 121
        7.2.2.1 Horizontal-Flow CWL.. 121
        7.2.2.2 Vertical-Flow CWL 121
        7.2.2.3 Hybrid CWL 122
    7.2.3 Enhanced CWL 122
        7.2.3.1 Baffled Sub Surface-Flow CWL.. 122
        7.2.3.2 Aerated CWL.. 122
7.3 Design and Operation of CWLs 122
    7.3.1 Selection of Macrophytes 123
    7.3.2 Selection of Substrate Media 124
7.4 Design and Operational Parameters of CWLs 124
    7.4.1 Environmental Conditions... 124
    7.4.2 Depth of Water... 124
    7.4.3 Hydraulic Retention Time 125
    7.4.4 Feeding Mode 125
7.5 Pollutant Removal Mechanisms in CWLs.. 125
    7.5.1 Removal of Organic Pollutants.. 126
    7.5.2 Nitrogen Removal 126
    7.5.3 Total Phosphate (TP) Removal 126
7.6 Advantages and Limitations of CWLs 127
7.7 Treated Wastewater Reuses Opportunities in Agriculture 127
7.8 Integrated Microbial Fuel Cells with CWLs (CWL-MFCs) 129
7.9 Cost Analysis.. 129
7.10 Challenges and Future Recommendations 129
7.11 Conclusion 130
References... 130

DOI: 10.1201/9781003202431-7

7.1   INTRODUCTION

Global water shortage is one of the leading challenges of the 21st century (Fitton et al., 2019). The increasing population and prosperity are changing the curve of global water demand, water consumption behavior, and unprecedented production of wastewater (Khan et al., 2015; Gosling and Arnell, 2016). This problem of water scarcity is expected to be worse in the future as freshwater resources are limited, and water consumption is rising (Boretti and Rosa, 2019). Along with this, sanitation facilities are not accessible to everyone, particularly to people from developing countries, which further deteriorates the available surface water quality. Researchers are trying to mitigate this tiresome situation by introducing the concept of wastewater reuse for different purposes (Trulli et al., 2016; Angelakis and Snyder, 2015). However, the presence of a vast range of pollutants, e.g., organics, dyes, metals, pharmaceuticals, and emerging micropollutants make the reuse process more complicated and costly (Crini and Lichtfouse, 2019). There is a huge scope for reusing wastewater by introducing more frequent and cost-effective treatment facilities (Alam et al., 2021). In this regard, constructed wetlands are considered one of the cheaper, most efficient, and most sustainable wastewater treatment and reuse technologies (Rai et al., 2013).

Constructed wetlands (CWLs) are engineered facilities inspired by natural processes, which use wetland soil, plants, substrate, and associated microbes to utilize and decompose pollutants present in wastewater by physical, chemical, and biological means (Abou-Elela, 2019; Li et al., 2021). Because of their relatively lower cost, easy operation, and green technology, this technology has been adopted extensively to treat various wastewater effluent including municipal, industrial, polluted river, and stormwater (Zhao et al., 2020; Álvarez et al., 2017; Chen et al., 2006; Song et al., 2009). With all the advantages of CWLs, many disadvantages such as lower pollutant removal efficiency, substrate clogging, mosquito breeding, and ineffectiveness toward stubborn contaminants exist, which undermine their application (Rahman et al., 2020; Manuel, 2003). Still, treatment with CWLs is well documented with various contaminants, including nutrients, pharmaceuticals, pathogens, and heavy metals (Ghimire et al., 2019; He et al., 2018; Bôto et al., 2016). These days, CWLs are coupled with other treatment technologies to mitigate major drawbacks associated with this process (Oon et al., 2015; Saeed and Sun, 2011; Zhai et al., 2011; Yeh and Wu, 2009).

Major challenges in this field are to maintain suitable plant species along with substrate media as they could vary according to geological and climatic factors (Stottmeister et al., 2003; Wu et al., 2015a). Along with this, most of the plants are vulnerable to toxic pollutants which affect plant growth and their contaminant removal capacity severely (Wu et al., 2013). In addition, this process is associated with several operational factors such as feed quality, feeding mode, hydraulic retention time (HRT), water depth, pH, temperature (T), dissolved oxygen (DO), and organic carbon availability, which needs to be optimized to achieve high removal efficiency (Saeed and Sun, 2012; Faulwetter et al., 2009). However, these parameters could be optimized using process optimization software (Quispe et al., 2019). Additionally, a thorough study is required to report recent developments in this field. Thus, this book chapter discusses the recent developments, challenges, and future perspectives of wastewater treatment wetlands in brief.

7.2 CONSTRUCTED WETLAND (CWL) TYPES

7.2.1 FREE WATER SURFACE-FLOW CWL

Free water surface-flow CWL is the facility where wastewater flows over the surface through flooded planted channels. It is simply an imitation of a natural wetland, marshy land, or swamp (Wu et al., 2014). Along with wastewater treatment, this system also helps in flood prevention and erosion control in shoreline areas (Farooqi et al., 2008). Since this system is closer to natural wetlands, a wide range of plant and wildlife diversity can exist, which can improve the treatment process (Almuktar et al., 2018). However, requirement of a sizable area in this facility makes it less popular as it increases the land cost, diminishes the vicinity, and has a potential threat to human health by exposure to pathogens. An average pollutant removal efficiency is around 50%–60% depending upon the pollutant type and concentration (El-Sheikh et al., 2010).

7.2.2 SUBSURFACE-FLOW CWL

7.2.2.1 Horizontal-Flow CWL

A horizontal-flow CWL is a type of subsurface flow in which wastewater flows permeate horizontally through the plant roots, rhizomes, and existing substrate media under the surface (Vymazal, 2009). Here, the pollutant present in wastewater is degraded by a combination of physical, chemical, and biological means as it passes through different zones (aerobic, anaerobic, and anoxic) (Vymazal, 2014). The underneath substrate is composed of planted macrophyte roots and sand/gravel, which provide oxygen, facilitate system interaction and allow water to pass (Brix, 1987). Usually, reeds (tall- and grass-like macrophytes) are grown for horizontal CWL with a substrate depth of 30–80 cm and a slope of 1%–3% to provide the gravitational flow (Akratos and Tsihrintzis, 2007). The major advantages of this system are as follows: it requires a smaller land area and prevents mosquito breeding or pathogen exposure (Vymazal et al., 2006). However, because of a sophisticated engineering structure, the cost of the system eventually increases (Tsihrintzis et al., 2007). Also, it has been reported that horizontal-flow CWL limits the nitrification of ammonia-nitrogen and favors the denitrification of nitrate-nitrogen because of the availability of anoxic and anaerobic conditions beneath the soil (Tsihrintzis et al., 2007; Zhang et al., 2014). However, an aerobic condition exists in the root zone (Parde et al., 2021). Horizontal-flow CWLs can be used to treat various wastewater influents including municipal, agricultural, industrial, and mine for different pollutants, e.g., COD, BOD, nitrogen, phosphorous, metals, and dyes (Solano et al., 2004; El Hamouri et al., 2007; Vohla et al., 2007; Coban et al., 2015; Obarska-Pempkowiak and Klimkowska, 1999).

7.2.2.2 Vertical-Flow CWL

Vertical-flow CWL was established to mitigate the major drawback of horizontal-flow CWL which is the incapability of nitrification because of the limited availability of oxygen (Almuktar et al., 2018). Here, initially, wastewater is flooded in the wetland channel and then infiltrated vertically through the wetland media. This system allows the circulation of oxygen in the appropriate amount to facilitate aerobic conditions as it is required during nitrification (Fan et al., 2013; Li et al., 2015).

In vertical-flow CWL, wastewater feed is applied in cycles of filling and draining through the substrate, which allows a high volume of wastewater treatment per unit area of wetland (Li et al., 2015). Also, comparatively higher pollutant removal efficiency is reported while using this type of wetland (Zhao et al., 2004). However, because of insufficient interaction between wastewater and substrate media, phosphorous removal efficiency is limited (Langergraber et al., 2007).

7.2.2.3   Hybrid CWL

To mitigate the drawbacks of horizontal- and vertical-flow CWL, a combination of both systems is considered a multistage treatment system in hybrid CWL. As of now, this hybrid CWL has been used extensively for various types of wastewater efficiently including winery (Serrano et al., 2011), pharmaceuticals (Reyes-Contreras et al., 2011), industrial (Vymazal, 2014), and gray wastewater (Comino et al., 2013). However, the pollutant removal efficiency and arrangement of hybrid CWL depends upon aerobic/anaerobic conditions which are required according to wastewater characteristics.

7.2.3   ENHANCED CWL

7.2.3.1   Baffled Sub Surface-Flow CWL

Baffled subsurface-flow CWLs include vertical baffles at regular intervals along the width of the horizontal-flow CWL to guide wastewater flow up and down through substrate media (Parde et al., 2021). This arrangement facilitates a longer pathway and provides more contact time to wastewater with substrate media, which is required for nitrogen removal (Gholipour and Stefanakis, 2021). The major advantage of this system is the efficient nitrogen removal, which is usually not achieved in horizontal/vertical-flow CWLs because of a shorter contact time. Recently, several modifications have been made to make this system more sustainable and efficient (Tee et al., 2012), (Chang et al., 2017; Zhao et al., 2016; Aalam and Khalil, 2019).

7.2.3.2   Aerated CWL

Aerated CWLs consist of several aerators to fulfill the oxygen requirement during wastewater treatment. Because of the lower availability of oxygen in conventional CWLs, the pollutant removal rate is also lower. To enhance the decomposition and removal rate, an adequate oxygen supply is needed, which is maintained in this system by external aerators (Sánchez-Monedero et al., 2008). However, an extra operational cost is added for running aerators.

7.3   DESIGN AND OPERATION OF CWLs

The design of CWLs is based on several components and their choices, including the choice of plant, substrate, depth, and mode of CWLs depending on material availability, location, and requirements. Other design parameters, such as HRT, hydraulic

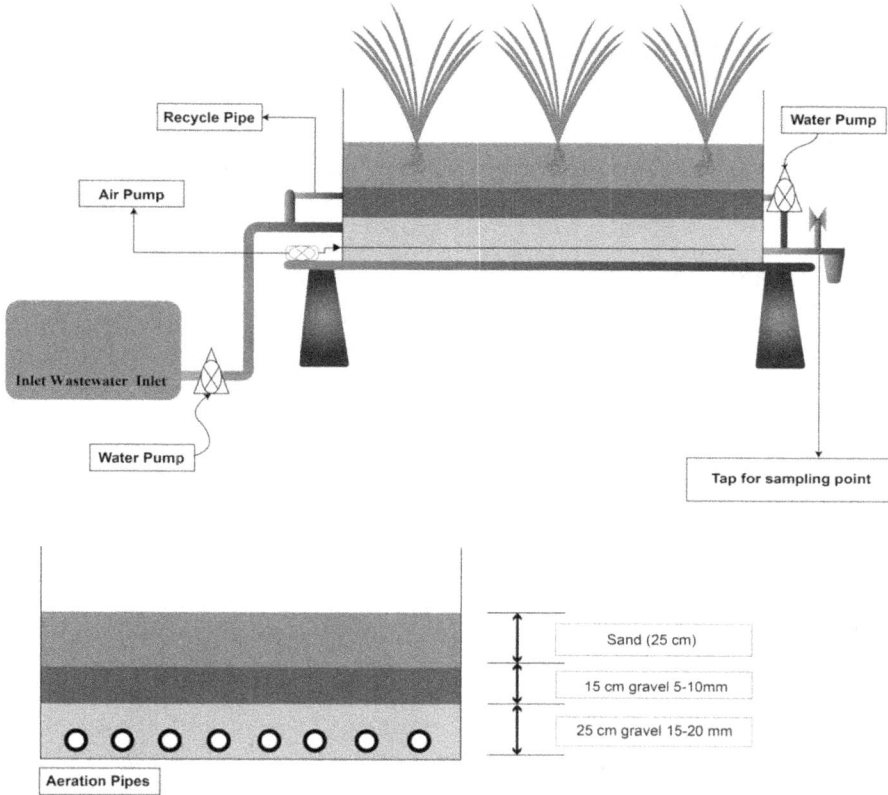

FIGURE 7.1 A typical design of CWLs.

loading rate (HLR), and plant and substrate quality, also need to be optimized for an efficient design (Akratos et al., 2009). A typical CWL design is shown in Figure 7.1.

7.3.1 SELECTION OF MACROPHYTES

Hundreds of macrophytes, including waterlogged and free-floating plants, have been used extensively in CWLs over the globe. However, very few species are used mostly in these facilities because of their rapid growth, their survival in extreme environments, and their excellent pollutant removal capacity (Vymazal, 2013). Few of them are Typhaceae, Phragmitesaustralis, Poaceae, Iridaceae, Hydrillaverticillata, Vallisnerianatans, Potamogetoncrispus, Nymphaea tetragona, Trapabispinosa, Marsilea quadrifolia, Eichhorniacrassipes, Lemna minor, Salvinianatans, and Hydrocharisdubia, mostly grown in free water surface-flow CWL (Rahman et al., 2020). However, Typhaceae and Phragmitesaustralis are frequently reported species (Vymazal, 2011).

The problem with macrophytes is that they have to survive in an extremely uncomfortable environment (Cong Manh et al., 2019). Pollutants present in wastewater reduce their life span, growth rate, and survival capacity. Along with this, other environmental uncertainties, such as eutrophication and high ammonia load, can disrupt plant growth (Xu et al., 2010). However, several species have the ability to survive an extreme pollutant load. For instance, *Typha* can survive for 20 days in a Cr (VI) solution with concentrations of up to 30 ppm (Ayele and Godeto, 2021). Similarly, *Arundodonax* along with Sarcocorniafruticose can survive in a highly saline environment (Calheiros et al., 2012). Therefore, more research on a suitable selection of the plant is needed for different pollutants, geography, and environment.

7.3.2 Selection of Substrate Media

Pollutant absorption potential and hydraulic conductivity play vital roles in the selection of substrate media (Wu et al., 2015b). Substrate media should be selected so that it can provide sufficient hydraulic conductivity to prevent major blockages which affect process capacity (Ry et al., 2010). Along with this, substrate media must provide enough biosorption to achieve the required pollutant removal efficiency. Generally, natural substances, industrial byproducts, and artificial media are used as a substrate, e.g., clay, sand, gravel, marble, calcite, fly ash, slag, recycled concrete, dolomite, calcite, zeolite, and activated carbon (Saeed et al., 2012; Yan and Xu, 2014; Cao et al., 2021). However, natural materials such as clay, sand, and gravel do not work well for long-term phosphate removal and can be replaced by artificial or industrial byproducts. Recently, several novel substrate media have been implemented successfully such as maerl bed (Gray et al., 2000), activated alumina (Tan et al., 2019), alum sludge (Zhao et al., 2011), modified carbon (Guo et al., 2020), and sludge ceramsite (Wu et al., 2016) for efficient removal of nutrients and organics. The mechanism of contaminant removal by substrate is comprised of several processes such as exchange, adsorption, complex formation, precipitation, or their combinations, which mainly depend on materials and their hydraulic conductivity. For phosphorous and ammonia adsorption a typical sorption capacity of sand is 0.13–0.29 g/kg and of zeolite is 11.6 g/kg for phosphorous and ammonia is reported (Xu et al., 2006; Rahman et al., 2020).

7.4 DESIGN AND OPERATIONAL PARAMETERS OF CWLs

7.4.1 Environmental Conditions

Environmental conditions of wetlands include physiochemical variables such as pH, temperature (T), moisture, dissolve oxygen (DO), oxidation–reduction potential (ORP), initial biological and chemical oxygen demand (BOD/COD), solids, and so on. A typical range of pH, T, moisture, and DO for the appropriate growth of an ordinary plant are 7–8, 25°C–28°C, and 5.8–7.9 ppm, respectively (Titah et al., 2014; OECD, 2006).

7.4.2 Depth of Water

Water depth in CWLs plays an important role in the selection of plant and pollutant removal as it influences biochemical reactions along with the DO rate. A typical shallow depth (around 27 cm) is reported for efficient removal of COD, BOD, ammonia, and phosphorous (Aguirre et al., 2005; García et al., 2005). Studies also suggest that metabolic pathways vary with water depth (García et al., 2005).

7.4.3 HYDRAULIC RETENTION TIME

HRT significantly influences the removal efficiency of pollutants in any treatment facility (Lee et al., 2009). For the removal of contaminants, appropriate contact time is needed between microbiological population and contaminants (Saeed et al., 2012; Yan and Xu, 2014). Studies have been conducted by varying HRT, and it is found that higher HRT eliminates ammonia and total nitrogen efficiently (Toet et al., 2005). Additionally, a typical HRT of 19.2 hours is reported for the successful elimination of nitrogen (Lee et al., 2009).

7.4.4 FEEDING MODE

The pollutant removal mechanism greatly depends on the oxygen transfer rate along with the oxidation–reduction reaction. Feeding modes of CWLs, such as continuous, intermittent, and batch modes, influence diffusion, and oxidation–reduction reactions of the system and, hence, alter the pollutant removal efficiency (Rahman et al., 2020). Several studies have been conducted based on the influence on the system efficiency by changing the feeding mode (Herrera-Melián et al., 2018; Bassani et al., 2021; Zhang et al., 2012; Sasikala et al., 2010; Zhang and Zou, 2010). These studies support that, because of the improved oxygen supply in the batch mode, it is found to be more efficient than the continuous mode. However, for the nitrogenous pollutant, the intermittent mode is found to be efficient. For instance, Caselles-Osorio and García (2007) studied ammonium removal efficiency in both batch and intermittent modes in the subsurface-flow CWL and found that ammonium removal efficiency was higher during the intermittent mode. However, according to the study conducted by Jia et al. (2010) removal efficiency in intermittent mode was found to be slower in the case of COD and phosphorous. Therefore, the feeding mode must be selected by considering wastewater characteristics and CWL types.

7.5 POLLUTANT REMOVAL MECHANISMS IN CWLs

Mechanisms of pollutant removal are typically based on separation and biogeochemical transformation, e.g., adsorption/absorption, leaching, stripping, filtration, acid–base reactions, oxidation–reduction, flocculation precipitation, and biochemical reactions (Choudhary et al., 2011). There are three major constituents, macrophytes, substrate media, and associated microorganisms that greatly influence removal mechanisms (Kumar and Dutta, 2019). Macrophytes provide sufficient oxygen and dissolve organic matter, which is required for the survival of microorganisms. Along with this, they increase the porosity and permeability of the substrate, support catalytic activities, and facilitate microbiological reactions (Meng et al., 2014; Noureddine and Ouakouak, 2018). On the other hand, substrate media provide sufficient hydraulic conductivity and enhance adsorption ability to support adsorption/absorption activity, filtration, and gravity separation (Ry et al., 2010). Finally, microorganisms associated with wetland systems, e.g., bacteria, protozoa, fungi, yeasts, and algae play a vital role in pollutant removal. These microorganisms get attached to plant roots, leaves, and substrate media by forming a biofilm and decomposing almost all the contaminants present in wastewater into insoluble simpler substances

(Faulwetter et al., 2009). Studies have found that the rhizosphere region of CWLs facilitates the microbial biofilm community by providing oxygen and greatly influencing the pollutant removal efficiency of the system (Zhang et al., 2016).

7.5.1 REMOVAL OF ORGANIC POLLUTANTS

Both aerobic and anaerobic microorganisms contribute to the degradation of organic pollutants depending upon oxygen availability. Naturally, oxygen is transferred through convection and macrophyte roots (Begg et al., 2001; Rehman et al., 2017). However, to facilitate aerobic degradation, oxygen can be supplied by external aeration (Zhang et al., 2010; Ouellet-Plamondon et al., 2006). On the other hand, small pores of substrate provide active sites for anaerobic degradation of organic pollutants. Insoluble organics are settled down by means of gravity and are filtered out, while soluble organics are eliminated by attached/suspended microbial growth. Aerobic degradation is mainly supported by chemoheterotrophic bacterial oxidation in which organic compounds are oxidized by chemoheterotrophic bacteria into simpler substances such as carbon dioxide, ammonia, and other stable compounds (García et al., 2010). Anaerobic degradation is maintained by anaerobic heterotrophic bacteria in two pathways, commonly known as methanogenesis and fermentation (Ali Shah et al., 2014). In the methanogenesis process, methane-producing bacteria are responsible for converting complex organic substances into methane and carbon dioxide along with new bacterial cells, whereas, in the fermentation process, acid-forming bacteria degrade organic pollutants into organic acids and alcohols (Amin et al., 2021).

7.5.2 NITROGEN REMOVAL

Nitrogen removal pathways generally follow nitrification, denitrification, ammonification, ammonia volatilization, and adsorption in CWLs (Saeed and Sun, 2012). However, novel nitrogen removal pathways also have been suggested recently (Wang et al., 2018). For nitrification, oxygen is required, which is provided by convection and plant roots. Along with this, the plant root system provides sufficient carbon for denitrification. External carbon sources such as biochar also can be added to improve nitrogen removal efficiency (Wang et al., 2018). Different plants supply oxygen and provide active surfaces in different amounts which greatly affects the extent of nitrification. Ammonification occurs in the aerobic and facultative zone of CWL because of proper oxygen availability. In these regards, pH should be maintained at 6.5–8.5 during this process as ammonification and ammonia volatilization are pH-dependent (Saeed and Sun, 2012). Out of several macrophytes, *Typha latifolia*, *Canna indica*, and *Phragmites australis* are found to be more efficient for nitrogen removal (Jesus et al., 2018).

7.5.3 TOTAL PHOSPHATE (TP) REMOVAL

A mixture of phosphates is available in the wastewater coming from different sources. However, orthophosphates (PO_4^{3-}) are more common. Studies suggest that CWLs do not work efficiently for the removal of phosphate (Bus and Karczmarczyk, 2017; Poor et al., 2020). However, removal efficiency can be optimized by a careful selection of macrophytes and loading rates. According to Guo et al. (2017), high water depth and low flow velocity enhance the phosphate removal rate. A typical phosphate removal efficiency ranges from 4.8% to 74.87% depending on different macrophytes as they have different phosphorous uptake (Jesus et al., 2018). The mechanism of removal mainly follows immobilization by microorganisms and adherence to substrate media. At present, several substrate media are being used to improve the phosphate removal efficiency, such as slag, zeolite, dolomite, and bauxite. Recently, other non-conventional materials such as biochar, magnesia, magnesite, zirconium oxide nanoparticle (ZON), and iron oxide coated granular activated carbon (Fe-GAC) are reported to be helpful in phosphate removal (Okochi and McMartin, 2011; Lan et al., 2018). However, these novel materials are expensive and, hence, increase the cost of treatment (Park et al., 2017). Therefore, the selection of an appropriate substrate medium is necessary for better performance in terms of phosphate removal.

7.6 ADVANTAGES AND LIMITATIONS OF CWLs

CWLs are considered to be one of the cheapest wastewater treatment facilities. The capital expenditure (CapEx) is lower than other facilities as it does not involve sophisticated structures and electricity requirements (Barbera et al., 2009). Along with this, operational cost (OpEx) is negligible (1%–2% of total cost) (Parde et al., 2021). However, the cost of the land and novel substrate material can increase the total cost. Another advantage is CWLs have the ability to treat various wastewater influents including those from municipal (Chung et al., 2008), agricultural (Kantawanichkul et al., 2003; Nan et al., 2020), diary (Schierano et al., 2020; O'Neill et al., 2011), tannery (Calheiros et al., 2007; Saeed et al., 2012), textile (Bulc and Ojstršek, 2008; Noonpui and Thiravetyan, 2011), mine (Nguyen et al., 2019; Singh and Chakraborty, 2020), and pulp and paper wastewater (Arivoli et al., 2015; Choudhary et al., 2012). Along with this, treated wastewater by CWLs can be used easily for various purposes, including in agriculture, gardening, industries, and flushing. However, some CWLs could be advantageous over other types depending upon the situation. Some advantages and limitations of different types of CWLs are summarized in Table 7.1.

Regardless of the advantages, there are many limitations of CWLs uncontrolled growth of biomass, varying removal efficiency of different species, higher treatment duration, disposal of contaminated plants, mosquito breeding, and degradation of ambiance. Nonetheless, this technology is suitable for various wastewater treatments.

7.7 TREATED WASTEWATER REUSES OPPORTUNITIES IN AGRICULTURE

Several attempts have been made to reuse recycled wastewater by CWLs for various purposes. Mainly studies on the reuse of wastewater are conducted for agricultural purposes. For instance, Cui et al. (2003) conducted an experiment on the treatment and reuse of septic tank effluent by vertical-flow constructed wetland. The authors used treated water for cultivation of lettuce and water spinach and they found that nitrogen level was high in those cultivated crops. Another study conducted by Morari and Giardini (2009) based on recycling and reuse used vertically constructed

TABLE 7.1

Advantages and Disadvantages of HSSF, VSFF, and FWS CWs

Advantages	Disadvantages
HSSF CWs	
Long flowing distance possible; nutrient gradients can establish	Higher area demand
Nitrification and denitrification possible	Careful calculation of hydraulics necessary for optimal O_2 supply
Formation of humic acids for N and P removal	Equal wastewater supply is complicated
Longer life cycle	
VSSF CWs	
Smaller area demand	Short flow distances
Good oxygen supply, good nitrification	Poor denitrification
Simple hydraulics	Higher technical demands
High purification performance from the beginning	Loss of performance especially in P removal (saturation)
FWS CWs	
Addition to the "green space" in a community	High area demand
BOD, TSS, COD, metals, and organic material removal in a reasonable detention time	Anoxic environment, poor nitrification
N and P removal in a significantly lower detention time	Mosquito production
Minimization of mechanical equipment, energy, and skilled operator requirement	

wetlands for the treatment of domestic wastewater. The authors found that suspended solids and phosphorous were high after the treatment making it unfit for reuse for irrigational purposes. Similarly, Marecos de Monte and Albuquerque (2010) studied wastewater treatment and their reuse by horizontal subsurface-flow CWLs. The authors found that all relevant water characteristics parameters such as COD, BOD, TN, TP, K, Ca, Mg, and Cl_2 were suitable for reuse in irrigation according to international standards. However, they suggested the implementation of wastewater disinfection to reduce pathogens before using it for irrigational purposes.

As recycled wastewater by CWLs contains nutrients and several trace minerals, they provide higher yields to crops and reduce fertilizer demands (Almuktar et al., 2017). However, reuse of wastewater in irrigation has several disadvantages, including potential public health impact, degradation of soil, crop, and groundwater quality, demonetization of property value, and other social and environmental impacts (Almuktar and Scholz, 2016). The main culprit is the presence of heavy metals and pathogens in recycled wastewater, which greatly affects human and ecological health. Along with this, chemicals present in recycled wastewater from different industries get accumulated in plant tissues and enter into the food chain and finally end up in the human body. These chemicals also affect the productivity and fertility of the soil.

The common problem with the use of recycled wastewater in irrigation is the increase of alkalinity, and the decrease in soil permeability because of the accumulation of toxic chemicals and nutrients. Also, the leaching of these contaminants into groundwater is a potential threat while using this water. Thus, while using recycled wastewater for irrigational purposes, groundwater should be evaluated to prevent water contamination (Almuktar et al., 2018).

7.8 INTEGRATED MICROBIAL FUEL CELLS WITH CWLs (CWL-MFCs)

Recently, a novel concept of formation of CWL-MFCs by inserting microbial fuel cells (MFCs) into CWLs is studied extensively (Zhao et al., 2013; Tang et al., 2019; Yang et al., 2020). The reason for using CWLs and MFCs as one system is because of the ubiquitous presence of anaerobic and aerobic zones in both systems. Along with this, this system facilitates a sustainable treatment approach as it generates electricity while treating wastewater (Vymazal et al., 2021). Several studies have been conducted around the globe on various aspects of CWL-MFC including electrode materials (Ge et al., 2020), system configuration (Aguirre-Sierra et al., 2020), increased biodegradation (Li et al., 2019), removal of emerging pollutants (Ji et al., 2020), biosensor development (Corbella et al., 2019), and maximum electricity generation (Xu et al., 2019). However, further research is needed to evaluate microbial behavior, electron transfer, and the development of different biosensors for a thorough understanding of CW-MFC technology.

7.9 COST ANALYSIS

The capital investment cost of CWLs includes the cost of the land, design cost, earthwork, liners, filtration, substrate media, plantation, hydraulic structures, and other miscellaneous costs (Wallace, 2015). As land, material, and labor cost varies around the world, the estimated total cost of CWLs also varies from place to place. For instance, the total cost/m^2 of CWL is estimated to be 29 USD in India and 290 USD in Belgium (Rousseau et al., 2004). Usually, the cost of subsurface-flow CWLs is higher than free water surface-flow CWLs as a subsurface-flow CWL needs a more sophisticated design. According to economic studies conducted by Rousseau et al. (2004), 392, 507, and 1258 euros are required for the construction of CWLs for a capacity of 201, 158, and 251 m^3, respectively.

7.10 CHALLENGES AND FUTURE RECOMMENDATIONS

Several studies have been conducted to improve and modify constructed wetland facilities in the past ten years. These modifications include hydraulic design, loading rate, selection of plant and novel substrates, different configurations, and mode of operation. However, most of these studies are conducted on the lab scale. Still, scaling up to a full-scale level is one of the major challenges. On a full-scale system, the scenario of maintenance, quality control, and environmental factors are different. Another main challenge is lack of long-term performance evaluation of these systems. Most of the studies have been done for a shorter span of time. Long-term system monitoring, like a decade or two, is needed as wastewater treatment systems should last for at least 20–30 years. Additionally, more studies should be conducted using real wastewater instead of synthetic ones to simulate field conditions.

There are recommendations for system design, maintenance, operation, and enhanced technologies. Since macrophytes and substrates greatly influence system performance, they must be chosen carefully. For the operation of the plant, the hydraulic condition needs to be optimized. Similarly, system maintenance on a regular interval must be done to avoid choking and clogging. In addition, the use of advanced technologies such as artificial aeration and microbial augmentation should be adopted to make the system more intelligent and efficient.

7.11 CONCLUSION

Constructed wetlands are considered a reliable option for wastewater treatment and reuse based on their performance. They treat almost all types of wastewater up to international standards so that it can be further discharged into the ecosystem or used for other purposes. Additionally, it can provide flood mitigation along with a reduction of carbon footprint. Different types of CWLs and their advantages/limitations are discussed here. Along with this, other advancements such as integrated CWL-MFCs and hybrid technologies are discussed. It is found that the latest integrated CWLs with novel substrate media are more efficient in nutrient and organic removal.

REFERENCES

Aalam, T., & Khalil, N. (2019). Performance of horizontal sub-surface flow constructed wetlands with different flow patterns using dual media for low-strength municipal wastewater: A case of pilot scale experiment in a tropical climate region. *Journal of Environmental Science and Health: Part A Toxic/Hazardous Substances and Environmental Engineering.* doi: 10.1080/10934529.2019.1635857.

Abou-Elela, S. I. (2019). Constructed Wetlands: The green technology for municipal wastewater treatment and reuse in agriculture. In: *Unconventional Water Resources and Agriculture in Egypt* (A. M. Negm (Ed.); pp. 189–239). Springer International Publishing: Cham. doi: 10.1007/698_2017_69.

Aguirre, P., Ojeda, E., García, J., Barragán, J., & Mujeriego, R. (2005). Effect of water depth on the removal of organic matter in horizontal subsurface flow constructed Wetlands. *Journal of Environmental Science and Health, Part A, 40*(6–7), 1457–1466. doi: 10.1081/ESE-200055886.

Aguirre-Sierra, A., Bacchetti-De Gregoris, T., Salas, J. J., De Deus, A., & Esteve-Núñez, A. (2020). A new concept in constructed wetlands: Assessment of aerobic electroconductive biofilters. *Environmental Science: Water Research and Technology.* doi: 10.1039/c9ew00696f.

Akratos, C. S., Papaspyros, J. N. E., & Tsihrintzis, V. A. (2009). Total nitrogen and ammonia removal prediction in horizontal subsurface flow constructed wetlands: Use of artificial neural networks and development of a design equation. *Bioresource Technology.* doi: 10.1016/j.biortech.2008.06.071.

Akratos, C. S., & Tsihrintzis, V. A. (2007). Effect of temperature, HRT, vegetation and porous media on removal efficiency of pilot-scale horizontal subsurface flow constructed wetlands. *Ecological Engineering.* doi: 10.1016/j.ecoleng.2006.06.013.

Alam, R., Khan, S. U., Basheer, F., & Farooqi, I. H.(2021). Nutrients and organics removal from slaughterhouse wastewater using phytoremediation: A comparative study on different aquatic plant species. *IOP Conference Series: Materials Science and Engineering, 1058*(1), 012068. IOP Publishing. doi: 10.1088/1757-899X/1058/1/012068.

Ali Shah, F., Mahmood, Q., Maroof Shah, M., Pervez, A., & Ahmad Asad, S. (2014). Microbial ecology of anaerobic digesters: The key players of anaerobiosis. *The Scientific World Journal*, 2014, 183752. doi: 10.1155/2014/183752.

Almuktar, S. A. A. A. N., Abed, S. N., & Scholz, M. (2017). Recycling of domestic wastewater treated by vertical-flow wetlands for irrigation of two consecutive Capsicum annuum generations. *Ecological Engineering*. doi: 10.1016/j.ecoleng.2017.07.002.

Almuktar, S. A. A. A. N., Abed, S. N., & Scholz, M. (2018). Wetlands for wastewater treatment and subsequent recycling of treated effluent: A review. *Environmental Science and Pollution Research*. doi: 10.1007/s11356-018-2629-3.

Almuktar, S. A. A. A. N., & Scholz, M. (2016). Mineral and biological contamination of soil and Capsicum annuum irrigated with recycled domestic wastewater. *Agricultural Water Management*. doi: 10.1016/j.agwat.2016.01.008.

Álvarez, J. A., Ávila, C., Otter, P., Kilian, R., Istenič, D., Rolletschek, M., Molle, P., Khalil, N., Ameršek, I., Mishra, V. K., Jorgensen, C., Garfi, A., Carvalho, P., Brix, H., & Arias, C. A. (2017). Constructed wetlands and solar-driven disinfection technologies for sustainable wastewater treatment and reclamation in rural India: SWINGS project. *Water Science and Technology*. doi: 10.2166/wst.2017.329.

Amin, F. R., Khalid, H., El-Mashad, H. M., Chen, C., Liu, G., & Zhang, R. (2021). Functions of bacteria and archaea participating in the bioconversion of organic waste for methane production. *Science of the Total Environment*. doi: 10.1016/j.scitotenv.2020.143007.

Angelakis, A. N., & Snyder, S. A. (2015). Wastewater treatment and reuse: Past, present, and future. *Water*, 7(9). doi: 10.3390/w7094887.

Arivoli, A., Mohanraj, R., & Seenivasan, R. (2015). Application of vertical flow constructed wetland in treatment of heavy metals from pulp and paper industry wastewater. *Environmental Science and Pollution Research*. doi: 10.1007/s11356-015-4594-4.

Ayele, A., & Godeto, Y. G. (2021). Bioremediation of chromium by microorganisms and its mechanisms related to functional groups. *Journal of Chemistry*. doi: 10.1155/2021/7694157.

Barbera, A. C., Cirelli, G. L., Cavallaro, V., Di Silvestro, I., Pacifici, P., Castiglione, V., Toscano, A., & Milani, M. (2009). Growth and biomass production of different plant species in two different constructed wetland systems in Sicily. *Desalination*. doi: 10.1016/j.desal.2008.03.046.

Bassani, L., Pelissari, C., da Silva, A. R., & Sezerino, P. H. (2021). Feeding mode influence on treatment performance of unsaturated and partially saturated vertical flow constructed wetland. *Science of the Total Environment*, 754, 142400. doi: doi: 10.1016/j.scitotenv.2020.142400.

Begg, J. S., Lavigne, R. L., & Veneman, P. L. M. (2001). Reed beds: Constructed wetlands for municipal wastewater treatment plant sludge dewatering. *Water Science and Technology*. doi: 10.2166/wst.2001.0857.

Boretti, A., & Rosa, L. (2019). Reassessing the projections of the world water development report. *NPJ Clean Water*. doi: 10.1038/s41545-019-0039-9.

Bôto, M., Almeida, C. M. R., & Mucha, A. P. (2016). Potential of constructed Wetlands for removal of antibiotics from saline aquaculture effluents. *Water*, 8(10). doi: 10.3390/w8100465.

Brix, H. (1987). The applicability of the wastewater treatment plant in othfresen as scientific documentation of the root-zone method. *Water Science and Technology*. doi: 10.2166/wst.1987.0093.

Bulc, T. G., & Ojstršek, A. (2008). The use of constructed wetland for dye-rich textile wastewater treatment. *Journal of Hazardous Materials*. doi: 10.1016/j.jhazmat.2007.11.068.

Bus, A., & Karczmarczyk, A. (2017). Supporting constructed wetlands in P removal efficiency from surface water. *Water Science and Technology*, 75(11), 2554–2561. doi: 10.2166/wst.2017.134.

Calheiros, C. S. C., Quitério, P. V. B., Silva, G., Crispim, L. F. C., Brix, H., Moura, S. C., & Castro, P. M. L. (2012). Use of constructed wetland systems with Arundo and Sarcocornia for polishing high salinity tannery wastewater. *Journal of Environmental Management*, 95(1), 66–71. doi: 10.1016/j.jenvman.2011.10.003.

Calheiros, C. S. C., Rangel, A. O. S. S., & Castro, P. M. L. (2007). Constructed wetland systems vegetated with different plants applied to the treatment of tannery wastewater. *Water Research*. doi: 10.1016/j.watres.2007.01.012.

Cao, Z., Zhou, L., Gao, Z., Huang, Z., Jiao, X., Zhang, Z., Ma, K., Di, Z., & Bai, Y. (2021). Comprehensive benefits assessment of using recycled concrete aggregates as the substrate in constructed wetland polishing effluent from wastewater treatment plant. *Journal of Cleaner Production*. doi: 10.1016/j.jclepro.2020.125551.

Caselles-Osorio, A., & García, J. (2007). Impact of different feeding strategies and plant presence on the performance of shallow horizontal subsurface-flow constructed wetlands. *Science of the Total Environment*. doi: 10.1016/j.scitotenv.2007.02.031.

Chang, J., Ma, L., Chen, J., Lu, Y., & Wang, X. (2017). Greenhouse wastewater treatment by baffled subsurface-flow constructed wetlands supplemented with flower straws as carbon source in different modes. *Environmental Science and Pollution Research*. doi: 10.1007/s11356-016-7922-4.

Chen, T. Y., Kao, C. M., Yeh, T. Y., Chien, H. Y., & Chao, A. C. (2006). Application of a constructed wetland for industrial wastewater treatment: A pilot-scale study. *Chemosphere*. doi: 10.1016/j.chemosphere.2005.11.069.

Choudhary, A. K., Kumar, S., & Sharma, C. (2011). Constructed wetlands : An approach for wastewater treatment. *Elixir Pollution*, 37, 3666–3672.

Choudhary, A. K., Kumar, S., & Sharma, C. (2012). Removal of chlorophenolics from pulp and paper mill wastewater through constructed Wetland. *Water Environment Research*. doi: 10.2175/106143012x13415215907419.

Chung, A. K. C., Wu, Y., Tam, N. F. Y., & Wong, M. H. (2008). Nitrogen and phosphate mass balance in a sub-surface flow constructed wetland for treating municipal wastewater. *Ecological Engineering*. doi: 10.1016/j.ecoleng.2007.09.007. Coban, O., Kuschk, P., Kappelmeyer, U., Spott, O., Martienssen, M., Jetten, M. S. M., & Knoeller, K. (2015). Nitrogen transforming community in a horizontal subsurface-flow constructed wetland. *Water Research*. doi: 10.1016/j.watres.2015.02.018.

Comino, E., Riggio, V., & Rosso, M. (2013). Grey water treated by an hybrid constructed wetland pilot plant under several stress conditions. *Ecological Engineering*. doi: 10.1016/j.ecoleng.2012.11.014.

Cong Manh, N., Van Minh, P., Tri Quang Hung, N., Thai Son, P., & Minh Ky, N. (2019). A study to assess the effectiveness of constructed wetland technology for polluted surface water treatment. *VNU Journal of Science: Earth and Environmental Sciences*. doi: 10.25073/2588-1094/vnuees.4372.

Corbella, C., Hartl, M., Fernandez-Gatell, M., & Puigagut, J. (2019). MFC-based biosensor for domestic wastewater COD assessment in constructed wetlands. *The Science of the Total Environment*, 660, 218–226. doi: 10.1016/j.scitotenv.2018.12.347.

Crini, G., & Lichtfouse, E. (2019). Advantages and disadvantages of techniques used for wastewater treatment. *Environmental Chemistry Letters*. doi: 10.1007/s10311-018-0785-9.

Cui, L.-H., Luo, S.-M., Zhu, X.-Z., & Liu, Y.-H. (2003). Treatment and utilization of septic tank effluent using vertical-flow constructed wetlands and vegetable hydroponics. *Journal of Environmental Sciences (China)*, 15(1), 75–82.

De Monte, H. M., & Albuquerque, A. (2010). Analysis of constructed wetland performance for irrigation reuse. *Water Science and Technology*. doi: 10.2166/wst.2010.063.

El Hamouri, B., Nazih, J., & Lahjouj, J. (2007). Subsurface-horizontal flow constructed wetland for sewage treatment under Moroccan climate conditions. *Desalination*. doi: 10.1016/j.desal.2006.11.018.

El-Sheikh, M. A., Saleh, H. I., El-Quosy, D. E., & Mahmoud, A. A. (2010). Improving water quality in polluated drains with free water surface constructed wetlands. *Ecological Engineering.* doi: 10.1016/j.ecoleng.2010.06.030.

Fan, J., Wang, W., Zhang, B., Guo, Y., Ngo, H. H., Guo, W., Zhang, J., & Wu, H. (2013). Nitrogen removal in intermittently aerated vertical flow constructed wetlands: Impact of influent COD/N ratios. *Bioresource Technology.* doi: 10.1016/j.biortech.2013.06.038.

Farooqi, I. H., Basheer, F., & Chaudhari, R. J. (2008). Constructed Wetland System (CWS) for wastewater treatment. In *Proceedings of Taal2007: The 12th World Lake Conference*, Vol. 1004, p. 1009.

Faulwetter, J. L., Gagnon, V., Sundberg, C., Chazarenc, F., Burr, M. D., Brisson, J., Camper, A. K., & Stein, O. R. (2009). Microbial processes influencing performance of treatment wetlands: A review. *Ecological Engineering.* doi: 10.1016/j.ecoleng.2008.12.030.

Fitton, N., Alexander, P., Arnell, N., Bajzelj, B., Calvin, K., Doelman, J., Gerber, J. S., Havlik, P., Hasegawa, T., Herrero, M., Krisztin, T., van Meijl, H., Powell, T., Sands, R., Stehfest, E., West, P. C., & Smith, P. (2019). The vulnerabilities of agricultural land and food production to future water scarcity. *Global Environmental Change.* doi: 10.1016/j.gloenvcha.2019.101944.

García, J., Aguirre, P., Barragán, J., Mujeriego, R., Matamoros, V., & Bayona, J. M. (2005). Effect of key design parameters on the efficiency of horizontal subsurface flow constructed wetlands. *Ecological Engineering.* doi: 10.1016/j.ecoleng.2005.06.010.

García, J., Rousseau, D. P. L., Morató, J., Lesage, E. L. S., Matamoros, V., & Bayona, J. M. (2010). Contaminant removal processes in subsurface-flow constructed wetlands: A review. *Critical Reviews in Environmental Science and Technology*, 40(7), 561–661. doi: 10.1080/10643380802471076.

Ge, X., Cao, X., Song, X., Wang, Y., Si, Z., Zhao, Y., Wang, W., & Tesfahunegn, A. A. (2020). Bioenergy generation and simultaneous nitrate and phosphorus removal in a pyrite-based constructed wetland-microbial fuel cell. *Bioresource Technology.* doi: 10.1016/j.biortech.2019.122350.

Ghimire, U., Nandimandalam, H., Martinez-Guerra, E., & Gude, V. G. (2019). Wetlands for wastewater treatment. *Water Environment Research*, 91(10), 1378–1389. doi: doi: 10.1002/wer.1232.

Gholipour, A., & Stefanakis, A. I. (2021). A full-scale anaerobic baffled reactor and hybrid constructed wetland for university dormitory wastewater treatment and reuse in an arid and warm climate. *Ecological Engineering.* doi: 10.1016/j.ecoleng.2021.106360.

Gosling, S. N., & Arnell, N. W. (2016). A global assessment of the impact of climate change on water scarcity. *Climatic Change.* doi: 10.1007/s10584-013-0853-x.

Gray, S., Kinross, J., Read, P., & Marland, A. (2000). The nutrient assimilative capacity of maerl as a substrate in constructed wetland systems for waste treatment. *Water Research.* doi: 10.1016/S0043-1354(99)00414-5.

Guo, C., Cui, Y., Dong, B., Luo, Y., Liu, F., Zhao, S., & Wu, H. (2017). Test study of the optimal design for hydraulic performance and treatment performance of free water surface flow constructed wetland. *Bioresource Technology*, 238, 461–471. doi: 10.1016/j.biortech.2017.03.163.

Guo, Z., Kang, Y., Hu, Z., Liang, S., Xie, H., Ngo, H. H., & Zhang, J. (2020). Removal pathways of benzofluoranthene in a constructed wetland amended with metallic ions embedded carbon. *Bioresource Technology.* doi: 10.1016/j.biortech.2020.123481.

He, Y., Nurul, S., Schmitt, H., Sutton, N. B., Murk, T. A. J., Blokland, M. H., Rijnaarts, H. H. M., & Langenhoff, A. A. M. (2018). Evaluation of attenuation of pharmaceuticals, toxic potency, and antibiotic resistance genes in constructed wetlands treating wastewater effluents. *Science of the Total Environment.* doi: 10.1016/j.scitotenv.2018.03.083.

Herrera-Melián, J. A., Borreguero-Fabelo, A., Araña, J., Peñate-Castellano, N., & Ortega-Méndez, J. A. (2018). Effect of substrate, feeding mode and number of stages on the performance of hybrid constructed wetland systems. *Water*, 10(1). doi: 10.3390/w10010039.

Jesus, J. M., Danko, A. S., Fiúza, A., & Borges, M.-T. (2018). Effect of plants in constructed wetlands for organic carbon and nutrient removal: A review of experimental factors contributing to higher impact and suggestions for future guidelines. *Environmental Science and Pollution Research International*, 25(5), 4149–4164. doi: 10.1007/s11356-017-0982-2.

Ji, B., Kang, P., Wei, T., & Zhao, Y. (2020). Challenges of aqueous per- and polyfluoroalkyl substances (PFASs) and their foreseeable removal strategies. *Chemosphere*. doi: 10.1016/j.chemosphere.2020.126316.

Jia, W., Zhang, J., Wu, J., Xie, H., & Zhang, B. (2010). Effect of intermittent operation on contaminant removal and plant growth in vertical flow constructed wetlands: A microcosm experiment. *Desalination*. doi: 10.1016/j.desal.2010.06.012.

Kantawanichkul, S., Somprasert, S., Aekasin, U., & Shutes, R. B. E. (2003). Treatment of agricultural wastewater in two experimental combined constructed wetland systems in a tropical climate. *Water Science and Technology*. doi: 10.2166/wst.2003.0319.

Khan, S. U., Noor, A., & Farooqi, I. H. (2015). GIS application for groundwater management and quality mapping in rural areas of District Agra, India. *International Journal of Water Resources & Arid Environments (IJWRAE)*, 4(1), 89–96.

Kumar, S., & Dutta, V. (2019). Constructed wetland microcosms as sustainable technology for domestic wastewater treatment: An overview. *Environmental Science and Pollution Research*. doi: 10.1007/s11356-019-04816-9.

Lan, W., Zhang, J., Hu, Z., Ji, M., Zhang, X., Zhang, J., Li, F., & Yao, G. (2018). Phosphorus removal enhancement of magnesium modified constructed wetland microcosm and its mechanism study. *Chemical Engineering Journal*. doi: 10.1016/j.cej.2017.10.150.

Langergraber, G., Prandtstetten, C., Pressl, A., Rohrhofer, R., & Haberl, R. (2007). Optimization of subsurface vertical flow constructed wetlands for wastewater treatment. *Water Science and Technology*. doi: 10.2166/wst.2007.129.

Lee, C. G., Fletcher, T. D., & Sun, G. (2009). Nitrogen removal in constructed wetland systems. *Engineering in Life Sciences*. doi: 10.1002/elsc.200800049.

Li, C., Wu, S., & Dong, R. (2015). Dynamics of organic matter, nitrogen and phosphorus removal and their interactions in a tidal operated constructed wetland. *Journal of Environmental Management*. doi: 10.1016/j.jenvman.2015.01.011.

Li, H., Zhang, S., Yang, X. L., Yang, Y. L., Xu, H., Li, X. N., & Song, H. L. (2019). Enhanced degradation of bisphenol A and ibuprofen by an up-flow microbial fuel cell-coupled constructed wetland and analysis of bacterial community structure. *Chemosphere*. doi: 10.1016/j.chemosphere.2018.11.022.

Li, Q., Long, Z., Wang, H., & Zhang, G. (2021). Functions of constructed wetland animals in water environment protection: A critical review. *Science of the Total Environment*. doi: 10.1016/j.scitotenv.2020.144038.

Manuel, P. M. (2003). Cultural perceptions of small urban wetlands: Cases from the Halifax Regional Municipality, Nova Scotia, Canada. *Wetlands*. doi: 10.1672/0277-5212(2003)023[0921:CPOSUW]2.0.CO;2.

Meng, P., Pei, H., Hu, W., Shao, Y., & Li, Z. (2014). How to increase microbial degradation in constructed wetlands: influencing factors and improvement measures. *Bioresource Technology*, 157, 316–326. doi: 10.1016/j.biortech.2014.01.095.

Morari, F., & Giardini, L. (2009). Municipal wastewater treatment with vertical flow constructed wetlands for irrigation reuse. *Ecological Engineering*. doi: 10.1016/j.ecoleng.2008.10.014.

Nan, X., Lavrnić, S., & Toscano, A. (2020). Potential of constructed wetland treatment systems for agricultural wastewater reuse under the EU framework. *Journal of Environmental Management*. doi: 10.1016/j.jenvman.2020.111219.

Nguyen, H. T. H., Nguyen, B. Q., Duong, T. T., Bui, A. T. K., Nguyen, H. T. A., Cao, H. T., Mai, N. T., Nguyen, K. M., Pham, T. T., & Kim, K.-W. (2019). Pilot-scale removal of arsenic and heavy metals from mining wastewater using adsorption combined with constructed wetland. *Minerals*, 9(6). doi: 10.3390/min9060379.

Noonpui, S., & Thiravetyan, P. (2011). Treatment of reactive azo dye from textile wastewater by burhead (Echinodorus cordifolius L.) in constructed wetland: Effect of molecular size. *Journal of Environmental Science and Health, Part A, 46*(7), 709–714. doi: 10.1080/10934529.2011.571577.

Noureddine, R., & Ouakouak, A. (2018). Study of domestic wastewater treatment by macrophyte plant in arid region of south-east Algeria (case of El Oued region). *Journal of Fundamental and Applied Sciences.* doi: 10.4314/jfas.v10i2.19.

O'Neill, A., Foy, R. H., & Phillips, D. H. (2011). Phosphorus retention in a constructed wetland system used to treat dairy wastewater. *Bioresource Technology.* doi: 10.1016/j.biortech.2011.01.075.

Obarska-Pempkowiak, H., & Klimkowska, K. (1999). Distribution of nutrients and heavy metals in a constructed wetland system. *Chemosphere.* doi: 10.1016/S0045-6535(99)00111-3.

OECD. (2006). Test No. 208: Terrestrial Plant Test: Seedling Emergence and Seedling Growth Test. doi: 10.1787/9789264070066-en.

Okochi, N. C., & McMartin, D. W. (2011). Laboratory investigations of stormwater remediation via slag: Effects of metals on phosphorus removal. *Journal of Hazardous Materials, 187*(1–3), 250–257. doi: 10.1016/j.jhazmat.2011.01.015.

Oon, Y. L., Ong, S. A., Ho, L. N., Wong, Y. S., Oon, Y. S., Lehl, H. K., & Thung, W. E. (2015). Hybrid system up-flow constructed wetland integrated with microbial fuel cell for simultaneous wastewater treatment and electricity generation. *Bioresource Technology.* doi: 10.1016/j.biortech.2015.03.014.

Ouellet-Plamondon, C., Chazarenc, F., Comeau, Y., & Brisson, J. (2006). Artificial aeration to increase pollutant removal efficiency of constructed wetlands in cold climate. Ecological Engineering. doi: 10.1016/j.ecoleng.2006.03.006.

Parde, D., Patwa, A., Shukla, A., Vijay, R., Killedar, D. J., & Kumar, R. (2021). A review of constructed wetland on type, treatment and technology of wastewater. *Environmental Technology and Innovation.* doi: 10.1016/j.eti.2020.101261.

Park, J.-H., Wang, J. J., Kim, S.-H., Cho, J.-S., Kang, S.-W., Delaune, R. D., & Seo, D.-C. (2017). Phosphate removal in constructed wetland with rapid cooled basic oxygen furnace slag. *Chemical Engineering Journal, 327*, 713–724. doi: doi: 10.1016/j.cej.2017.06.155.

Poor, C., Burrill, K., & Jarvis, M. (2020). Efficiency of constructed wetlands for nutrient removal. *World Environmental and Water Resources Congress 2020: Groundwater, Sustainability, Hydro-Climate/Climate Change, and Environmental Engineering - Selected Papers from the Proceedings of the World Environmental and Water Resources Congress 2020.* doi: 10.1061/9780784482964.001.

Quispe, R., Soto, M., Ingaruca, E., Bulege, W., & Custodio, M. (2019). Optimization of the operation of a municipal wastewater treatment plant with hydrocotyle ranunculoides. *Journal of Ecological Engineering.* doi: 10.12911/22998993/112486.

Rahman, M. E., Halmi, M. I. E. Bin, Samad, M. Y. B. A., Uddin, M. K., Mahmud, K., Shukor, M. Y. A., Abdullah, S. R. S., & Shamsuzzaman, S. M. (2020). Design, operation and optimization of constructed wetland for removal of pollutant. *International Journal of Environmental Research and Public Health.* doi: 10.3390/ijerph17228339.

Rai, U. N., Tripathi, R. D., Singh, N. K., Upadhyay, A. K., Dwivedi, S., Shukla, M. K., Mallick, S., Singh, S. N., & Nautiyal, C. S. (2013). Constructed wetland as an ecotechnological tool for pollution treatment for conservation of Ganga river. *Bioresource Technology.* doi: 10.1016/j.biortech.2013.09.005.

Rehman, F., Pervez, A., Khattak, B. N., & Ahmad, R. (2017). Constructed wetlands: Perspectives of the oxygen released in the rhizosphere of macrophytes. *Clean: Soil, Air, Water.* doi: 10.1002/clen.201600054.

Reyes-Contreras, C., Matamoros, V., Ruiz, I., Soto, M., & Bayona, J. M. (2011). Evaluation of PPCPs removal in a combined anaerobic digester-constructed wetland pilot plant treating urban wastewater. *Chemosphere.* doi: 10.1016/j.chemosphere.2011.06.003.

Rousseau, D. P. L., Vanrolleghem, P. A., & De Pauw, N. (2004). Constructed wetlands in Flanders: A performance analysis. *Ecological Engineering.* doi: 10.1016/j.ecoleng.2004.08.001.

Ry, W., Korboulewsky, N., Prudent, P., Domeizel, M., Rolando, C., & Bonin, G. (2010). Feasibility of using an organic substrate in a wetland system treating sewage sludge: Impact of plant species. *Bioresource Technology, 101*(1), 51–57. doi: 10.1016/j.biortech.2009.07.080.

Saeed, T., Afrin, R., Muyeed, A. Al, & Sun, G. (2012). Treatment of tannery wastewater in a pilot-scale hybrid constructed wetland system in Bangladesh. *Chemosphere.* doi: 10.1016/j.chemosphere.2012.04.055.

Saeed, T., & Sun, G. (2011). Enhanced denitrification and organics removal in hybrid wetland columns: Comparative experiments. *Bioresource Technology.* doi: 10.1016/j.biortech.2010.09.056.

Saeed, T., & Sun, G. (2012). A review on nitrogen and organics removal mechanisms in subsurface flow constructed wetlands: Dependency on environmental parameters, operating conditions and supporting media. *Journal of Environmental Management.* doi: 10.1016/j.jenvman.2012.08.011.

Sánchez-Monedero, M. A., Aguilar, M. I., Fenoll, R., & Roig, A. (2008). Effect of the aeration system on the levels of airborne microorganisms generated at wastewater treatment plants. *Water Research.* doi: 10.1016/j.watres.2008.06.028.

Sasikala, S., Tanaka, N., Jinadasa, K., & Mowjood, M. (2010). Comparison study of pulsing and continuous flow for improving effluent water quality and plant growth of a constructed wetland to treat domestic wastewater. *Tropical Agricultural Research.* doi: 10.4038/tar.v21i2.2596.

Schierano, M. C., Panigatti, M. C., Maine, M. A., Griffa, C. A., & Boglione, R. (2020). Horizontal subsurface flow constructed wetland for tertiary treatment of dairy wastewater: Removal efficiencies and plant uptake. *Journal of Environmental Management.* doi: 10.1016/j.jenvman.2020.111094.

Serrano, L., de la Varga, D., Ruiz, I., & Soto, M. (2011). Winery wastewater treatment in a hybrid constructed wetland. *Ecological Engineering.* doi: 10.1016/j.ecoleng.2010.06.038.

Singh, S., & Chakraborty, S. (2020). Performance of organic substrate amended constructed wetland treating acid mine drainage (AMD) of North-Eastern India. *Journal of Hazardous Materials.* doi: 10.1016/j.jhazmat.2020.122719.

Solano, M. L., Soriano, P., & Ciria, M. P. (2004). Constructed wetlands as a sustainable solution for wastewater treatment in small villages. *Biosystems Engineering.* doi: 10.1016/j.biosystemseng.2003.10.005.

Song, H. L., Nakano, K., Taniguchi, T., Nomura, M., & Nishimura, O. (2009). Estrogen removal from treated municipal effluent in small-scale constructed wetland with different depth. *Bioresource Technology.* doi: 10.1016/j.biortech.2009.01.045.

Stottmeister, U., Wießner, A., Kuschk, P., Kappelmeyer, U., Kästner, M., Bederski, O., Müller, R. A., & Moormann, H. (2003). Effects of plants and microorganisms in constructed wetlands for wastewater treatment. *Biotechnology Advances.* doi: 10.1016/j.biotechadv.2003.08.010.

Tan, X., Yang, Y., Liu, Y., Li, X., Fan, X., Zhou, Z., Liu, C., & Yin, W. (2019). Enhanced simultaneous organics and nutrients removal in tidal flow constructed wetland using activated alumina as substrate treating domestic wastewater. *Bioresource Technology.* doi: 10.1016/j.biortech.2019.02.036.

Tang, C., Zhao, Y., Kang, C., Yang, Y., Morgan, D., & Xu, L. (2019). Towards concurrent pollutants removal and high energy harvesting in a pilot-scale CW-MFC: Insight into the cathode conditions and electrodes connection. *Chemical Engineering Journal.* doi: 10.1016/j.cej.2019.05.035.

Tee, H. C., Lim, P. E., Seng, C. E., & Nawi, M. A. M. (2012). Newly developed baffled subsurface-flow constructed wetland for the enhancement of nitrogen removal. *Bioresource Technology.* doi: 10.1016/j.biortech.2011.11.032.

Titah, H. S., Abdullah, S. R. S., Mushrifah, I., Anuar, N., Basri, H., & Mukhlisin, M. (2014). Phytotoxicity and uptake of arsenic by ludwigia octovalvis in a pilot reed bed system. *Environmental Engineering Science.* doi: 10.1089/ees.2013.0207.

Toet, S., Van Logtestijn, R. S. P., Kampf, R., Schreijer, M., & Verhoeven, J. T. A. (2005). The effect of hydraulic retention time on the removal of pollutants from sewage treatment plant effluent in a surface-flow wetland system. *Wetlands, 25*(2), 375–391. doi: 10.1672/13.

Trulli, E., Torretta, V., & Rada, E. C. (2016). Water restoration of an urbanized karst stream by free-water-surface constructed wetlands as municipal wastewater post-treatment. *UPB Scientific Bulletin, 78*, 163–174.

Tsihrintzis, V. A., Akratos, C. S., Gikas, G. D., Karamouzis, D., & Angelakis, A. N. (2007). Performance and cost comparison of a FWS and a VSF constructed wetland system. *Environmental Technology.* doi: 10.1080/09593332808618820.

Vohla, C., Alas, R., Nurk, K., Baatz, S., & Mander, Ü. (2007). Dynamics of phosphorus, nitrogen and carbon removal in a horizontal subsurface flow constructed wetland. *Science of the Total Environment.* doi: 10.1016/j.scitotenv.2006.09.012.

Vymazal, J. (2009). The use constructed wetlands with horizontal sub-surface flow for various types of wastewater. *Ecological Engineering.* doi: 10.1016/j.ecoleng.2008.08.016.

Vymazal, J. (2011). Plants used in constructed wetlands with horizontal subsurface flow: A review. *Hydrobiologia.* doi: 10.1007/s10750-011-0738-9.

Vymazal, J. (2013). Emergent plants used in free water surface constructed wetlands: A review. *Ecological Engineering.* doi: 10.1016/j.ecoleng.2013.06.023.

Vymazal, J. (2014). Constructed wetlands for treatment of industrial wastewaters: A review. *Ecological Engineering.* doi: 10.1016/j.ecoleng.2014.09.034.

Vymazal, J., Greenway, M., Tonderski, K., Brix, H., & Mander, Ü. (2006). Constructed wetlands for wastewater treatment. In: *Wetlands and Natural Resource Management* (J. T. A. Verhoeven, B. Beltman, R. Bobbink, & D. F. Whigham (Eds.); pp. 69–96). Springer: Berlin Heidelberg. doi: 10.1007/978-3-540-33187-2_5.

Vymazal, J., Zhao, Y., & Mander, Ü. (2021). Recent research challenges in constructed wetlands for wastewater treatment: A review. *Ecological Engineering.* doi: 10.1016/j.ecoleng.2021.106318.

Wallace, S. (2015). Feasibility, design criteria, and O&M requirements for small scale constructed wetland wastewater treatment systems. *Water Intelligence Online.* doi: 10.2166/9781780403991.

Wang, M., Zhang, D., Dong, J., & Tan, S. K. (2018). Application of constructed wetlands for treating agricultural runoff and agro-industrial wastewater: A review. *Hydrobiologia.* doi: 10.1007/s10750-017-3315-z.

Wu, H., Fan, J., Zhang, J., Ngo, H. H., Guo, W., Liang, S., Lv, J., Lu, S., Wu, W., & Wu, S. (2016). Intensified organics and nitrogen removal in the intermittent-aerated constructed wetland using a novel sludge-ceramsite as substrate. *Bioresource Technology.* doi: 10.1016/j.biortech.2016.01.051.

Wu, H., Zhang, J., Ngo, H. H., Guo, W., Hu, Z., Liang, S., Fan, J., & Liu, H. (2015a). A review on the sustainability of constructed wetlands for wastewater treatment: Design and operation. *Bioresource Technology.* doi: 10.1016/j.biortech.2014.10.068.

Wu, J., Xu, D., He, F., He, J., & Wu, Z. (2015b). Comprehensive evaluation of substrates in vertical-flow constructed wetlands for domestic wastewater treatment. *Water Practice and Technology.* doi: 10.2166/wpt.2015.077.

Wu, S., Kuschk, P., Brix, H., Vymazal, J., & Dong, R. (2014). Development of constructed wetlands inperformance intensifications for wastewater treatment: A nitrogen and organic matter targeted review. *Water Research.* doi: 10.1016/j.watres.2014.03.020.

Wu, S., Kuschk, P., Wiessner, A., Müller, J., Saad, R. A. B., & Dong, R. (2013). Sulphur transformations in constructed wetlands for wastewater treatment: A review. *Ecological Engineering.* doi: 10.1016/j.ecoleng.2012.11.003.

Xu, D., Xu, J., Wu, J., & Muhammad, A. (2006). Studies on the phosphorus sorption capacity of substrates used in constructed wetland systems. *Chemosphere.* doi: 10.1016/j. chemosphere.2005.08.036.

Xu, F., Ouyang, D., Rene, E. R., Ng, H. Y., Guo, L., Zhu, Y., Zhou, L., Yuan, Q., Miao, M., Wang, Q., & Kong, Q. (2019). Electricity production enhancement in a constructed wetland-microbial fuel cell system for treating saline wastewater. *Bioresource Technology, 288,* 121462. doi: doi: 10.1016/j.biortech.2019.121462.

Xu, J., Zhang, J., Xie, H., Li, C., Bao, N., Zhang, C., & Shi, Q. (2010). Physiological responses of Phragmites australis to wastewater with different chemical oxygen demands. *Ecological Engineering.* doi: 10.1016/j.ecoleng.2010.06.010.

Yan, Y., & Xu, J. (2014). Improving winter performance of constructed wetlands for wastewater treatment in northern china: A Review. *Wetlands.* doi: 10.1007/s13157-013-0444-7.

Yang, Y., Zhao, Y., Tang, C., Xu, L., Morgan, D., & Liu, R. (2020). Role of macrophyte species in constructed wetland-microbial fuel cell for simultaneous wastewater treatment and bioenergy generation. *Chemical Engineering Journal.* doi: 10.1016/j.cej.2019.123708.

Yeh, T. Y., & Wu, C. H. (2009). Pollutant removal within hybrid constructed wetland systems in tropical regions. *Water Science and Technology.* doi: 10.2166/wst.2009.846.

Zhai, J., Xiao, H. W., Kujawa-Roeleveld, K., He, Q., & Kerstens, S. M. (2011). Experimental study of a novel hybrid constructed wetland for water reuse and its application in Southern China. *Water Science and Technology.* doi: 10.2166/wst.2011.790.

Zhang, D. Q., Gersberg, R. M., Zhu, J., Hua, T., Jinadasa, K. B. S. N., & Tan, S. K. (2012). Batch versus continuous feeding strategies for pharmaceutical removal by subsurface flow constructed wetland. *Environmental Pollution.* doi: 10.1016/j.envpol.2012.04.004.

Zhang, D. Q., Jinadasa, K. B. S. N., Gersberg, R. M., Liu, Y., Ng, W. J., & Tan, S. K. (2014). Application of constructed wetlands for wastewater treatment in developing countries: A review of recent developments (2000–2013). *Journal of Environmental Management, 141,* 116–131. doi: 10.1016/j.jenvman.2014.03.015.

Zhang, L., Zhang, L., Liu, Y., Shen, Y., Liu, H., & Xiong, Y. (2010). Effect of limited artificial aeration on constructed wetland treatment of domestic wastewater. *Desalination, 250*(3), 915–920. doi: doi: 10.1016/j.desal.2008.04.062.

Zhang, S., & Zou, Q. (2010). Vertical-flow constructed wetlands applied in a recirculating aquaculture system for channel catfish culture: Effects on water quality and zooplankton. *Polish Journal of Environmental Studies, 19*(5), 1063–1070. http://www.pjoes. com/Vertical-Flow-Constructed-Wetlands-Applied-r-nin-a-Recirculating-Aquaculture-System,88483,0,2.html.

Zhang, Y., Carvalho, P. N., Lv, T., Arias, C., Brix, H., & Chen, Z. (2016). Microbial density and diversity in constructed wetland systems and the relation to pollutant removal efficiency. *Water Science and Technology.* doi: 10.2166/wst.2015.542.

Zhao, J., Zhao, Y., Xu, Z., Doherty, L., & Liu, R. (2016). Highway runoff treatment by hybrid adsorptive media-baffled subsurface flow constructed wetland. *Ecological Engineering.* doi: 10.1016/j.ecoleng.2016.02.020.

Zhao, Y., Collum, S., Phelan, M., Goodbody, T., Doherty, L., & Hu, Y. (2013). Preliminary investigation of constructed wetland incorporating microbial fuel cell: Batch and continuous flow trials. *Chemical Engineering Journal.* doi: 10.1016/j.cej.2013.06.023.

Zhao, Y., Ji, B., Liu, R., Ren, B., & Wei, T. (2020). Constructed treatment wetland: Glance of development and future perspectives. *Water Cycle.* doi: 10.1016/j.watcyc.2020.07.002.

Zhao, Y. Q., Babatunde, A. O., Hu, Y. S., Kumar, J. L. G., & Zhao, X. H. (2011). Pilot field-scale demonstration of a novel alum sludge-based constructed wetland system for enhanced wastewater treatment. *Process Biochemistry.* doi: 10.1016/j.procbio.2010.08.023.

Zhao, Y. Q., Sun, G., & Allen, S. J. (2004). Anti-sized reed bed system for animal wastewater treatment: A comparative study. *Water Research.* doi: 10.1016/j.watres.2004.03.038.

8 Introduction to Micropollutants and Their Sources

Mohd Azfar Shaida
Aligarh Muslim University

Soumita Talukdar
G. D. Goenka University

Mohd Salim Mahtab and Izharul Haq Farooqi
Aligarh Muslim University

CONTENTS

8.1 Introduction to Micropollutants... 139
    8.1.1 Micropollutants in Human Health... 142
8.2 Sources of Micropollutants... 144
    8.2.1 Point Source Pollution 144
    8.2.2 Diffuse Source Pollution 145
    8.2.3 Occurrence of Micropollutants.. 146
8.3 Conclusion 147
Acknowledgments.. 147
References.. 148

8.1 INTRODUCTION TO MICROPOLLUTANTS

Micropollutants (MPs) can be described as the manmade organic contaminants of pharmaceuticals, pesticides, plasticizers, consumable products, soaps, etc. that are present in very low concentrations ranging from µg/L to ng/L (Kumar et al., 2020; Dubey et al., 2021). Over 70,000 varieties of chemicals are estimated to be produced by the chemical industry, and their uncontrolled release poses a serious threat to aquatic life (Rogowska et al., 2020). Also, frequent access to and long-term use of these chemicals have significantly increased the incoming of these trace compounds (Chavoshani et al., 2020). Most of the MPs are not regulated by law; therefore, their routine monitoring is not performed for surface or drinking waters; therefore, significant concentrations of MPs are present in the treated waters (Luo et al., 2014). The small fractions of the contaminants that are regulated by law are schematically

DOI: 10.1201/9781003202431-8

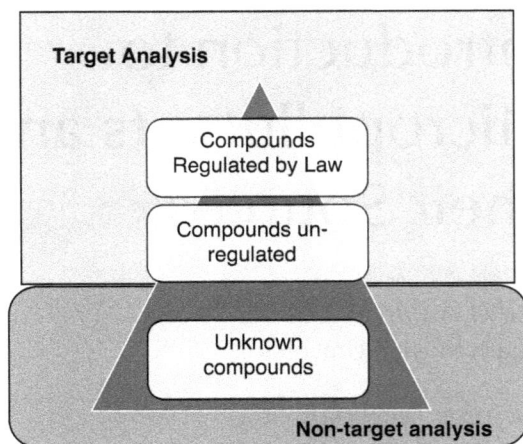

FIGURE 8.1 Schematic of contaminants in aquatic systems.

shown in Figure 8.1. Moreover, the degradation products and byproducts of the regulated contaminants may also generate secondary contamination, which may impose enhanced ecotoxicity on the aquatic environment (Rogowska et al., 2020). The removal of these MPs is also challenging due to their biological and physio-chemical stability (Zdarta et al., 2022). Prolonged exposure to such contaminants may induce harmful effects such as mutations and serious damage to the ecosystem (Zdarta et al., 2022).

The main constituent of these MPs includes pharmaceutical drugs, i.e., pain relief, anti-inflammatory drugs, personal care products, fertilizers, etc., which mainly come from domestic wastewater (Oturan et al., 2015). Drugs such as ofloxacin, ibuprofen, furosemide, ciprofloxacin, and sulfamethoxazole were found to intoxicate Italian rivers and induce harmful effects on the human embryonic cells (Zdarta et al., 2022). Diclofenac on the other hand was reported to alter the gene transcription in *Mytilusgallo provincialis* upon exposure to 1 μg/L in water. Estrogen compounds such as 17-β estradiol of concentration 11.2 ng/L and 17α-ethinylestradiol of concentration 0.5 ng/L induced feminine characteristics in fish (Zdarta et al., 2022). Ethylhexyl-methoxycinnamate, one of the major constituents in sunscreens, shampoos, creams, perfumes, etc., exhibits harmful effects on aquatic life predominantly on snails such as *Melanoides tuberculata* and *Potamopyrgus antipodarum* (Zdarta et al., 2022).

Industrial effluents such as dyes and other harmful products exhibit mutagenic, teratogenic, and carcinogenic properties. Moreover, these dyes are complex to degrade and inhibit sunlight penetration into the water, retarding the rate of photosynthesis (Lado Ribeiro et al., 2019). Phenols and halogenated phenols also impose significant threats to the environment. Tetrabromobisphenol A (TBBPA) is associated with endocrine disruption and cytotoxicity (Ngweme et al., 2021). Pharmaceutical ethynylestradiol affects the F1 generation of zebrafish and renders them infertile, whereas no such toxic effects have been observed in the F0 generation (Nash et al., 2004).

TABLE 8.1

Toxicity of Various MP Compounds and Their Harmful Effects on Various Species

Compound	Measured Parameter	Toxicity (mg/L)	Species
Citalopram	EC_{50}	3.300	*Pseudokirchneriella subcapitata*
Gabapentin	LD_{50}	50	*Oncorhynchus mykiss*
Tramadol	LC_{50}	130	Unspecified fish
Lamivudine	EC_{50}	49.06	*Pseudokirchneriella subcapitata*
Efavirenz	EC_{50}	0.012	*Pseudokirchneriella subcapitata*
N-acetyl-4-aminoantipyrine	LC_{20}	10	*Daphnia magna*
Diclofenac	EC_{50}	1,950	*Raoultella* sp., strain DD4
1H-benzotriazole	EC_{10}	1.18	*Desmodesmus subspicatus*
Irbesartan	EC_{50}	460	*Pseudok*
Iopromide	EC_{50}	10,000	*Pseudokirch*
Darunavir	EC_{50}	43–100	Green algae
Naproxen	EC_{50}	174	*Daphnia magna*
Acetaminophen	EC_{50}	>160	*Oryzias latipes*
Caffeine	EC_{50}	290.2	*Desmodesmus subspicatus*
Metronidazole-OH	IC_{50}	>100	Aerobic bacteria
Metformin	EC_{50}	64	*Daphnia magna*

Researchers have found that the concentration of diazinon was as high as 684 ng/L in treated wastewater. The pesticides that are reported to have the most harmful effects are diuron and diazinon. Other harmful compounds include simazine, atrazine, chlortoluron, malathion, isoproturon, and terbuthylazine (Zdarta et al., 2022). Dichlorodiphenyltrichloroethane (DDT) causes thinning of eggs shells in black ducks on exposure to 0.6 mg/kg of body weight (Kumar et al., 2020). Heptachlor, dieldrin, and organochlorine also cause toxic effects on aquatic organisms. The list of highly concentrated MPs found in wastewater and their toxicity is summarized in Table 8.1 (Rogowska et al., 2020). EC_{50} is the effective concentration that has a significant effect on 50% of the tested population, LC_{50}/LD_{50} is the lethal concentration or dosage that causes death in 50% of the tested population and IC_{50} is the concentration that causes inhibition of growth in 50% of the population (Rogowska et al., 2020).

Biofilms are one of the major constituents in the streams; they are a complex combination of algae, fungi, bacteria, and other microorganisms commonly known as periphyton (Desiante, Minas, and Fenner, 2021). These biofilms play a significant role in the respiration of aquatic systems and geochemical cycles. MPs affect different species to different extents and ways, and an array of MPs are reported to induce significant biotransformation in various biofilm communities (Desiante, Minas, and Fenner, 2021). The transformation rate follows pseudo-first order rate kinetics; the biotransformation rates of BUP, BDown, HUP, and Hdown biofilms, by various MPs, are listed in Table 8.2 (Desiante, Minas, and Fenner, 2021). There is also a significant effect on root exudation of aquatic plants such as *Lemna* and *Salvinia* in the presence

TABLE 8.2

Biotransformations of Four Microbial Biofilm Communities BUP, BDown, HUP, and Hdown in the Presence of Various MP Compounds

MP	k_{bio}, BUP (day^{-1})	k_{bio}, Bdown (day^{-1})	k_{bio}, HUP (day^{-1})	k_{bio}, Hdown (day^{-1})
Atenolol	0.185	0.191	1.718	1.604
Capecitabine	0.083	0.130	2.555	1.194
Cilastatin	0.386	0.482	14.864	32.951
Fenhexamid	0.170	0.265	0.348	0.288
Oxcarbazepine	1.357	2.548	6.327	7.999
Propachlor	0.817	2.696	1.888	1.147
Ranitidine	0.121	0.122	0.382	0.296
Sulfadiazine	0.689	0.794	0.692	0.271
Sulfamethazine	0.365	0.637	0.420	0.310
Sulfamethoxazole	0.738	1.140	0.585	0.384
Sulfapyridine	0.399	0.892	0.577	0.372
Sulfathiazole	0.576	0.943	0.497	0.429
Trinexapac ethyl	0.226	0.394	0.916	0.666

of various MPs (Escolà Casas and Matamoros, 2022). Upon exposure to various MPs, a decrease in fatty acids and sugars exudation was observed in *Lemna*, whereas in *Salvinia*, long-chain compounds and germacrene increased upon MP exposure (Escolà Casas and Matamoros, 2022).

8.1.1 Micropollutants in Human Health

Human beings are exposed around >90% to persistent bioaccumulative and toxic contaminants (PBT) through their food, especially via food samples derived from animal origin, although significant contaminants are also derived from vegetables as the consumption of vegetables in many countries is as high as 100 kg/capita (Lü et al., 2014). Persistent organic pollutants in vegetables and food products induce severe health risks such as neurotoxicity, cancer, endocrine disruption, leukemia, reproductive disorders, asthma, and birth defects (Ngweme et al., 2021). The daily intake of polyaromatic hydrocarbons (PAHs) from the vegetables was calculated:

$$\text{Intake} = \left(C_{PAH} \times Y_{factor} \times V_{intake} \right) / BW$$

where C_{PAH} is the concentration of PAH in vegetables (μg/kg), Y_{factor} is the fresh weight to dry weight conversion factor, V_{intake} is the vegetable intake per day, and BW is the average body weight.

Antibiotics are also reported to get 90% excreted back after their administration; thereby, their accumulation occurs in wastewater, surface water, soil, etc. (Patel et al., 2019). Continuous accumulation of these MPs (such as erythromycin) may render the microorganism resistant and induce risk to human health. Organophosphates were also reported to induce neurotoxic and genotoxic effects on humans, whereas methyl parathion causes the exchange of chromatids in human lymphocytes (Kumar et al., 2020). 17β-Ethinylestradiol (EE2) is also toxic for humans and is also reported to induce prostate cancer (Kumar et al., 2020). With recent exponential growth in nanotechnology, the accumulation of nanoparticles also imposes health risks on wildlife. Studies have exhibited that exposure to single-walled carbon nanotubes (SWCNT) and multi-walled carbon nanotubes (MWCNT) results in sustained chronic disease and pulmonary inflammatory disease (Dong and Ma, 2018). Bottini *et al.* reported cell death of human T cells upon exposure to oxidized multi-walled CNTs (Girardello et al., 2017). The harmful effects on various species by MPs are listed in Table 8.3.

TABLE 8.3
Harmful Effects of Various Micropollutants

Micropollutant	Concentration	Harmful Effect	References
Naproxen, diclofenac, benzotriazole, and carbamazepine	10, 100, 1,000 µg/L	Reduction of fatty acids and sugar in root exudation of Lemna minor	Escolà Casas and Matamoros (2022)
Naproxen, diclofenac, benzotriazole, and carbamazepine	10, 100, 1,000 µg/L	Long-chain compounds release in *Salvinianatans*	Escolà Casas and Matamoros (2022)
Tetrabromobisphenol A (TBBPA)	–	Endocrine disruption	Ngweme et al. (2021)
CNTs	400 µg/L	Cell death in human T cells	Girardello et al. (2017)
Pharmaceutical ethynylestradiol	5 ng/L	Infertility in the F1 generation of zebrafish	Girardello et al. (2017)
17-α-ethinylestradiol	5–6 ng/L	Feminization of male fish (*Pimephales promelas*) near extinction	Kidd et al. (2007)
DDT	0.6 mg/kg	Thinning of eggshells of black duck	Kumar et al. (2020)
DDT	113–118 mg/kg	Toxic to mammals	Kumar et al. (2020)
Aldrin	1–200 mg/L	Toxic to aquatic insects	Kumar et al. (2020)
Aldrin	2.2–53 mg/L	Toxic to fish	Kumar et al. (2020)
Dichlorofluorescin	5 mg/L	Histopathological changes in rainbow trout's liver and kidney	Kumar et al. (2020)
17-β-Ethinylestradiol	0.005–0.09 mg/kg	Carcinogenic effects on female mice	Kumar et al. (2020)
Diclofenac	1 µg/L	Alters gene transcription in *Mytilusgallo provincialis*	Zdarta et al. (2022)
Ethylhexyl-methoxycinnamate	0.4 mg/kg	Reproductive toxic effect on *Potamopyrgus antipodarum*	Kaiser et al. (2012)
Ethylhexyl-methoxycinnamate	10 mg/kg	Reproductive toxic effects on *Melanoides tuberculata*	Kaiser et al. (2012)

8.2 SOURCES OF MICROPOLLUTANTS

The presence of MPs in both aquatic and terrestrial environments nowadays has become a worldwide issue. Because of the rising concern, it is important first to understand and explore the major sources and pathways from which they enter into the environment. For many MPs, there may be many pathways to groundwater and other associated receptors (see Figure 8.2). Sources of MPs are generally classified into point source pollution and diffusion pollution (Lapworth et al., 2012). Table 8.4 summarizes the major sources of MPs in the environment.

8.2.1 POINT SOURCE POLLUTION

This type of pollution originates from a specific location whose inputs into aquatic systems can often be defined in a spatially discrete manner, and its environmental loading is more constrained (Lapworth et al., 2012). Important examples include industrial effluents (pharmaceuticals industry, textile industry, etc.), hospital effluents, food processing plants, municipal sewage treatment plants, waste disposal sites like landfill sites, and domestic septic tanks (Tran, Reinhard, and Gin, 2018). A large number of studies investigated that artificial recharge or infiltration of wastewater is a very important source of MPs in groundwater (Menger et al., 2020; McCance et al., 2018). It has been also found that not proper functioning of wastewater treatment plants may cause the release of MPs into the environment (Margot et al., 2015). However, some potential pollutants entered the environment by passing sophisticated treatment processes. For example, metaldehyde (an important ingredient of slug pellets) widely occurred in treated drinking water sources (Castle et al., 2017). For instance, some antibiotics are not easily removed by sewage treatment plants

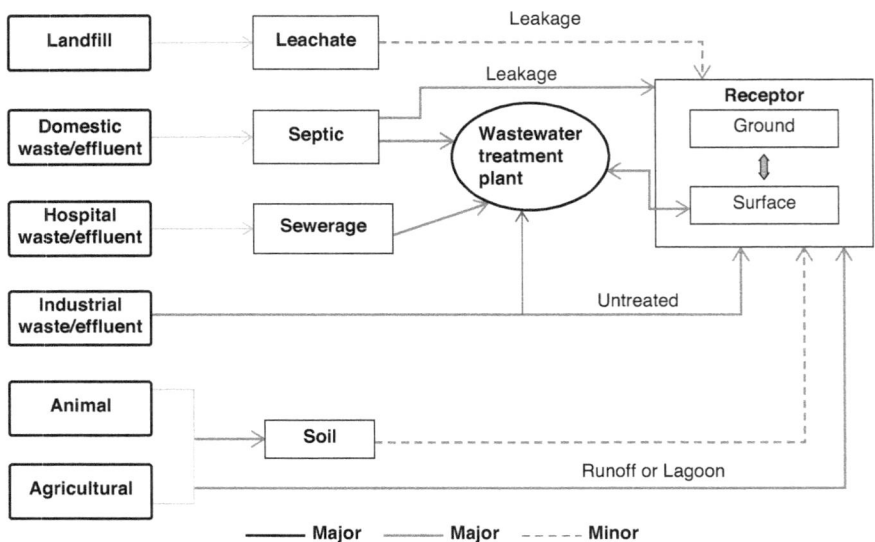

FIGURE 8.2   Major sources and pathways of MPs in soil and water.

TABLE 8.4

Major Sources of Micropollutants

Category	Sub-Class	Major Sources	
		Point Source	Diffuse Source
Pharmaceuticals	Antibiotics, β-blockers, stimulants, and lipid regulators	Hospital effluents, domestic wastewater (human excretion), municipal landfill leachate	Biosolids (sewage sludge)
Pesticides	Herbicides, insecticides, and fungicides	Domestic waster (runoff gardens and lawns)	Agricultural runoff
Personal care products	Fragrances, disinfectants, and insect repellants	Landfill leachate(improper disposal), domestic wastewater (from laundry, bathing, shaving)	Biosolids (sewage sludge)
Industrial chemicals	Plasticizers and fire retardants	Industrial effluents, wastewater (improper treatment)	
Steroid hormones	Estrogen	Aquaculture	Runoff from animal feeding operations

(removal efficiency only around 34%–72%), and therefore, the soil zone is adversely affected by the wastewater discharge (Sabri et al., 2020).

Among many point sources, municipal landfill leachate is the primary source of MPs in groundwater as it contains a wide variety of contaminants. According to many studies, a variety of organic pollutants are detected in landfill leachates worldwide (Oturan et al., 2015). These compounds are basically chlorinated aliphatics, higher fatty acids, pesticides, phenolic compounds, aromatic compounds, phenolic compounds, polychlorinated biphenyls, polyaromatic hydrocarbons (PAHs), phthalates, pharmaceuticals, and personal care products (PPCPs) and some emerging contaminants such as perfluorinated compounds (PFCs) are also found in high concentration (Eggen, Moeder, and Arukwe, 2010; Tijani et al., 2016). Many reported compounds in leachate belong to the 126 priority pollutants list defined by the USEPA (Boonyaroj et al., 2012). Domestic septic tanks also remain an important source of groundwater pollution by MPs mainly PPCPs, particularly where shallow groundwater tables and high aquifers transmissivity are observed (Li, 2014). Furthermore, it is very difficult to effectively monitor and regulate contamination from septic tanks. One case study carried out in Canada revealed that septic tank wastewater samples contained elevated concentrations of many pharmaceutical compounds, and the leakage of these septic tanks may become an important source of MPs (Carrara et al., 2008). Some studies reported that the concentration of pollutants detected in groundwater impacted by septic tanks ranges between 10 and 10^3 ng/L (Lapworth et al., 2012).

8.2.2 Diffuse Source Pollution

In contrast to point source pollution, diffuse pollution generally originated from not very properly defined locations, and these locations are typically spread over large geographical scales (Li, 2014). Compared to point source pollution, diffuse pollution has generally low environmental loading and has a higher potential for natural attenuation in the soil and subsurface (Luo et al., 2014). Agricultural runoff from manures and biosolid sources are the paramount diffuse sources of MPs. Both manure and biosolids (from sewage sludge) have important applications in enhancing soil nutrient levels (Oun et al., 2014). Incomplete removal of pollutants during wastewater treatment may cause residual concentrations of MPs that are stuck with biosolids, and due to their high solubility, they are easily leached out into the soil and contaminate groundwater. In this regard, two compounds, perfluorochemicals, and polychlorinated alkanes, are important groundwater contaminants (Clarke and Smith, 2011). One study reviewed that a range of antimicrobial compounds (concentration between 5 and 42 ng/L) were found in groundwater sources in China which were attributed to soil manure applications during agricultural practices (Hu, Zhou, and Luo, 2010).

Managed aquifer recharge is also considered diffuse source pollution as it uses surface water sources (mainly treated wastewater) to artificially recharge an aquifer and make them natural temporary water storage system; this is basically a management tool in water-scarce areas (Alam et al., 2021). However, this artificial recharge in some instances disturbed the water balance in the soil and subsurface that caused potential long-term contamination of groundwater sources.

8.2.3 OCCURRENCE OF MICROPOLLUTANTS

The occurrence of MPs in water depends upon the source of water and the season. Concentration levels were much higher in winters than that in summers or monsoons (Luo et al., 2014). This may be due to enhanced degradation in warmer climates or dilution by rainfall (Luo et al., 2014; Rogowska et al., 2020). According to the available literature, MP concentration generally follows an order: Industrial effluent > wastewater treatment plants > surface water > ground water > drinking water (Patel et al., 2019; Rogowska et al., 2020). In wastewater treatment plants (WWTPs), the influent and effluent show significant variations in MP concentration which may be due to various factors like the WWTP size and efficacy of wastewater treatment processes, water consumption per person per day, excretion rate, etc. (Kosek et al., 2020). Research showed that climatic conditions could also affect MP concentration in the influent; for example, the use of pesticides is seasonal, which depends on the pest prevalence in different climatic conditions (Kiefer et al., 2019). Then, after the release of WWTPs, effluents in surface water could be considered the main source of MPs in surface water. Following WWTP treatment processes, MPs undergo natural attenuation of varying degrees (e.g., sorption on solids and suspended solids, aerobic biodegradation). Research revealed that the natural attenuation of PCPs is more likely due to the river water dilution or sorption of solids; therefore, in dry weather conditions, the occurrence levels of PCPs in surface water enhanced in comparison to wet weather conditions (Ebele, Abou-Elwafa Abdallah, and Harrad, 2017). Contrary to PCPs, pharmaceuticals showed lower occurrence levels in summer water samples in comparison to winter water samples (Patel et al., 2019). Some studies revealed that chemicals like bisphenol A and biocides which are mainly used as pavement materials and in roof paintings were leached out during precipitation and accumulated in roof runoff up to a remarkable level and entered the surface water (Durak et al., 2021; Frankowski et al., 2021). Now in drinking water, the concentration of MPs is found to be the least. The maximum concentration of common MPs found in drinking water is <1,000 ng/L of carbamazepine (Canada), <100 ng/L of ibuprofen (US), >10 ng/L of atenolol (US), <10 ng/L of sulfamethoxazole (US), 100 ng/L of nonylphenol (US), 100 ng/L of bisphenol A (Canada), and >100 ng/L of caffeine (US), which are much below their predicted non-effective concentration (PNEC) (Rogowska et al., 2020). Similarly, in groundwater, the concentration of most of the MPs was much below their PNEC levels, except that the concentration of carbamazepine in Germany (<50, 2,325 ng/L) and sulfamethoxazole in the US (1,110, 160 ng/L) exceeded their PNEC levels of 25,000 and 20,000 ng/L, respectively (Rogowska et al., 2020). As compared to groundwater and drinking water, surface water contains the maximum concentration of MPs. This may be due to the direct release of the effluent of treated wastewater into the surface water (Kasprzyk-Hordern, Dinsdale, and Guwy, 2009). However, dilution of concentration occurs with the course of natural processes such as sorption on solids, dilution by water bodies, or rainfall. Among the common MPs, ibuprofen and caffeine concentration (Costa Rica) were as high as 36, 788, and 5,000 ng/L, respectively, which was much higher than that of their PNECs (Spongberg et al., 2011). It is to be noted that the PNEC values were standardized with respect to individual MPs, and mixtures are not taken into consideration (Luo et al., 2014).

8.3   CONCLUSION

The extensive use of consumer products, pharmaceuticals, plasticizers, pesticides, fertilizers, industrial waste, etc., in daily life leads to the accumulation and increase in the concentration of MPs which impose harmful effects on the environment, especially aquatic life. The concentration of these MPs further accumulates exponentially due to unregulated waste management, the generation of toxic byproducts, subsequent reactions, and also the leaching of the landfills. MPs have exhibited diverse harmful effects such as mutagenic, teratogenic, and evolutionary disorders in many aquatic organisms such as snails, fish, biofilms, etc. The sources of these MPs are mainly categorized into point sources and diffused sources depending upon the concentration and spread of the MPs. In point source pollution, the loading of the MP concentration is much more than the diffuse source as in the former, the effluence of the MPs is direct and localized, whereas, in the latter, the pollution is indirect and is spread over a diffused geographical stretch, thereby diluting the harmful impact.

ACKNOWLEDGMENTS

Mohd Azfar Shaida acknowledges the University Grants Commission for granting Dr. D S Kothari Post-Doctoral Fellowship (Letter No. F. 4-2/2006 (BSR)/ CH/20-21/0058). Soumita Talukdar would like to express their gratitude to the School of Science and Engineering (G. D. Goenka University) for providing technical assistance with this chapter.

REFERENCES

Alam, S. et al. (2021) 'Managed aquifer recharge implementation criteria to achieve water sustainability', *Science of the Total Environment*, 768, p. 144992. doi: 10.1016/j.scitotenv.2021.144992.

Boonyaroj, V. et al. (2012) 'Toxic organic micro-pollutants removal mechanisms in long-term operated membrane bioreactor treating municipal solid waste leachate', *Bioresource Technology*, 113, pp. 174–180. doi: 10.1016/j.biortech.2011.12.127.

Carrara, C. et al. (2008) 'Fate of pharmaceutical and trace organic compounds in three septic system plumes, Ontario, Canada', *Environmental Science and Technology*, 42(8), pp. 2805–2811. doi: 10.1021/es070344q.

Castle, G. D. et al. (2017) 'Review of the molluscicide metaldehyde in the environment', *Environmental Science: Water Research and Technology*, 3(3), pp. 415–428. doi: 10.1039/c7ew00039a.

Chavoshani, A. et al. (2020) *Introduction, Micropollutants and Challenges*. INC. doi: 10.1016/b978-0-12-818612-1.00001-5.

Clarke, B. O. and Smith, S. R. (2011) 'Review of "emerging" organic contaminants in biosolids and assessment of international research priorities for the agricultural use of biosolids', *Environment International*, 37(1), pp. 226–247. doi: 10.1016/j.envint.2010.06.004.

Desiante, W. L., Minas, N. S. and Fenner, K. (2021) 'Micropollutant biotransformation and bioaccumulation in natural stream biofilms', *Water Research*, 193. doi: 10.1016/j.watres.2021.116846.

Dong, J. and Ma, Q. (2018) 'Macrophage polarization and activation at the interface of multi-walled carbon nanotube-induced pulmonary inflammation and fibrosis', *Nanotoxicology*, 12(2), pp. 153–168. doi: 10.1080/17435390.2018.1425501.

Dubey, M. et al. (2021) 'Occurrence, fate, and persistence of emerging micropollutants in sewage sludge treatment', *Environmental Pollution*, 273, p. 116515. doi: 10.1016/j.envpol.2021.116515.

Durak, J. et al. (2021) 'Environmental risk assessment of priority biocidal substances on Polish surface water sample', *Environmental Science and Pollution Research*, 28(1), pp. 1254–1266. doi: 10.1007/s11356-020-11581-7.

Ebele, A. J., Abou-Elwafa Abdallah, M. and Harrad, S. (2017) 'Pharmaceuticals and personal care products (PPCPs) in the freshwater aquatic environment', *Emerging Contaminants*, 3(1), pp. 1–16. doi: 10.1016/j.emcon.2016.12.004.

Eggen, T., Moeder, M. and Arukwe, A. (2010) 'Municipal landfill leachates: A significant source for new and emerging pollutants', *Science of the Total Environment*, 408(21), pp. 5147–5157. doi: 10.1016/j.scitotenv.2010.07.049.

Escolà Casas, M. and Matamoros, V. (2022) 'Linking plant-root exudate changes to micropollutant exposure in aquatic plants (Lemna minor and Salvinia natans). A prospective metabolomic study', *Chemosphere*, 287. doi: 10.1016/j.chemosphere.2021.132056.

Frankowski, R. et al. (2021) 'Biodegradation and photo-Fenton degradation of bisphenol A, bisphenol S and fluconazole in water', *Environmental Pollution*, 289(August). doi: 10.1016/j.envpol.2021.117947.

Girardello, R. et al. (2017) 'Cellular responses induced by multi-walled carbon nanotubes: In vivo and in vitro studies on the medicinal leech macrophages', *Scientific Reports*, 7(1), pp. 1–12. doi: 10.1038/s41598-017-09011-9.

Hu, X., Zhou, Q. and Luo, Y. (2010) 'Occurrence and source analysis of typical veterinary antibiotics in manure, soil, vegetables and groundwater from organic vegetable bases, northern China', *Environmental Pollution*, 158(9), pp. 2992–2998. doi: 10.1016/j.envpol.2010.05.023.

Kaiser, D. et al. (2012) 'Ecotoxicological effect characterisation of widely used organic UV filters', *Environmental Pollution*, 163, pp. 84–90. doi: 10.1016/j.envpol.2011.12.014.

Kasprzyk-Hordern, B., Dinsdale, R. M. and Guwy, A. J. (2009) 'The removal of pharmaceuticals, personal care products, endocrine disruptors and illicit drugs during wastewater treatment and its impact on the quality of receiving waters', *Water Research*, 43(2), pp. 363–380. doi: 10.1016/j.watres.2008.10.047.

Kidd, K. A. et al. (2007) 'Collapse of a fish population after exposure to a synthetic estrogen', *Proceedings of the National Academy of Sciences of the United States of America*, 104(21), pp. 8897–8901. doi: 10.1073/pnas.0609568104.

Kiefer, K. et al. (2019) 'New relevant pesticide transformation products in groundwater detected using target and suspect screening for agricultural and urban micropollutants with LC-HRMS', *Water Research*, 165, p. 114972. doi: 10.1016/j.watres.2019.114972.

Kosek, K. et al. (2020) 'Implementation of advanced micropollutants removal technologies in wastewater treatment plants (WWTPs): Examples and challenges based on selected EU countries', *Environmental Science and Policy*, 112(June), pp. 213–226. doi: 10.1016/j.envsci.2020.06.011.

Kumar, N. M. et al. (2020) 'Micro-pollutants in surface water: Impacts on the aquatic environment and treatment technologies', *Current Developments in Biotechnology and Bioengineering*, pp. 41–62. doi: 10.1016/b978-0-12-819594-9.00003-6.

Lado Ribeiro, A. R. et al. (2019) 'Impact of water matrix on the removal of micropollutants by advanced oxidation technologies', *Chemical Engineering Journal*, 363(January), pp. 155–173. doi: 10.1016/j.cej.2019.01.080.

Lapworth, D. J. et al. (2012) 'Emerging organic contaminants in groundwater: A review of sources, fate and occurrence', *Environmental Pollution*, 163, pp. 287–303. doi: 10.1016/j.envpol.2011.12.034.

Li, W. C. (2014) 'Occurrence, sources, and fate of pharmaceuticals in aquatic environment and soil', *Environmental Pollution*, 187, pp. 193–201. doi: 10.1016/j.envpol.2014.01.015.

Lü, H. et al. (2014) 'Levels of organic pollutants in vegetables and human exposure through diet: A review', *Critical Reviews in Environmental Science and Technology*, 44(1), pp. 1–33. doi: 10.1080/10643389.2012.710428.

Luo, Y. et al. (2014) 'A review on the occurrence of micropollutants in the aquatic environment and their fate and removal during wastewater treatment', *Science of the Total Environment*, 473–474, pp. 619–641. doi: 10.1016/j.scitotenv.2013.12.065.

Margot, J. et al. (2015) 'A review of the fate of micropollutants in wastewater treatment plants', *WIREs Water*, 2(5), pp. 457–487. doi: 10.1002/wat2.1090.

McCance, W. et al. (2018) 'Contaminants of Emerging Concern as novel groundwater tracers for delineating wastewater impacts in urban and peri-urban areas', *Water Research*, 146, pp. 118–133. doi: 10.1016/j.watres.2018.09.013.

Menger, F. et al. (2020) 'Wide-scope screening of polar contaminants of concern in water: A critical review of liquid chromatography-high resolution mass spectrometry-based strategies', *Trends in Environmental Analytical Chemistry*, 28, p. e00102. doi: 10.1016/j.teac.2020.e00102.

Nash, J. P. et al. (2004) 'Long-term exposure to environmental concentrations of the pharmaceutical ethynylestradiol causes reproductive failure in fish', *Environmental Health Perspectives*, 112(17), pp. 1725–1733. doi: 10.1289/ehp.7209.

Ngweme, G. N. et al. (2021) 'Occurrence of organic micropollutants and human health risk assessment based on consumption of Amaranthus viridis, Kinshasa in the Democratic Republic of the Congo', *Science of the Total Environment*, 754. doi: 10.1016/j.scitotenv.2020.142175.

Oturan, N. et al. (2015) 'Occurrence and removal of organic micropollutants in landfill leachates treated by electrochemical advanced oxidation processes', *Environmental Science and Technology*, 49(20), pp. 12187–12196. doi: 10.1021/acs.est.5b02809.

Oun, A. et al. (2014) 'Effects of biosolids and manure application on microbial water quality in rural areas in the US', *Water (Switzerland)*, 6(12), pp. 3701–3723. doi: 10.3390/w6123701.

Patel, M. et al. (2019) 'Pharmaceuticals of emerging concern in aquatic systems: Chemistry, occurrence, effects, and removal methods', *Chemical Reviews*, pp. 3510–3673. doi: 10.1021/acs.chemrev.8b00299.

Rogowska, J. et al. (2020) 'Micropollutants in treated wastewater', *Ambio*, 49(2), pp. 487–503. doi: 10.1007/s13280-019-01219-5.

Sabri, N. A. et al. (2020) 'Fate of antibiotics and antibiotic resistance genes during conventional and additional treatment technologies in wastewater treatment plants', *Science of the Total Environment*, 741, p. 140199. doi: 10.1016/j.scitotenv.2020.140199.

Spongberg, A. L. et al. (2011) 'Reconnaissance of selected PPCP compounds in Costa Rican surface waters', *Water Research*, 45(20), pp. 6709–6717. doi: 10.1016/j. watres.2011.10.004.

Tijani, J. O. et al. (2016) 'Pharmaceuticals, endocrine disruptors, personal care products, nanomaterials and perfluorinated pollutants: A review', *Environmental Chemistry Letters*, 14(1), pp. 27–49. doi: 10.1007/s10311-015-0537-z.

Tran, N. H., Reinhard, M. and Gin, K. Y. H. (2018) 'Occurrence and fate of emerging contaminants in municipal wastewater treatment plants from different geographical regions-a review', *Water Research*, 133, pp. 182–207. doi: 10.1016/j.watres.2017.12.029.

Zdarta, J. et al. (2022) 'Free and immobilized biocatalysts for removing micropollutants from water and wastewater: Recent progress and challenges', *Bioresource Technology*, 344(PB), p. 126201. doi: 10.1016/j.biortech.2021.126201.

9 Effects of Micropollutants on Human Health

Areeba Aziz
Aligarh Muslim University

Asad Aziz
State University of New York at Buffalo

CONTENTS

9.1 Introduction 152
    9.1.1 What Are the Micropollutants? 152
    9.1.2 Sources and Pathways to Human Beings 152
9.2 Effects of Micropollutants on Human Health 153
    9.2.1 Gastrointestinal Effects 153
    9.2.2 Cardiovascular Effects 154
    9.2.3 Neurological Effect 155
    9.2.4 Incident of Toroku Arsenic Pollution 156
    9.2.5 The Minimata Disaster and the Disease That Followed:
         Mercury Poisoning Sickened an Entire Japanese Town 157
    9.2.6 Reproductive Effects 157
    9.2.7 Lead: A Silent Killer in Nigeria 159
    9.2.8 Carcinogenic Effects 159
         9.2.8.1 Agrochemicals 160
         9.2.8.2 Weedkiller 'Raises the Risk of Non-Hodgkin
                 Lymphoma by 41%' 160
         9.2.8.3 POPs 161
9.3 Future Perspectives and Need for Public Awareness 162
    9.3.1 Measures to be Taken 162
    9.3.2 Why Is Environmental Awareness Important? 163
9.4 Convention and Regulation 163
9.5 Conclusion 163
Acknowledgment 165
References 165

DOI: 10.1201/9781003202431-9

9.1 INTRODUCTION

9.1.1 WHAT ARE THE MICROPOLLUTANTS?

Micropollutants, which are also called emerging contaminants, consist of an extensive assemblage of manmade and natural substances. They are found almost everywhere on earth including water bodies, soil, and food. The main groups of micropollutants are pharmaceuticals, personal care products, steroid hormones, surfactants, pesticides, and industrial chemicals. They are generally present in trace amounts, i.e., nanogram/liter to microgram/liter. Released by industry, household, or agriculture, they enter the environment and spread throughout the water ecosystem. After being taken up by aquatic organisms or humans via contaminated water or food, micropollutants are transported to different tissues within the organism. Depending on the properties of micropollutants (MPs) and the biology of target species, they may bioaccumulate, metabolize or cause adverse effects (Burkhardt-Holm, 2011).

Over the past few decades, the occurrence of micropollutants has become a worldwide issue of increasing environmental concern. Albeit MPs are present in trace amounts, their persistent exposure causes adverse effects on human health such as cancer and diabetes. Here, in this chapter, we have summarized major classes of MPs and their impact on human health. The chapter also focuses on future perspectives, the need for public awareness, and conventions and regulations regarding the mitigation of micropollutants (Luo et al., 2014).

9.1.2 SOURCES AND PATHWAYS TO HUMAN BEINGS

The main sources of micropollutants include effluents from sewage treatment plants, discharge from hospitals, agricultural runoffs, and industrial outflow of chemicals (Table 9.1). Drugs disposed at dumping sites also lead to groundwater contamination (Luo et al., 2014).

TABLE 9.1
Sources of Micropollutants

Category	Important Subclass	Major Sources
Pharmaceuticals	Lipid regulators, antibiotics, beta blockers and stimulants	Domestic wastewater Hospital effluents A dumping ground of undisposed drugs Pharmaceutical industry
Personal care products	Fragrances, disinfectants, insect repellents	Domestic wastewater from bathing, shaving, swimming, etc.
Steroid hormones	Estrogen	Cattle and livestock wastewater Domestic wastewater from excretion
Surfactants	Non-ionic surfactants	Industrial wastewater Wastewater from laundry
Industrial chemicals	Plasticizers, fire retardants	Domestic wastewater (by leaching out of material) Runoff from gardens, lawns, and roadways
Pesticides	Herbicides, insecticides, and fungicides	Agricultural runoff

Source: Luo et al. (2014).

9.2 EFFECTS OF MICROPOLLUTANTS ON HUMAN HEALTH

9.2.1 GASTROINTESTINAL EFFECTS

Many trace elements are vital for the proper functioning of the human body (copper, zinc, and iron), but some of them are harmful (lead, arsenic, and mercury) depending upon their concentration and chemical forms. Since the main source of exposure is oral, as they pass through the gastrointestinal tract, they bring about damaging effects (Vázquez et al., 2015).

I. **Mercury and its compounds**: Mercury is a rare element found in the earth's crust, having an average abundance by mass of only 0.08 parts per million. Anthropogenic sources of mercury in the environment include mining, industrial processes, combustion of fossil fuels, production of cement, and incineration of medical waste. Dental amalgam fillings are also a primary source for the general population. Mercury exists in its elemental state as well as in the form of its mercuric and mercurous salts (Fisher, 2003).

The acute ingestion of Hg (II) and Ch_3Hg causes irritation to tissues of the gastrointestinal (GI) tract. In humans, there is evidence of extensive precipitation of intestinal mucosal proteins, colicky abdominal pain, diarrhea, oropharyngeal pain, ulceration, and hemorrhages throughout the entire length of the GI tract after accidental ingestion of HgCl (Vázquez et al., 2015). It has been observed that a high dose of mercury salts causes burning chest pain, mercurial stomatitis, and discoloration of the oral mucous membrane. People who are occupationally exposed to mercury vapors develop gingivitis, excessive salvation, difficulty in swallowing, and stomatitis (Park and Zheng, 2012).

Mercury produced abdominal pain, sore gums, ulceration, and diarrhea in five out of nine individuals in a thermometer manufacturing plant. Blisters on the lips and tongue as well as vomiting were reported in the case of a 19-month-old boy who ingested an unknown amount of mercuric chloride powder. Several patients who were hypersensitive to mercury developed stomatitis at the point of contact with the amalgam filling (Fisher, 2003).

II. **Lead**: Short-term exposure to lead causes GI disturbances in adults at a blood concentration level of 100–200 μg/dL, although consequences have been reported at a concentration as low as 40–60 μg/dL also. Chronic exposure at work causes nausea, anorexia, constipation, and abdominal cramps (Vázquez et al., 2015). Inorganic lead compounds are classified as carcinogenic by the International Agency for Research on Cancer. Several studies have shown that lead compounds are associated with GI cancers in workers exposed to them.

III. **Arsenic**: Nausea, vomiting, and diarrhea are very common symptoms in humans following oral exposure to arsenic and its compounds due to irritation in GI mucosa. The cause of death in acute exposure to high doses is massive fluid loss from the GI tract. A study carried out in Taiwan in a population exposed to a high concentration of arsenic in drinking water (0.35–1.14 mg/L) also demonstrated the occurrence of colon cancer (Vázquez et al., 2015).

9.2.2 Cardiovascular Effects

Cardiovascular diseases (CVDs) are a group of disorders of the heart and blood vessels. They are the leading cause of death globally. An estimated 17.9 million people died from CVDs in 2019, representing 32% of all global deaths. Out of these 85% were due to heart attack and stroke. One of the major contributors to this is exposure to micropollutants such as persistent organic pollutants (POPs) and mercury.

I. **POPs**: POPs are also known as silent killers because of their bioaccumulative and tenacious nature. These consist of pesticides, industrial chemicals such as PCBs, polychlorinated biphenyls, PBDEs, and byproducts of industrial processes such as dioxins and furans. These toxins get incorporated and biomagnified in the food chain and cause notorious health hazards and environmental effects. Most POPs are highly lipid-soluble with semi-volatile (Pv between 10^{-4} and 10^{-10}) properties (Alharbi et al., 2018).

The amalgamation of social, physiological, and metabolic factors along with the pollution of POPs leads to the development of cardiac diseases (Figure 9.1). According to a study conducted in New York, there was a significant elevation of 19.2% in hypertension discharge rate in areas exposed to POPs as compared to clean sites. Thus, proving that living near hazardous waste sites, particularly those containing POPs, may constitute a risk of developing hypertension (Huang et al., 2006).

In a study conducted in Sweden, it was suggested that circulating levels of POPs are related to the stultification of left systolic and diastolic function which possibly caused heart failure. It has been experimentally proven that exposure to POPs induces oxidative stress and inflammation, which are two presumed players of myocyte dysfunction (Sjöberg Lind et al., 2013).

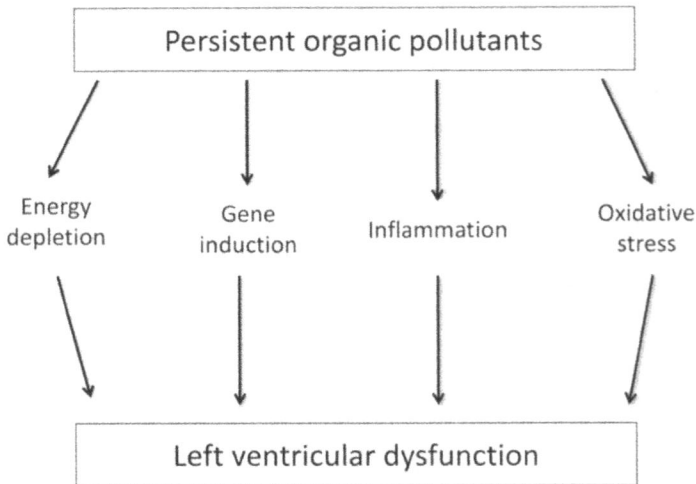

FIGURE 9.1 Possible mechanisms by which POPs might affect the function of the left ventricle (Lind et al., 2013).

II. **Mercury and its compounds**: Acute and chronic exposure to mercury and its compounds causes cardiovascular anomalies. Short-term inhalation of elemental mercury vapors can cause increased blood pressure and heart palpitations (Fisher, 2003). Mercurous chloride also known as calomel was used as a purgative as well as for the treatment of intestinal worms. It was also used as a teething powder up until 1954 in Great Britain. Tachycardia and elevated blood pressure were reported in children who were being treated with mercurous chloride tablets for worm infection or who were given the same for teething discomfort (Fisher, 2003). These medicinal uses were later discontinued when the toxicity of the compound was confirmed.

It has been experimentally proved that statistically there is an increase of approximately 5 mmHg in both systolic and diastolic pressure in 50 volunteers who had a dental amalgam filling as compared to those who did not (Fisher, 2003).

9.2.3 NEUROLOGICAL EFFECT

Neurological disorders are defined in medicine as disorders that affect the brain and nerves present throughout the body (Table 9.2). Symptoms may include paralysis, muscle weakness, poor coordination, loss of sensation, seizures, and altered level of consciousness. Neurotoxicity may be associated with exposure to an increased level of neurotoxic agents that pollute the environment. Approximately 3% of neuronal disorders are caused by exposure to neurotoxins. The most vulnerable age groups are infants as their brain is in the developing stage and older adults because they show neurological degeneration (Vargas and Ponce-Canchihuamán, 2017).

Polychlorinated biphenyls: PCBs have widely distributed contaminants and persist for a long period because of their resistance to biological degradation. Their

TABLE 9.2
Substances Associated with Neurotoxic Damage and Their Effects on Human Health

Toxic Agents	Pathologies	Source
Mercury	Acute: Nausea, tremors, and headache	Scientific instruments
	Chronic: Peripheral neuropathy, encephalopathy	Dental amalgam
		Mining
Insecticides	Acute: Cholinergic poisoning	Agriculture industry
	Chronic: Ataxia, peripheral neuropathy, and paralysis	
Arsenic	Acute encephalopathy	Pesticides seafood
	Peripheral neuropathy	
Lead	Encephalopathy	Welding services
		Insecticides
Tetrachloroethylene	Acute: Necrosis	Paint removers, textile
	Chronic: Encephalopathy and neuropathy	industry, etc.

Source: Vargas and Ponce-Canchihuamán (2017).

half-lives lie between several months to years. People who are occupationally exposed to PCBs have reported symptoms such as headache, dizziness, depression, fatigue, and a tingling sensation in their hands. Studies on humans and animals have proved that they have a lasting impact on developing brains as these accumulate in adipose tissue and are mobilized during lactation (Chance, 2001).

The studies above proved that Newborn children from mothers who ate contaminated fish were more likely to exhibit symptoms like hypoactive reflexes, more motor immaturity, and a greater amount of startle. Prenatal exposure was also associated with poorer performance on verbal and memory scales, poorer reading word comprehension, less muscle tone, and less activity (Faroon and Ruiz, 2016).

Arsenic: Arsenic is widely distributed in the earth's crust. Its compound exists in both organic and inorganic forms either in a trivalent or pentavalent oxidation state. Environmental sources of arsenic are mining, pesticides, medicines, and drinking water aquifers contaminated with heavy metals. Groundwater contamination by arsenic is a common phenomenon in countries like India, Myanmar, Bangladesh, etc. Water containing an arsenic concentration of 10–50 ppb causes peripheral neuropathy. The toxicity generally affects sensory nerve fibers more than motor nerve fibers.

Several mechanisms that play a role in As-induced neurotoxicity are as follows:

i. **Mitochondrial dysfunction**: Arsenic suppresses the activity of mitochondria and leads to the accumulation of reactive oxygen species (ROS) that causes neurodegeneration.

ii. **Lipid peroxidation**: Arsenic causes oxidative stress that in turn causes lipid peroxidation, which leads to DNA damage and brain cell death and also causes deterioration of the central nervous system.

iii. **Apoptosis**: Apoptosis is a form of programmed cell death that occurs in multicellular organisms. Arsenic neurotoxicity causes apoptosis by triggering p-38 mitogen and JNK3 pathways.

iv. **Thiamine deficiency**: The deficiency of thiamine causes neuronal issues. Axonal neuropathy that is similar to Wernicke's encephalopathy may be induced by it.

v. **Decreased acetylcholinesterase activity**: It is one of the essential enzymes that are needed for the correct functioning of the central nervous system (CNS). Its decreased activity leads to a cholinergic crisis that may cause peripheral neuropathy or CNS damage (Mochizuki, 2019).

9.2.4 INCIDENT OF TOROKU ARSENIC POLLUTION

Toroku is a small village on Kyushu island in Japan. Before the Second World War, the production of arsenopyrite was done there, and workers used to mold small pieces of it with their bare hands before taking them to burn in a kiln. Kilns also produced smoke and ashes containing arsenic. These ashes were disposed of in the river, causing contamination of water bodies. The livestock started dying mysteriously, and dozens of people died in areas living near the mine. Symptoms such as sensory disturbances, skin lesions, dizziness, and neurological abnormalities were observed on examining the residents.

In 1990, the Supreme Court ordered the mining company to compensate the victims of arsenic exposure under pollution-related health damage compensation (PRDCL). It was a victory, except that many had already died.

Mercury and its compounds: The CNS is probably the most prone target for mercury exposure. Several cases of cognitive, personality, sensory, and motor defects have been reported over the years. Well-known symptoms of mercury poisoning include tremors, emotional liability, insomnia, neuromuscular changes, memory loss polyneuropathy (stocking-glove sensory loss, reduced sensory, and motor nerve conduction velocities). A case study reported dementia and irritability in two women who ingested a tablet of a laxative that contained 120 mg of mercurous chloride (Fisher, 2003).

9.2.5 THE MINIMATA DISASTER AND THE DISEASE THAT FOLLOWED: MERCURY POISONING SICKENED AN ENTIRE JAPANESE TOWN

Minimata disease was first discovered about 40 years ago in Minimata Bay, Japan. A study group from the Kumamoto University of Medicine discovered that the source of illness is human ingestion of a large amount of methylmercury-contaminated fish and shellfish from Minimata Bay. The mercury came from a nearby petrochemical industry that dumped 27 tons of mercury compounds into Minimata Bay.

People suffering from the syndrome showed symptoms such as sensory disturbances in the extremities followed by ataxia, impairment of gait and speech, muscle weakness, tremors, abnormal eye movement, hearing impairment, disequilibrium, and disturbed sense of smell and taste (Eto, 1997).

The company finally quit poisoning Minimata's water in 1968 and financially compensated the victims. After 22 years, the plaintiffs achieved their goal of making those responsible for Japan's worst case of industrial pollution pay for their negligence (Kugler, 2016).

9.2.6 REPRODUCTIVE EFFECTS

Reproductive health is a state of complete physical, mental, and social well-being and not merely the absence of disease or infirmity in all matters relating to the reproductive system and its functions and processes. These systems are made of organs and hormone-producing glands, including the pituitary gland in the brain. Ovaries in females and testicles in males are reproductive organs, or gonads, which maintain the health of their respective systems.

The reproductive problems involve improper functioning of the male or female reproductive system that can lead to birth defects, preterm birth, developmental disorders, low birth weight, impotence, infertility, and menstrual disorders (Alharbi et al., 2018).

Under this heading, we will be discussing two major compounds that cause reproductive issues: EDCs and lead.

I. **Endocrine-disrupting compound**: According to the definition adopted by European Union, "an endocrine disruptor is an exogenous substance that alters the function of the endocrine system, causing adverse effects on the health of an organism or its progeny". EDCs are generally present in low

TABLE 9.3
Effects of EDCs on the Reproductive System

Study Conducted On	EDCs	Effect
Humans	BPA phthalate esters	Anogenital distance alteration
Humans	Phthalates	Alteration of sex steroid hormones
Humans	BPA	Alteration of sexual functions: erectile function, orgasmic function
Humans	BPA phthalates	Influence on sperm motility, sperm concentration, DNA damage
Animals	BPA	Effect on the hypothalamic-pituitary-ovarian axis
Animals	BPA parabens	Effect on oocyte
Humans	BPA phthalates	Endometriosis
Humans	BPA phthalates	PCOS

Source: Giulivo et al. (2016).

concentrations, but they are sufficient to cause health problems. Also, we cannot prevent our exposure to EDCs since these are present in our daily use products such as food, flame retardants, plastic bottles, food cans, detergents, toys, cosmetics, pesticides, etc. (Table 9.3). EDCs can alter the normal activities carried out by estrogens and androgens that either disrupt the metabolism of sex steroids or inhibit their synthesis. Endocrine disruption may lead to an irregular menstrual cycle, reduced fertility, endometriosis, a polycystic ovarian syndrome in females, while in males, it can cause low sperm motility, erectile dysfunction, and infertility (Kabir et al., 2015).

Phthalates and bisphenols are two major classes of EDCs. Phthalates are associated with infertility, especially in men. Recently, it has been observed that exposure to phthalates causes DNA damage and low sperm motility and concentration.

Studies conducted on men working in BPA and epoxy resin manufacturing companies in China demonstrated lower sexual function as compared to non-exposed ones. Furthermore, the effects of BPA on oocytes such as reduced ovarian weight and disturbance in prophase are well known (Giulivo et al., 2016).

II. **Lead**: Lead is a heavy metal. It is present in contaminated drinking water, paints, leaded gas, hair dyes, pesticides, pencil, tobacco smoke, and ceramics (Amadi et al., 2017). The developing fetus and young children are more vulnerable to the toxicity of lead. Exposure to lead may affect libido, semen quality by reducing sperm count, motility, viability, and sperm DNA integrity. These alterations lead to reduced fertility potential, increasing chances of miscarriages and preterm birth. Lead exposure impairs hormonal synthesis and regulations in both sexes. It affects female reproduction by impairing menstruation, reducing fertility potential, delaying conception time, and altering the hormonal production and circulation that affects pregnancy and its outcome. A combination of genetic, environmental, occupational, and lifestyle factors contributes to adverse effects on the reproductive health of men also. Most of the studies have confirmed that even moderate to low-level exposure to lead affects certain reproductive parameters (Kumar, 2018).

9.2.7 LEAD: A SILENT KILLER IN NIGERIA

Nigeria recorded the world's worst case of lead poisoning in 2010 in Zamfara State where an estimated 400 children died from lead toxicity following exposures to mining sites. Later, a study was conducted and lead levels of maternal blood and umbilical blood were recorded (Ladele et al., 2019).

A blood lead level of more than 25 µg/dL during pregnancy has been recognized to cause miscarriages. Since the concentration of umbilical and maternal blood lead levels are almost the same, lead can easily pass from the mother to the fetus, causing various developmental and neurological disorders in infants (Amadi et al., 2017). It is evident that environmental lead contamination remains a major challenge in Nigeria and calls for policy on environmental lead contamination and its management (Figure 9.2).

9.2.8 CARCINOGENIC EFFECTS

Cancer is a large group of diseases that can start in almost any organ or tissue of the body when abnormal cells grow uncontrollably, go beyond their usual boundaries to invade adjoining parts of the body, and/or spread to other organs. The latter process

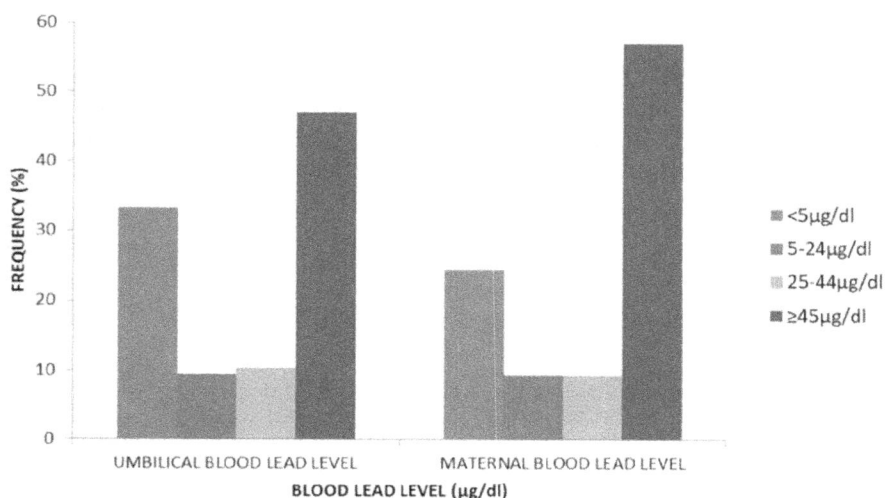

FIGURE 9.2 Demonstrating lead levels in the umbilical blood and maternal blood. (Ladele et al., 2019.)

is called metastasizing and is a major cause of death from cancer. Cancer is the second leading cause of death globally, accounting for an estimated 9.6 million deaths, or one in six deaths, in 2018. Lung, prostate, colorectal, stomach, and liver cancers are the most common types of cancer in men, while breast, colorectal, lung, cervical, and thyroid cancer are the most common among women.

Cancer is caused by mutations in the DNA. The DNA contains genes, each of which contains a set of instructions telling the cell what functions to perform, as well as how to grow and divide. Errors in the instructions can cause the cell to stop its normal function and may allow a cell to become cancerous. Mutations could be hereditary or acquired. Several forces can cause gene mutations, such as smoking, radiation, viruses, carcinogens, etc. (The and Fearon, 2020).

The main carcinogens that we will be discussing here are agrochemicals and POPs.

9.2.8.1 Agrochemicals

In recent years, there has been an increasing concern that long-term exposure to agrochemicals leads to cancer. Certain types of cancer, especially those associated with the hematopoietic system, have been observed in people who are occupationally exposed to pesticides. The International Agency for Research on Cancer recently stated that 43% of pesticides that they observed were carcinogenic to humans.

Organophosphate, thiocarbamate insecticides, and several fungicides and herbicides are genotoxic; that is, they are chemical mutagens. Other than these, nongenotoxic pesticides may also cause cancer via their epigenetic properties such as tumor promotion, inhibition of intercellular communication, or induction of peroxisome proliferation (Table 9.4).

9.2.8.2 Weedkiller 'Raises the Risk of Non-Hodgkin Lymphoma by 41%'

This was the headline of the *Guardian*, a leading newspaper in the UK. The article mentioned that a broad new scientific analysis of the cancer-causing potential of glyphosate herbicides has found that people with high exposures to the popular pesticides have a 41% increased risk of developing a type of cancer called non-Hodgkin lymphoma.

The evidence "supports a compelling link" between exposures to glyphosate-based herbicides and increased risk for non-Hodgkin lymphoma, concluded the authors. This finding contradicted the US environmental protection agency's assurance of safety over the weed killer.

TABLE 9.4

Types of Cancer Associated with Various Pesticides

Cancer	Pesticides
Soft tissue sarcoma and lymphoma	Phenoxy acetic acid herbicides (DDT)
Non-Hodgkin's lymphoma	Organophosphate and organochloride insecticides, fumigants
Leukemia	DDT, ethylene oxide, chlordane, methoxychlor, etc.
Lung carcinoma	DDT, arsenicals
Ovarian carcinoma	Triazines

Source: Weisenburger (1993).

Phenoxy herbicides include 2,4-D-2,4,5 T-2-methyl 4 chlorophenoxyacetic acid and other related compounds. These compounds are extensively used in agricultural fields. A case study performed in Kansas concluded a two-fold excess of Non Hodgkin's lymphoma (NHL) in farmers who used phenoxy herbicides. Three other case studies were performed in Washington, Sweden, and Italy to associate herbicides and NHL. A Swedish study stated an odds ratio of 4.9%, a Washington study stated a 1.3 odds ratio for farmers and 4.8 for people who sprayed forests with herbicides. Thus, establishing the fact that exposure to pesticides leads to an increased risk of developing cancer (Zahm and Blair, 1992).

9.2.8.3 POPs

Acute exposure to POPs has recently been linked to immunosenescence, an essential mechanism in carcinogenesis. Based on epidemiological evidence, PCBs were concluded as carcinogenic. The large intestine is a potential target because its mucosa is frequently exposed to POPs as they are mainly excreted via feces.

The change in gut microbiota could be one possible reason linking POPs and colorectal cancer because the gut microbiota is strongly associated with colorectal carcinogenesis by multiple mechanisms. Also, several *in vitro* studies reported that DDT or DDE can induce colorectal adenocarcinoma and cell proliferation by oxidative stress or beta-catenin signaling (Lee et al., 2018).

Several studies have examined the relationship between serum levels of PCBs with the risk of breast and pancreatic cancer, and have successfully established the proof. Many PCBs are known endocrine disrupters and may modulate steroid sex hormones as agonists, antagonists, or mixed agonist–antagonists, particularly concerning estrogen or testosterone activity. As a result, it is plausible that these chemicals contribute to the development of prostate cancer through hormone-mediated effects (Figure 9.3).

POPs get accumulated in adipose tissues due to their lipophilic properties; therefore, they are responsible for causing hormone-dependent breast cancer. POPs present in adipose tissue promote carcinogenesis and progression of mammary cancer. They can interfere with estrogen synthesis, interact with transcription factors, induce genotoxic enzymes and cytochrome 450, leading to an increased level of ROS, and

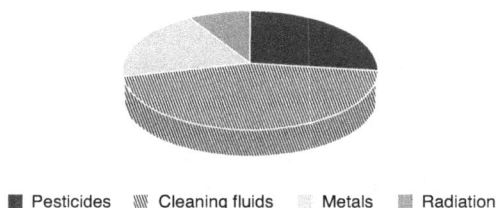

SELF REPORTED CHEMICAL EXPOSURE IN
PROSTRATE CANCER CASES

■ Pesticides ▨ Cleaning fluids ▨ Metals ▨ Radiation

FIGURE 9.3 Demonstrating common causes of prostate cancer. (Ritchie et al., 2003.)

induce transgenerational phenotypic changes by altering the epigenome. Numerous *in vitro* studies have proven that some POPs stimulate the growth of estrogen receptor-positive breast cancer cells (Ennour-Idrissi et al., 2019).

9.3 FUTURE PERSPECTIVES AND NEED FOR PUBLIC AWARENESS

9.3.1 MEASURES TO BE TAKEN

Over time, we are reviving our ways of dealing with pollutants. Improved methods have enabled us to detect even the tiniest concentration of pollutants in water at early stages. Early discovery helps us to counteract them before it is too late. This requires an efficient set of measures that have been evaluated in terms of their effectiveness, cost, and feasibility. We need to properly implement already framed laws regarding the mitigation of hazardous substances and environmental pollution.

The following measures are necessary to be included more strongly in the future:

1. Introducing a monograph/master file system. It promises more consistent, up-to-date assessments as well as resource savings, and better availability of environmental data from each substance evaluation.
2. Developing and harmonizing effective reduction measures within the authorization process.
3. Boosting research into environmentally friendlier active pharmaceutical ingredients and dosage forms – "green pharmacy".
4. Educating and informing specific target groups on the environmentally friendly use of products.
5. Education about proper disposal.
6. Banning medicinal products with persistent, bioaccumulative, and toxic properties.
7. Creating permanently green riparian buffer strips: These strips permanently covered with vegetation are known for reducing the entry of pollutants into water.
8. Increasing the percentage of organically farmed areas.
9. Setting and enforcing better standards.
10. Limiting or preventing the use of plant protection products.
11. Prohibiting aerial spraying of biocidal products.
12. Systematically recording and monitoring environmental pollution.
13. Education and communication: actively sensitizing the population about the proper and sustainable use of biocidal products.
14. Introducing a ban on using antifouling products in sensitive areas.
15. Using a more realistic dilution factor for treatment plants in the exposure assessment of industrial chemicals.
16. Enhancing the criteria for eco-labels for detergents.
17. Information campaigns for sustainable handling of detergents.
18. Upgrading treatment plants: In practice, two methods for advanced wastewater treatment have proven technically feasible on a large scale: oxidation with ozone and adsorption onto activated carbon or a combination of the two methods.

19. Initiating projects in the departmental research plan for analyzing the entry of micropollutants into the environment from industries and businesses and possible measures at the source.
20. Voluntary initiatives for phasing out certain chemicals.
21. Using potential synergies between EU directives that provide for measures to reduce the emission of micropollutants (HazBREF).
22. Separate collection/disposal of radiocontrast agents.
23. Improved infrastructure and facilities in the health sector (Ahting et al., 2018).

9.3.2   Why Is Environmental Awareness Important?

We have a moral obligation to protect the environment and promote sustainable development for generations yet to come. We must take responsibility for our actions and understand their impact. We also need to take measures to protect the planet and hopefully undo some of the damage already caused. Since the industrial revolution, the concentration of pollutants has drastically increased, severely damaging the health of animals and humans. We need public awareness to instill empathy toward the planet, especially in children so that they can be well equipped with the tools necessary to behave in a responsible and informed way toward the environment.

Awareness can be created using newspapers, pamphlets, fundraising events, organizing educational events, social media, etc. The goal is to spread the message so that more and more people can participate in environmental protection. We all need to play our part however small it may be.

9.4   CONVENTION AND REGULATION

The recognition that the environment requires legal protection did not occur until the 1960s. At that time, various influences such as growing awareness of the unity and fragility of the biosphere increased public concern over the impact of industrial activity on natural resources and human health, and the success of various movements to protect the environment led to a collection of laws in a relatively short period (Table 9.5).

9.5   CONCLUSION

Since the past decade or so, there has been a continuous increase in the concentration of micropollutants in the environment. These so-called micropollutants are traces of medicinal products, sewage wastewater, industrial effluents, and other chemicals. These are the substances that have toxic effects on humans as well as the environment. They are persistent, bioaccumulative, and endocrine active.

Prominent micropollutants that are known to cause human health disorders are heavy metals such as mercury, arsenic, lead, POPs, polychlorinated biphenyls, endocrine-disrupting compounds, and agrochemicals. Mercury exposure, lead, and arsenic causes gastrointestinal, cardiovascular, reproductive, and neurological disorders. POPs and agrochemicals are genotoxic substances having epigenetic properties; they have recently been associated with immunosenescence, which is an essential

TABLE 9.5
Regulations and Their Objectives

S. No.	Regulation	Objective
1.	National Environmental Policy Act, 1970	To encourage harmony between humans and their environment
2.	Occupational Safety and Health Act, 1971	To consolidate and amend the laws regulating the occupational safety, health, and working conditions of employees
3.	Federal Water Pollution Control Act, 1972	Its objective is to restore and maintain the chemical, physical, and biological integrity of the nation's waters; recognize the responsibilities of the states in addressing pollution and providing assistance to states to do so, for the improvement of wastewater treatment; and maintain the integrity of wetlands
4.	Clean Water Act, 1972	To eliminate the discharge of pollutants into navigable waters
5.	Safe Drinking Water Act, 1974	To protect public health by regulating the nation's public drinking water supply
6.	The Toxic Substance Control Act, 1976	Address production, implementation, use, and disposal of PCBs, asbestos, lead-based paints, etc.
7.	The Emergency Planning and Community Right to Know Act, 1986	Companies must report inventories of specific chemicals kept in the workplace and the annual release of hazardous substances
8.	Pollution Prevention Act, 1990	Strategize to promote source reduction
9.	Public Liability Insurance Act, 1991	To provide for damages to victims of an accident that occurs as a result of handling hazardous substances
10.	Federal Insecticide, Fungicide, and Rodenticide Act (FIFRA), 1947/1996	Non-registered chemicals may not be sold or distributed in the US
11.	Biomedical Waste Management Rule, 1998	For disposal, segregation, and transportation of infectious waste
12.	Regulation on Detergents, 2004	Regulate complete aerobic biodegradation of surfactants and derogations for placing them on market
13.	REACH, 2006	Registration, evaluation, authorization, and restriction of chemicals
14.	Hazardous Waste Management, Handling, and Transboundary Rules, 2008	Guide for manufacturing, storage, import of hazardous chemicals, and management of waste
15.	Plant Protection Product Regulation, 2009	Authorization, placing on the market, use, and control of plant protection products
16.	Directive on Sustainable Use of Pesticides, 2009	Commitment to sustainable and environmentally friendly use of pesticides
17.	Directive for Industrial Emission, 2010	Sets of requirements for constructing, operating, and cessation of industrial installations
18.	Regulation of Biocidal Products, 2012	Authorization of biocidal products based on an environmental risk assessment of active biocidal substances
19.	Directive on Environmental Quality Standards, 2008/2013	Quality standards for priority substances. The list is updated every 6 years.
20.	Pesticide Registration Improvement Act, 2018	Amends FIFRA to revise registration and maintenance fee requirements for pesticides

Source: Speight (2016).

process in carcinogenicity. PCBs cause neurological defects, mainly affecting developing brains, and EDCs lead to reproductive disorders as they disrupt the normal course of action of androgen and estrogen.

Furthermore, the process of degradation of some substances can yield transformed metabolites that are more harmful than the original substance. The substance properties such as resistance to biodegradation and bioaccumulation call for a precautionary approach in handling micropollutants to amply protect ecology and health. We need a combination of precautionary measures at the source and during product use, the implementation of the best available technologies, and adherence to environmental quality standards.

ACKNOWLEDGMENT

The authors would like to express their gratitude to the Department of Civil Engineering (Aligarh Muslim University) for providing technical assistance with this chapter.

REFERENCES

Ahting, M., Brauer, F., Duffek, A., Ebert, I., Eckhardt, A., Hassold, E., Helmecke, M., Kirst, I., Krause, B., Lepom, P. and Leuthold, S., 2018. *Recommendations for Reducing Micropollutants in Waters.* German Environment Agency, Germany.

Alharbi, O.M., Khattab, R.A. and Ali, I., 2018. Health and environmental effects of persistent organic pollutants. *Journal of Molecular Liquids, 263*, pp. 442–453.

Amadi, C.N., Igweze, Z.N. and Orisakwe, O.E., 2017. Heavy metals in miscarriages and still-births in developing nations. *Middle East Fertility Society Journal, 22*(2), pp. 91–100.

Burkhardt-Holm, P., 2011. Linking water quality to human health and environment: The fate of micropollutants. Institute of Water Policy (IWP) – NUS, Singapore, pp. 1–62.

Chance, G.W., 2001. Environmental contaminants and children's health: Cause for concern, time for action. *Paediatrics & Child Health, 6*(10), pp. 731–743.

Ennour-Idrissi, K., Ayotte, P. and Diorio, C., 2019. Persistent organic pollutants and breast cancer: A systematic review and critical appraisal of the literature. *Cancers, 11*(8), p. 1063.

Eto, K., 1997. Pathology of Minamata disease. *Toxicologic Pathology, 25*(6), pp. 614–623.

Faroon, O. and Ruiz, P., 2016. Polychlorinated biphenyls: New evidence from the last decade. *Toxicology and Industrial Health, 32*(11), pp. 1825–1847.

Fisher, J.F. and World Health Organization, 2003. *Elemental Mercury and Inorganic Mercury Compounds: Human Health Aspects.* World Health Organization, Geneva, Switzerland.

Giulivo, M., de Alda, M.L., Capri, E. and Barceló, D., 2016. Human exposure to endocrine disrupting compounds: Their role in reproductive systems, metabolic syndrome and breast cancer. A review. *Environmental Research, 151*, pp. 251–264.

Huang, X., Lessner, L. and Carpenter, D.O., 2006. Exposure to persistent organic pollutants and hypertensive disease. *Environmental Research, 102*(1), pp. 101–106.

Kabir, E. R., Rahman, M. S. and Rahman, I. 2015. A review on endocrine disruptors and their possible impacts on human health. *Environmental Toxicology and Pharmacology, 40*(1), pp. 241–258.

Kugler, M., 2016. The Minamata disaster and the disease that followed: Mercury poisoning that sickened an entire Japanese town. *Verywell*, pp. 1–3.

Kumar, S., 2018. Occupational and environmental exposure to lead and reproductive health impairment: An overview. *Indian Journal of Occupational and Environmental Medicine, 22*(3), p. 128.

Ladele, J.I., Fajolu, I.B. and Ezeaka, V.C., 2019. Determination of lead levels in maternal and umbilical cord blood at birth at the Lagos University Teaching Hospital, Lagos. *PLoS One, 14*(2), p. e0211535.

Lee, Y.M., Kim, S.A., Choi, G.S., Park, S.Y., Jeon, S.W., Lee, H.S., Lee, S.J., Heo, S. and Lee, D.H., 2018. Association of colorectal polyps and cancer with low-dose persistent organic pollutants: A case-control study. *PLoS One, 13*(12), p. e0208546.

Lind, Y.S., Lind, P.M., Salihovic, S., van Bavel, B. and Lind, L., 2013. Circulating levels of persistent organic pollutants (POPs) are associated with left ventricular systolic and diastolic dysfunction in the elderly. *Environmental Research, 123*, pp. 39–45.

Luo, Y., Guo, W., Ngo, H.H., Nghiem, L.D., Hai, F.I., Zhang, J., Liang, S. and Wang, X.C., 2014. A review on the occurrence of micropollutants in the aquatic environment and their fate and removal during wastewater treatment. *Science of the Total Environment, 473*, pp. 619–641.

Mochizuki, H., 2019. Arsenic neurotoxicity in humans. *International Journal of Molecular Sciences, 20*(14), p. 3418.

Park, J.D. and Zheng, W., 2012. Human exposure and health effects of inorganic and elemental mercury. *Journal of Preventive Medicine and Public Health, 45*(6), p. 344.

Ritchie, J.M., Vial, S.L., Fuortes, L.J., Guo, H., Reedy, V.E. and Smith, E.M., 2003. Organochlorines and risk of prostate cancer. *Journal of Occupational and Environmental Medicine*, pp. 692–702.

Speight, J.G., 2016. *Environmental Organic Chemistry for Engineers*. Butterworth-Heinemann, Oxford.

Teh, B.T. and Fearon, E.R., 2020. Genetic and epigenetic alterations in cancer. In: J.E. Niederhuber, et al. (Eds), *Abeloff's Clinical Oncology* (pp. 209–224). Elsevier, Amsterdam, Netherlands.

Vargas, R. and Ponce-Canchihuamán, J., 2017. Emerging various environmental threats to brain and overview of surveillance system with zebrafish model. *Toxicology Reports, 4*, pp. 467–473.

Vázquez, M., Calatayud, M., Piedra, C.J., Chiocchetti, G.M., Vélez, D. and Devesa, V., 2015. Toxic trace elements at gastrointestinal level. *Food and Chemical Toxicology, 86*, pp. 163–175.

Weisenburger, D.D., 1993. Human health effects of agrichemical use. *Human Pathology, 24*(6), pp. 571–576.

Zahm, S.H. and Blair, A., 1992. Pesticides and non-Hodgkin's lymphoma. *Cancer Research, 52*(19 Supplement), pp. 5485s–5488s.

10 Biodegradability of Micropollutants in Wastewater and Natural Systems

Monika Dubey, Bhanu Prakash Vellanki,
and Absar Ahmad Kazmi
Indian Institute of Technology Roorkee

CONTENTS

10.1 Introduction 167
10.2 Removal Mechanisms 169
    10.2.1 Volatilization 169
    10.2.2 Adsorption 170
    10.2.3 Biodegradation 170
    10.2.4 Photolysis 170
10.3 Factors Affecting Biodegradation of ECs in Wastewater Treatment 171
    10.3.1 SRT 171
    10.3.2 HRT 171
    10.3.3 pH 172
    10.3.4 Redox Condition 172
    10.3.5 Temperature 172
    10.3.6 Microbial Community 173
    10.3.7 Suspended vs. the Attached Growth Process 173
10.4 Factors Affecting Biodegradation of ECs in Natural Systems 174
10.5 Biotransformation 174
10.6 Conclusion 175
10.7 Research Scope 176
References 176

10.1 INTRODUCTION

As per the United States Geological Survey, Emerging contaminants (ECs) are any synthetic or naturally occurring chemical or any microorganism that are not commonly monitored in the environment but have the potential to enter the environment and cause known or suspected adverse ecological and or human health effects

DOI: 10.1201/9781003202431-10

(USGS, 2016). ECs have been increasingly being identified in wastewater for the past two decades due to recent development and an increase in people's standard of living, leading to the frequent use of pharmaceuticals and personal care products (PCPs) collectively called PPCPs. Other contaminants commonly observed are illicit drugs, per- and polyfluorinated chemicals (PFCs), flame retardants, steroids, endocrine disruptors (EDCs), and their metabolites.

According to Philip, Aravind, and Aravindakumar (2018), a total of 166 compounds belonging to 36 classes of ECs have been detected in India with categories in terms of abundance as pharmaceuticals (95 nos.), followed by Per- and polyfluoroalkyl substances (PFASs) (35 nos.), EDCs (20 nos.), PCPs (11 nos.), artificial sweeteners (ASWs) (4 nos.) and flame retardants (1 nos.). Most of these compounds are usually partially metabolized in the human body and ultimately get into municipal wastewater. The percentage excretion of a certain drug from the body of the host varies from drug to drug. It could be as high as 90% for antibiotics excreted from human and animal bodies, unchanged (Hirsch et al., 1999) or as low as 5%–10% for azithromycin (Grujić et al., 2009). Most pharmaceutical compounds are small, organic, moderately hydrophobic, or lipophilic with less than 500 Da molecular weight (Lipinski, Lombardo, Dominy, and Feeney, 2001; Radjenovic, Petrović, and Barceló, 2007). For pharmaceuticals to work properly, they must remain stable in harsh conditions inside the human body. Because of this, the compounds in many pharmaceuticals do not break down when excreted or disposed of in water where they are bioavailable and biologically active (Halling-Sorensen et al., 1998). This is why a number of pharmaceutical compounds have been detected in wastewater. Once taken, pharmaceuticals are absorbed, distributed, metabolized, and excreted in their original form or as conjugates, dissolved in urine, or biodegraded in feces (Tambosi et al., 2010a). These micropollutants are generally found at a concentration from low ng/L to μg/L in the solid and liquid phases in the sewage. A review of analgesics and non-steroidal anti-inflammatory drug (NSAID) concentration in the influent wastewater suggests it to be in the range from ng/L to mg/L (Grandclément et al., 2017).

ECs in the water environment have proven to be hazardous for the health of human beings and particularly aquatic life. This is because of their constant and direct contact with contaminated water. The effect can be chronic or acute depending on the degradability of the compound and its bioavailability in the environment (Jjemba, 2006). Pharmaceuticals continuously enter the environment with variable removal and transformation rates. Hence, they may have the same exposure potential as persistent organic pollutants (Gros et al., 2006). Adverse effects include toxicity, antibiotic-resistant pathogenic bacteria, genotoxicity, and endocrine disruption effects (Tambosi et al., 2010a). Some of the primary sources of ECs, common treatment options, and potential receptors of ECs in the environment are shown in Figure 10.1.

Wastewater treatment plants (WWTPs) have been observed to be not the only but the largest source of ECs entering the environment (Petrović, Hernando, Díaz-Cruz, and Barceló, 2005). They may act as a barrier to the spread of such compounds in the surface water. This explains the need to have an effective technique for removing ECs in the WWTPs. A high proportion of sewage-generating networks is not connected to the wastewater collection and treatment facilities in developing and underdeveloped countries. Since the natural degradation of these contaminants is

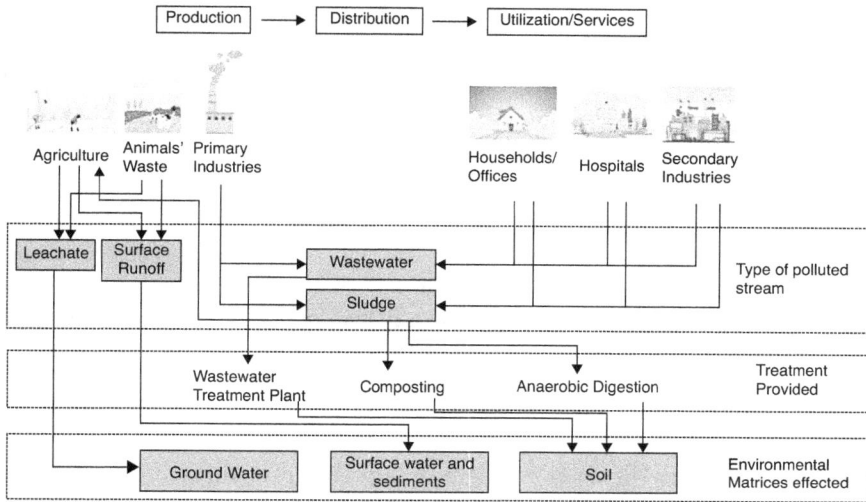

FIGURE 10.1   Sources, treatment, and potential receptors of ECs in the environment.

observed to be very low, it is essential to understand the removal mechanisms of ECs in wastewater and natural systems.

Reverse osmosis and advanced oxidation processes have shown promising results for removing ECs in wastewater, but high treatment costs and the possibility of toxic byproducts' formation limits their wide usage. There are various conventional treatment techniques such as activated sludge process (ASP), modified ludzack ettinger (MLE) technique, sequential batch reactors, waste stabilization ponds (WSP), up-flow anaerobic sludge blanket reactors (UASBR), and other hybrid techniques evolving with time with the hope to address the issue better. The major removal mechanisms of ECs in natural and engineered WWTPs are discussed in the following section.

10.2   REMOVAL MECHANISMS

ECs' removal in WWTPs occurs mainly by four mechanisms: (1) volatilization by aeration, (2) adsorption to sewage sludge and activated carbon or any such media, (3) biological degradation depending on specific conditions like solids retention time (SRT), hydraulic retention time (HRT), pH, redox condition, temperature (Ejhed et al., 2018), and (4) photodegradation.

10.2.1   VOLATILIZATION

The volatile compounds in the unionized state can be significantly removed by the air-stripping effect. Matamoros et al. (2016) reported more than 90% removal of 4-octylphenol, galaxolide, and tributyl phosphate in the aerated systems. However, when Henry's coefficient (K_H) value is less than 10^{-1} Pa m^3/mol, the removal of the compound by the volatilization pathway can be neglected (Joss et al., 2006). With K_H generally very low, volatilization is minimal for most of the ECs.

10.2.2 ADSORPTION

Adsorption occurs both by hydrophobic as well as electrostatic interaction of ECs with particulates and microbes. When the pk_a of the compound is less than the pH of wastewater to be treated, which is almost neutral in most cases, the compound exists in an ionic form. In such conditions, the compound remains dissolved in water and has less tendency for adsorption on the sludge. However, with the decrease in pH, adsorption tends to increase (Tambosi et al., 2010b). Compounds with a high log K_{ow} value also tend to adsorb onto the sludge, and thus the removal efficiency of such compounds is higher (Ejhed et al., 2018; Fan, Li, Zhang, and Feng, 2014). The study also found a strong correlation between the removal of caffeine, ibuprofen, estrone, naproxen, and estradiol with the removal of particles and sludge.

10.2.3 BIODEGRADATION

The ECs vary widely in terms of degradability. The recalcitrance of some compounds concerning biological degradation depends on the molecular properties of the compounds (Onesios et al., 2009). These properties define the degradation of certain compounds by a specific strain of microorganisms under favorable optimum conditions (Tahri Joutey et al., 2018). Studies have suggested that aromatic rings or elements such as chlorine in the molecular structure result in lesser susceptibility to biological degradation. Compounds like clofibric acid, diclofenac, and dichloroprop are some such compounds with complex structures (Kimura, Hara, & Watanabe, 2005). Kümmerer and Al-Ahmad (1997) observed low degradability of cytarabine, an anti-tumor agent in hospital wastewater due to its fluorinated structure requiring high redox potential for degradation.

Apart from the expensive and time-consuming experimental studies, the biodegradability of ECs can also be estimated based on the regression models-based structure-activity relationship, commonly called SAR models. The models are highly useful in predicting the biodegradability of the compound based on the structural fragments of the compound (functional groups, steric structure, and bonding type) and their biodegradability in WWTPs. These models are currently adopted by the United States Environmental Protection Agency as the EPA BIOWIN model and the Japanese Ministry of International Trade and Industry (MITI) for use in chemical screening (Tunkel et al., 2000).

10.2.4 PHOTOLYSIS

Yu-Chen and Martin (2009) found that several ECs such as ketoprofen, propranolol, naproxen, E2, EE2, gemfibrozil, and ibuprofen could be successfully removed from a contaminated river water sample by photolysis. The breakdown by photolysis depends on the chemical structure. For ketoprofen, the carbonyl moiety is in conjugation with two aromatic rings is attributed to making it susceptible to photolytic degradation. However, complete mineralization may not be possible, and thus, complete removal with the removal of the parent compound is not possible. The process can be seriously hindered by the high concentration of dissolved organic matter and suspended particles.

10.3 FACTORS AFFECTING BIODEGRADATION OF ECs IN WASTEWATER TREATMENT

The existing wastewater treatment infrastructure is designed to treat conventional pollutants such as organic carbon, present at much higher concentrations and not micropollutants. Therefore, the metabolism of the macropollutants acts as the major source of energy for the microbial community, whereas the major removal pathway for ECs is mostly through co-metabolism (Fischer and Majewsky, 2014). The rate of co-metabolism can be controlled with the concentration of carbon, nitrogen, and other nutrients (Fischer and Majewsky, 2014). There exists a vast difference in the biodegradability of ECs reported in the literature. This is primarily because of differences in the initial concentration of primary substrate and ECs, varying microbial populations, and inoculation time (Onesios et al., 2009). Some of the critical factors impacting the biodegradation of ECs in WWTPs are discussed below.

10.3.1   SRT

SRT is one of the critical parameters used to design a wastewater treatment facility. The SRT governs the mean cell residence time of microbes in the active biological reactor and therefore impacts the fate of ECs in WWTPs. This is because a more extended SRT aids in the development of slow-growing microbes, thereby increasing their diversity in the treatment system. Kimura et al. (2007) reported higher removal of six acidic pharmaceuticals at higher SRT (7, 15, and 65 days) mostly due to enhanced biodegradation. The authors also reported enhanced specific surface area of sludge in the bioreactor at higher SRT, leading to higher sorption of ECs. Comparing the membrane bioreactors with the ASP system to remove several ECs, including pharmaceuticals, musk fragrances, and EDCs, Clara et al. (2005b) reported similar removal of ECs but favored membrane bioreactor as it can achieve similar or higher SRT in a compact design. Since washout is not a problem, an MBR provides a stable environment for slow-growing bacteria that can degrade bio-recalcitrant compounds.

Owing to enhanced heterotrophic biodegradation of estrogens in carbonaceous ASP-based systems, an increase in SRT is recommended for higher degradation of estrogens to >70% and >80% at SRT of 10 and 20 days, typically higher than the SRT of ASP systems (McAdam et al., 2010). The available literature indicates that a lower concentration of ECs in the effluent can be achieved at SRT > 10 days at 10°C (Clara et al., 2005a). Enhanced degradation of moderately biodegradable compounds can be achieved at higher internal recirculation of mixed liquor suspended solids (MLSS) (Suárez et al., 2012). However, for compounds with high sorption potential, such as ethinylestradiol, an increase in removal by 11% is reported by Suárez et al. (2012) at SRT >20 days.

10.3.2   HRT

Petrie et al. (2014) recommended an increase in the HRT of the CAS system for a generic removal of ECs. The removal rate is generally inversely proportional to the food-to-microorganism ratio, possibly due to the selective uptake of the substrate at a high F/M ratio (Joss et al., 2004). The % removal of 17α-ethinylestradiol increased to 65% with an increase in HRT leading to a reduction in the F/M ratio (Petrie et al., 2014). Similarly, Escolà Casas and Bester (2015) reported an increase in the removal of recalcitrant ECs (diclofenac, iopromide, ioxehol, propranolol, propiconazole, tebuconazole, and iomeprol) with an increase in the HRT from four to 35 hours, predominantly by biodegradation in a column study.

10.3.3 pH

High deviation from the working pH ranges of the wastewater can change the microbial fauna, thereby decreasing the biological performance of the treatment plant. The change is reflected in the lower removal of organic carbon and micronutrients (N and P) and the removal of ECs (Tadkaew et al., 2010).

The pH of the mixed liquor or the reacting medium influences the speciation of ECs, thereby affecting their adsorption on suspended particles and sludge. The neutral/negatively charged compounds are likely to be present in the water matrix, available for biodegradation (Tadkaew et al., 2010). Whereas the positively charged ECs are likely to be adsorbed onto the solid matrix, unavailable for biodegradation. Sui et al. (2016) studied the effect of pH on the removal of bezafibrate through the ASP. At a pKa of 3.6, bezafibrate exist in anionic form. Thus, at a working pH range of 6–9, the compound is unaffected by pH. The authors also reported high adaption of the microbes to bezafibrate at a wide pH range (5–9).

10.3.4 REDOX CONDITION

Studies suggest that the removal of ECs depends on the redox condition. Though the most efficient configuration redox conditions to remove an array of ECs is unknown, high degradation is reported under aerobic regime due to high metabolization and availability of free energy (Xue et al., 2010). McAdam et al. (2010) reported higher removal of estrogens in nitrifying (91%) and nitrifying/denitrifying (89%) than in the conventional carbonaceous (heterotrophic) ASP (51%). The biomass activity was not found related to the removal of estrogens as the biomass activity for carbonaceous ASP was higher (80 μg/kg biomass/day) than under nitrifying (61 μg/kg biomass/day) and nitrifying/denitrifying (15 μg/kg biomass/day) conditions.

10.3.5 TEMPERATURE

In general, higher degradation of ECs is observed at elevated temperatures due to increased microbial activity. Every 10°C increase in the temperature doubles the microbial activity. Since the temperature affects the microbial population, the temperature is related to the SRT of the treatment process and can be calculated at a reference temperature according to (Clara et al., 2005a):

$$\text{SRT}_{T\text{ref}} = \text{SRT}_T \times 1.072^{(T\text{ref}-10)} \tag{10.1}$$

The effect of temperature on microbial activity is more pronounced at temperatures <20°C. Sui et al. (2016) reported a minimal increase in the removal rate of bezafibrate from 0.050 to 0.056 hours⁻¹ with an increase in temperature from 20°C to 30°C. However, with a decrease in temperature to 10°C, the reaction rate decreased dramatically from 0.050 to 0.010 hours^{-1}.

10.3.6 MICROBIAL COMMUNITY

A diverse microbial consortium can help increase the biodegradation of ECs. There exists a positive relationship between the diversity of microbial population and the biotransformation of specific but not all ECs in WWTPs (Johnson et al., 2015). Studies point to increased removal of ECs by oxidase enzymes such as oxygenase Cytochromes P450 and laccases enzymes acting as biocatalysts in pure culture (Singhal and Perez-Garcia, 2016). Laccases are multi-copper oxidases derived from bacteria, plants, insects, and fungi and capable of transforming aminophenols, polyphenols, polyamines, and aryl diamines. Laccases act as electron acceptors to oxidize the substrate and are capable of oxidizing monophenols, hydroquinone-like structures, and in some cases, non-phenolic compounds (Gasser et al., 2014). However, such studies are at their developmental stage and have not been implemented in full-scale systems. The current WWTPs utilize prokaryotic microbes to treat the organic matter (Singhal and Perez-Garcia, 2016). The feast-famine regime (intermittent feeding with settled raw and effluent wastewater) in a MBBR-based WWTP was reported to increase the biotransformation of 24 ECs to as high as 66 times for propranolol and 10 times for atenolol (Liang et al., 2021). The method utilized preferential development of microbes favoring the utilization of recalcitrant carbon and suggested that the growth of potential degraders was lower than the growth rate of the non-degrading microbial population.

Similar to bacteria, several classes of fungi such as *Bjerkandera* sp. R1, *Bjerkanderaadusta*, and *Phanerochaetechrysosporium* are reported successful in completely mineralizing naproxen in a continuously stirred tank reactor (Rodarte-Morales et al., 2012).

10.3.7 SUSPENDED VS. THE ATTACHED GROWTH PROCESS

Biofilm-based treatment techniques are reported to be more efficient than conventional activated sludge-based systems (suspended growth process) in removing some recalcitrant ECs such as diclofenac (Liang et al., 2021). This could be possibly due to higher biomass concentration in the attached growth systems. The removal efficiency of ECs by attached growth slow sand filters, as studied by Escolà Casas and Bester (2015), is reported to be 41% and 85% for diclofenac and iomeprol, which are otherwise considered bio-recalcitrant compounds. Similarly, Falås et al. (2012) reported higher removal of diclofenac, ketoprofen, gemfibrozil, clofibric acid, and mefenamic acid per unit biomass in biofilm carrier aerobic treatment plant as compared to suspended growth system. Though a higher nitrification rate was observed in the suspended system, higher removal in the biomass carrier system indicates that the removal of ECs is not solely dependent on the activity of ammonia-oxidizing bacteria.

10.4 FACTORS AFFECTING BIODEGRADATION OF ECs IN NATURAL SYSTEMS

The microbial communities in the natural systems such as soil, surface water, and groundwater are key factors in maintaining the ecosystem and deciding the fate of ECs. The metabolic/co-metabolic action can degrade various classes of ECs (Barra Caracciolo et al., 2015) in different time scales. For example, naproxen, a common anti-inflammatory drug, is reported to be degraded within 30–40 days by α and γ-proteobacteria found in river water (Grenni et al., 2014). Similarly, naproxen can be mineralized by soil microcosms to $^{14}CO_2$ in 27 days by natural degradation under aerobic conditions at 30°C (Topp et al., 2008).

10.5 BIOTRANSFORMATION

Conventional technologies based on biological processes are not sufficient to altogether remove the ECs from wastewater. Studies suggest that some of these contaminants may hinder the biological processes taking place in conventional WWTPs. Metabolites formed in due course of time may also act as secondary pollutants (Yang et al., 2017).

The transformation products are of great concern in the biodegradation study due to their stability and toxicity. Phase I (oxidation and hydrolysis) and phase II (glucuronidation and sulfation) metabolites of the ECs are detected in the wastewater and biosolids. The enzymatic activity assisted glucuronidation as phase II transformation is observed for hydroxyl, carboxyl, −NHOH, amino, tertiary amine, thiol, and 1,3-dicarbonyl groups, whereas sulfonation is observed more frequently for hydroxyl and less frequently for amino groups (Sunkara and Wells, 2010). For example, a list of some pharmaceuticals, metabolites, and their conjugated form is provided in Table 10.1. The study

TABLE 10.1

Ratio of Excretion of Pharmaceuticals and Its Metabolite in Raw Sewage through Urine (Gurke et al., 2015)

Parent/Metabolite	Excretion in Non-Conjugated and Conjugated Form				Ratio Metabolite/ Parent
	Parent (%)	Conjugated Parent (%)	Metabolite (%)	Conjugated Metabolite (%)	Expected in Sewage
Citalopram/N-desmethyl citalopram	26	12	19	0	0.5–0.7
Clozapine/N-desmethyl clozapine	2.5	3.1	1.3–5.3	0	0.2–2
Mirtazapine/N-desmethyl mirtazapine	4	25	25–35	0	0.8–8.3
Oxcarbazepine/10,11-dihydro-10-hydroxy carbamazepine	0.6	9	26.9	41.1–44	2.8–119.1
Venlafaxine/O-desmethyl venlafaxine	4.7	0	29.4	26.4	5.9–11.3

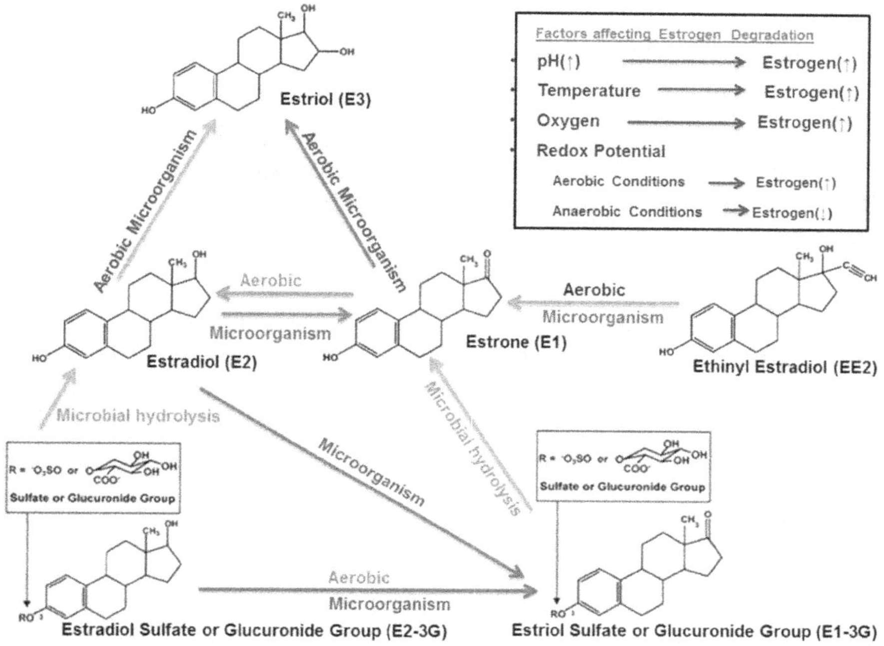

FIGURE 10.2 Conversion pathways for estrogens in wastewater. (Adapted with permission from Adeel et al. 2017.)

of metabolites is of particular interest as phase II metabolized glucuronate or sulfate metabolites can be back-transformed to parent compounds during the treatment.

Transformation products can also be formed by breaking down long and complex structures into simpler structures during biodegradation. For example- nonyl phenols with a long hydrophilic chain are hydrophobic. However, their biotransformation can result in smaller and short-chained intermediates (Petrie et al., 2014). Zwiener et al. (2002) studied the formation and degradation of transformation products of ibuprofen under aerobic and anoxic conditions in a lab-scale biofilm reactor and reported a positive relationship between an increase in HRT and the formation of hydroxy-ibuprofen under aerobic conditions. Estrogens have an interconnected metabolic pathway during biological degradation, as explained in Figure 10.2. Under aerobic conditions, E1 can be transformed to E3, and microbes such as *Sphingobacterium* sp. can degrade EE2 to E1 (Adeel et al., 2017). There also exist several microbial strains in anaerobic mode, capable of transforming estrogens to other forms. The study also indicated that high temperature and dissolved oxygen are favorable for enhanced degradation of estrogens in wastewater.

10.6 CONCLUSION

This chapter highlights the features of potential ECs in wastewater. The main factors affecting the biodegradation of ECs are summarized. Change in lifestyle, industrialization, and unnecessary dependence on pharmaceuticals are some of the causes of the introduction of ECs in wastewater. The ECs are not sufficiently treated (partially treated or metabolized) in the WWTPs and are discharged in the surface water bodies. There are proofs of endocrine disruption, teratogenic effects, and neurological disorders linked to the presence of ECs in water bodies. Biodegradation is reported as the main removal mechanism of ECs in the WWTPs. However, complete mineralization through biodegradation is rare. Several plant design parameters, including SRT, HRT, redox conditions, microbial diversity, characteristics of the wastewater such as pH, temperature, characteristics of the compounds such as pKa, hydrophobicity, functional groups, and climatic conditions such as temperature, can affect the biodegradation of ECs. An increase in HRT, SRT, and temperature is reported to impact the removal of ECs positively. In pure cultures, several specialized enzymes are reported as potential solutions for degrading ECs, including bio-recalcitrant ones. However, a sufficient understanding of biodegradation, including the metabolic pathways of biotransformation of ECs, is still lacking. Some of the research areas requiring further research are discussed in the following section.

10.7   RESEARCH SCOPE

The issue with the occurrence of ECs in the environment has attracted the attention of the scientific community in the past three decades. However, there still exist certain research gaps related to the formation of transformation products during biodegradation. Not only the proper mass balance can be estimated by incorporating transformation products (TPs), but also their ecological effects can be estimated. The microbial consortia can be further explored to understand the interaction of bacteria-fungi, and bacteria-archaea to increase the biodegradation of ECs. Oxygenase Cytochromes P450 (CYPs or Cyt P450), an efficient class of monooxygenases, are reported to degrade pharmaceuticals and toxins (Singhal and Perez-Garcia, 2016). However, studies on microbes expressing these strains are insufficient to understand the interactions in real wastewater fully.

Studies on the impact of ECs on microbial populations are scarce. Reports indicate the negative effect of ECs on the microbial population, impeding biodegradation (Gomes et al., 2020). An enhanced understanding of the above areas can help better understand the total removal, biodegradability, and related biotransformation in the WWTPs and allow for efficient application of biodegradation to remove ECs (parent and/or transformation product).

REFERENCES

Adeel, M., Song, X., Wang, Y., Francis, D., Yang, Y., 2017. Environmental impact of estrogens on human, animal and plant life: A critical review. *Environ. Int.* 99, 107–119. doi: 10.1016/j.envint.2016.12.010.

Barra Caracciolo, A., Topp, E., Grenni, P., 2015. Pharmaceuticals in the environment: Biodegradation and effects on natural microbial communities: A review. *J. Pharm. Biomed. Anal.* 106, 25–36. doi: 10.1016/j.jpba.2014.11.040.

Clara, M., Kreuzinger, N., Strenn, B., Gans, O., Kroiss, H., 2005a. The solids retention time: A suitable design parameter to evaluate the capacity of wastewater treatment plants to remove micropollutants. *Water Res.* 39, 97–106. doi: 10.1016/j.watres.2004.08.036.

Clara, M., Strenn, B., Gans, O., Martinez, E., Kreuzinger, N., Kroiss, H., 2005b. Removal of selected pharmaceuticals, fragrances and endocrine disrupting compounds in a membrane bioreactor and conventional wastewater treatment plants. *Water Res.* 39, 4797–4807. doi: 10.1016/j.watres.2005.09.015.

Ejhed, H., Fång, J., Hansen, K., Graae, L., Rahmberg, M., Magnér, J., Dorgeloh, E., Plaza, G., 2018. The effect of hydraulic retention time in onsite wastewater treatment and removal of pharmaceuticals, hormones and phenolic utility substances. *Sci. Total Environ.* 618, 250–261. doi: 10.1016/J.SCITOTENV.2017.11.011.

Escolà Casas, M., Bester, K., 2015. Can those organic micro-pollutants that are recalcitrant in activated sludge treatment be removed from wastewater by biofilm reactors (slow sand filters)? *Sci. Total Environ.* 506–507, 315–322. doi: 10.1016/j.scitotenv.2014.10.113.

Falås, P., Baillon-Dhumez, A., Andersen, H.R., Ledin, A., La Cour Jansen, J., 2012. Suspended biofilm carrier and activated sludge removal of acidic pharmaceuticals. *Water Res.* 46, 1167–1175. doi: 10.1016/j.watres.2011.12.003.

Fan, H., Li, J., Zhang, L., Feng, L., 2014. Contribution of sludge adorption and biodegradation to the removal of five pharmaceuticals in a submerged MBR.pdf. *Biochem. Eng. J.* 88, 101–107.

Fischer, K., Majewsky, M., 2014. Cometabolic degradation of organic wastewater micropollutants by activated sludge and sludge-inherent microorganisms. *Appl. Microbiol. Biotechnol.* 98, 6583–6597. doi: 10.1007/s00253-014-5826-0.

Gasser, C.A., Ammann, E.M., Shahgaldian, P., Corvini, P.F.X., 2014. Laccases to take on the challenge of emerging organic contaminants in wastewater. *Appl. Microbiol. Biotechnol.* 98, 9931–9952. doi: 10.1007/s00253-014-6177-6.

Gomes, I.B., Maillard, J.Y., Simões, L.C., Simões, M., 2020. Emerging contaminants affect the microbiome of water systems: Strategies for their mitigation. *NPJ Clean Water* 3. doi: 10.1038/s41545-020-00086-y.

Grandclément, C., Seyssiecq, I., Piram, A., Wong-Wah-Chung, P., Vanot, G., Tiliacos, N., Roche, N., Doumenq, P., 2017. From the conventional biological wastewater treatment to hybrid processes, the evaluation of organic micropollutant removal: A review. *Water Res.* 111, 297–317. doi: 10.1016/j.watres.2017.01.005.

Grenni, P., Patrolecco, L., Ademollo, N., Di Lenola, M., Barra Caracciolo, A., 2014. Capability of the natural microbial community in a river water ecosystem to degrade the drug naproxen. *Environ. Sci. Pollut. Res.* 21, 13470–13479. doi: 10.1007/s11356-014-3276-y.

Gros, M., Petrović, M., Barceló, D., 2006. Development of a multi-residue analytical methodology based on liquid chromatography-tandem mass spectrometry (LC-MS/MS) for screening and trace level determination of pharmaceuticals in surface and wastewaters. *Talanta* 70, 678–690. doi: 10.1016/j.talanta.2006.05.024.

Grujić, S., Vasiljević, T., Laušević, M., 2009. Determination of multiple pharmaceutical classes in surface and ground waters by liquid chromatography-ion trap-tandem mass spectrometry. *J. Chromatogr. A* 1216, 4989–5000. doi: 10.1016/j.chroma.2009.04.059.

Gurke, R., Rößler, M., Marx, C., Diamond, S., Schubert, S., Oertel, R., Fauler, J., 2015. Occurrence and removal of frequently prescribed pharmaceuticals and corresponding metabolites in wastewater of a sewage treatment plant. *Sci. Total Environ.* 532, 762–770. doi: 10.1016/j.scitotenv.2015.06.067.

Halling-Sorensen, B., Nors Nielsen, S., Lanzky, P.F., Ingerslev, F., Lutzhoft, H.C., Jorensen, S.E., 1998. Occurrence, fate and effects of pharmaceutical substances in the environment: A review. *Chemosphere* 36, 357–393. doi: 10.1016/j.jhazmat.2012.10.020.

Hirsch, R., Ternes, T., Haberer, K., Kratz, K.L., 1999. Occurrence of antibiotics in the aquatic environment. *Sci. Total Environ.* 225, 109–118. doi: 10.1016/S0048-9697(98)00337-4.

Jjemba, P.K., 2006. Excretion and ecotoxicity of pharmaceutical and personal care products in the environment. *Ecotoxicol. Environ. Saf.* 63, 113–130. doi: 10.1016/j.ecoenv.2004.11.011.

Johnson, D.R., Helbling, D.E., Lee, T.K., Park, J., Fenner, K., Kohler, H.P.E., Ackermann, M., 2015. Association of biodiversity with the rates of micropollutant biotransformations among full-scale wastewater treatment plant communities. *Appl. Environ. Microbiol.* 81, 666–675. doi: 10.1128/AEM.03286-14.

Joss, A., Andersen, H., Ternes, T., Richle, P.R., Siegrist, H., 2004. Removal of estrogens in municipal wastewater treatment under aerobic and anaerobic conditions: Consequences for plant optimization. *Environ. Sci. Technol.* 38, 3047–3055. doi: 10.1021/es0351488.

Joss, A., Zabczynski, S., Göbel, A., Hoffmann, B., Löffler, D., McArdell, C.S., Ternes, T.A., Thomsen, A., Siegrist, H., 2006. Biological degradation of pharmaceuticals in municipal wastewater treatment: Proposing a classification scheme. *Water Res.* 40, 1686–1696. doi: 10.1016/j.watres.2006.02.014.

Kimura, K., Hara, H., Watanabe, Y., 2005. Removal of pharmaceutical compounds by submerged membrane bioreactors (MBRs). *Desalination* 178, 135–140. doi: 10.1016/j.desal.2004.11.033.

Kimura, K., Hara, H., Watanabe, Y., 2007. Elimination of selected acidic pharmaceuticals from municipal wastewater by an activated sludge system and membrane bioreactors. *Environ. Sci. Technol.* 41, 3708–3714. doi: 10.1021/es061684z.

Kümmerer, K., Al-Ahmad, A., 1997. Biodegradability of the anti-tumour agents 5-fluorouracil, cytarabine, and gemcitabine: Impact of the chemical structure and synergistic toxicity with hospital effluent. *Acta Hydrochim. Hydrobiol.* 25, 166–172. doi: 10.1002/aheh.19970250402.

Liang, C., de Jonge, N., Carvalho, P.N., Nielsen, J.L., Bester, K., 2021. Biodegradation kinetics of organic micropollutants and microbial community dynamics in a moving bed biofilm reactor. *Chem. Eng. J.* 415. doi: 10.1016/j.cej.2021.128963.

Lipinski, C.A., Lombardo, F., Dominy, B.W., Feeney, P.J., 2001. Experimental and computational approaches to estimate solubility and permeability in drug discovery and development settings. *Adv. Drug Deliv. Rev.* 3–26. doi: 10.1021/ac0101454.

Matamoros, V., Uggetti, E., García, J., Bayona, J.M., 2016. Assessment of the mechanisms involved in the removal of emerging contaminants by microalgae from wastewater: A laboratory scale study. *J. Hazard. Mater.* 301, 197–205. doi: 10.1016/j.jhazmat.2015.08.050.

McAdam, E.J., Bagnall, J.P., Koh, Y.K.K., Chiu, T.Y., Pollard, S., Scrimshaw, M.D., Lester, J.N., Cartmell, E., 2010. Removal of steroid estrogens in carbonaceous and nitrifying activated sludge processes. *Chemosphere* 81, 1–6. doi: 10.1016/j.chemosphere.2010.07.057.

Onesios, K.M., Yu, J.T., Bouwer, E.J., 2009. Biodegradation and removal of pharmaceuticals and personal care products in treatment systems: A review. *Biodegradation* 20, 441–466. doi: 10.1007/s10532-008-9237-8.

Petrie, B., McAdam, E.J., Lester, J.N., Cartmell, E., 2014. Assessing potential modifications to the activated sludge process to improve simultaneous removal of a diverse range of micropollutants. *Water Res.* 62, 180–192. doi: 10.1016/j.watres.2014.05.036.

Petrović, M., Hernando, M.D., Díaz-Cruz, M.S., Barceló, D., 2005. Liquid chromatography-tandem mass spectrometry for the analysis of pharmaceutical residues in environmental samples: A review. *J. Chromatogr. A* 1067, 1–14. doi: 10.1016/j.chroma.2004.10.110.

Philip, J.M., Aravind, U.K., Aravindakumar, C.T., 2018. Emerging contaminants in Indian environmental matrices: A review. *Chemosphere* 190, 307–326. doi: 10.1016/j.chemosphere.2017.09.120.

Radjenovic, J., Petrovic, M., Barceló, D., 2007. Analysis of pharmaceuticals in wastewater and removal using a membrane bioreactor. *Anal. Bioanal. Chem.* 387, 1365–1377. doi: 10.1007/s00216-006-0883-6.

Rodarte-Morales, A.I., Feijoo, G., Moreira, M.T., Lema, J.M., 2012. Biotransformation of three pharmaceutical active compounds by the fungus Phanerochaete chrysosporium in a fed batch stirred reactor under air and oxygen supply. *Biodegradation* 23, 145–156. doi: 10.1007/s10532-011-9494-9.

Singhal, N., Perez-Garcia, O., 2016. Degrading organic micropollutants: The next challenge in the evolution of biological wastewater treatment processes. *Front. Environ. Sci.* 4, 1–5. doi: 10.3389/fenvs.2016.00036.

Suárez, S., Reif, R., Lema, J.M., Omil, F., 2012. Mass balance of pharmaceutical and personal care products in a pilot-scale single-sludge system: Influence of T, SRT and recirculation ratio. *Chemosphere* 89, 164–171. doi: 10.1016/j.chemosphere.2012.05.094.

Sui, Q., Yan, P., Cao, X., Lu, S., Zhao, W., Chen, M., 2016. Biodegradation of bezafibrate by the activated sludge under aerobic condition: Effect of initial concentration, temperature and pH. *Emerg. Contam.* 2, 173–177. doi: 10.1016/j.emcon.2016.09.001.

Sunkara, M., Wells, M.J.M., 2010. Phase II pharmaceutical metabolites acetaminophen glucuronide and acetaminophen sulfate in wastewater. *Environ. Chem.* 7, 111–122. doi: 10.1071/EN09098.

Tadkaew, N., Sivakumar, M., Khan, S.J., McDonald, J.A., Nghiem, L.D., 2010. Effect of mixed liquor pH on the removal of trace organic contaminants in a membrane bioreactor. *Bioresour. Technol.* 101, 1494–1500. doi: 10.1016/j.biortech.2009.09.082.

Tahri Joutey, N., Bahafid, W., Sayel, H., El Ghachtouli, N., 2018. Biodegradation: Involved microorganisms and genetically engineered microorganisms. In: R. Chamy (Ed.), *Biodegradation: Life of Science*, p. 13. Books on Demand: Norderstedt, Germany.

Tambosi, J.L., de Sena, R.F., Favier, M., Gebhardt, W., José, H.J., Schröder, H.F., de Fatima Peralta Muniz Moreira, R., 2010a. Removal of pharmaceutical compounds in membrane bioreactors (MBR) applying submerged membranes. *Desalination* 261, 148–156. doi: 10.1016/j.desal.2010.05.014.

Tambosi, J.L., Yamanaka, L.Y., José, H.J., De Fátima Peralta Muniz Moreira, R., Schröder, H.F., 2010b. Recent research data on the removal of pharmaceuticals from sewage treatment plants (STP). *Quim. Nova* 33, 411–420. doi: 10.1590/S0100-40422010000200032.

Topp, E., Hendel, J.G., Lapen, D.R., Chapman, R., 2008. Fate of the non-steroidal anti-inflammatory drug naproxen in agricultural soil receiving liquid municipal biosolids. *Environ. Toxicol. Chem.* 27, 2005–2010. doi: 10.1897/07-644.1.

Tunkel, J., Howard, P.H., Boethling, R.S., Stiteler, W., Loonen, H., 2000. Predicting ready biodegradability in the Japanese Ministry of International Trade and Industry test. *Environ. Toxicol. Chem.* 19, 2478–2485. doi: 10.1002/etc.5620191013.

USGS, 2016. Drinking Water Exposure to Chemical and Pathogenic Contaminants: Emerging Contaminants [WWW Document]. U.S. Geol. Surv. URL https://archive.usgs.gov/archive/sites/health.usgs.gov/dw_contaminants/emc.html (accessed 5.6.19).

Xue, W., Wu, C., Xiao, K., Huang, X., Zhou, H., Tsuno, H., Tanaka, H., 2010. Elimination and fate of selected micro-organic pollutants in a full-scale anaerobic/anoxic/aerobic process combined with membrane bioreactor for municipal wastewater reclamation. *Water Res.* 44, 5999–6010. doi: 10.1016/j.watres.2010.07.052Yang, Y., Ok, Y.S., Kim, K.H., Kwon, E.E., Tsang, Y.F., 2017. Occurrences and removal of pharmaceuticals and personal care products (PPCPs) in drinking water and water/sewage treatment plants: A review. *Sci. Total Environ.* 596–597, 303–320. doi: 10.1016/j.scitotenv.2017.04.102.

Yu-Chen, L.A., Martin, R., 2009. Photodegradation of common environmental pharmaceuticals and estrogens in river water. *Environ. Toxicol. Chem.* 24, 1303–1309. doi: 10.1897/04-236R.1.

Zwiener, C., Seeger, S., Glauner, T., Frimmel, F.H., 2002. Metabolites from the biodegradation of pharmaceutical residues of ibuprofen in biofilm reactors and batch experiments. *Anal. Bioanal. Chem.* 372, 569–575. doi: 10.1007/s00216-001-1210-x.

11 Biodegradation Technology for the Removal of Micropollutants
A Critical Review

Kaushar Hussain and Nadeem Ahmad Khan
Mewat Engineering College

Izharul Haq Farooqi
Aligarh Muslim University

Sirajuddin Ahmed
Jamia Millia Islamia

CONTENTS

11.1 Introduction 181
11.2 Physicochemical Treatment for Degradation 183
11.3 Photocatalysis 184
11.4 Sonochemical Methods and Nanoremediation 186
11.5 Biotechnological Approaches for Micropollutant Degradation 186
11.6 Microbial Electrochemical System 187
11.7 Immobilized Enzymes for Micropollutant Degradation 188
11.8 Metabolic Engineering Approaches for Pollutant Degradation 191
11.9 Invention of Novel Genes Involved in Bioremediation 192
11.10 Enhanced Bioremediation via Metabolic Engineering Processes 193
11.11 Conclusions 194
References 195

11.1 INTRODUCTION

Modern researchers are interested in environmental pollutants that persist in the soil and water. There is an ever-increasing amount of these toxins in the environment. A few examples of these long-lasting polluters are pharmaceuticals, pesticides, phthalate plasticizers, polycyclic aromatic compounds, chlorinated phenols, organic

DOI: 10.1201/9781003202431-11

solvents, and even some metal ions (Xin et al., 2019). In addition to previously reported pollutants whose effects are not fully understood. Urban, industrial, and agricultural lands are all at risk due to the widespread use of chemicals. On the other hand, persistent environmental pollutants are pesticides, chlorinated compounds, aromatic hydrocarbons, and antibiotics (Singh et al., 2018a). To put it another way, these pollutants are being released into the environment through various means, including pesticides sprayed on crops, antibiotics discarded in hospitals, industrial effluents, and the wastewater treatment process (Shekoohiyan et al., 2016). About 3,000 suspected sites in the United States have been identified by the European Environmental Agency (2007), while there are 3 million contaminated points in Europe. There are more polluted sites in emerging countries than in developed countries due to the widespread use of toxic chemicals in countries like China, Nigeria, Pakistan, Ethiopia, Vietnam, Indonesia, and South Africa. Some of these compounds have regional variations in their concentration (Wang et al., 2014). The dangerous concentration of persistent environmental pollutants varies from chemical to chemical, even though many are present in low levels in soil and water. From 0.6 to 921 ng/L of pesticides in the US, China, and European water. Even in trace amounts, they pose a serious threat to life and are called micropollutants. There are four types of pesticides: organochlorines (organophosphorus), organophosphates (carbamates), and pyrethroids (pyrethroids) (Ali Zomorodian et al., 2017). Because they are nonbiodegradable and persistent, these pesticides harm the environment irreparably. It is possible for pesticides used on farmland or waterways to build up and enter the food chain, posing serious risks to the health of farm animals and humans. Asthma and fetal death are all possible long-term effects of cancer and endocrine disruption. Pesticides accumulate in food chains, where they end up wreaking havoc on all living things they encounter. Another issue with synthetic hormones like progestins in the water system is the potential for bioconcentration in fish, leading to developmental abnormalities (Perullini et al., 2014). These harmful chemicals accumulate in the body's tissues, harming the human body's organs and systems. Pesticides, for example, have been linked to cardiovascular problems, endocrine system dysfunction, liver dysfunction, and hypertension when used at high concentrations. Bird populations have been reported to decline due to the accumulation of chemicals like DDT in their food chains, linked to the decline of amphibians and other wildlife populations (Hou et al., 2019). Children, environmental pollutants that persist over time, and agricultural workers are just a few of the negatively affected by these chemicals. In addition to inhalation, ingestion, or skin penetration, many persistent environmental pollutants are found in contaminated food (Li et al., 2013). Studies show that patients with cancer-causing diseases have higher blood levels of these pollutants than healthy people. According to numerous studies, the accumulation of pesticides is a major factor in the development of lymphoma and other cancers of the brain, breast, testes, and ovaries (Li et al., 2019). It is impossible to ignore the negative effects of xenobiotics on plants and animals when used to control pests and increase crop yields. Therefore, reduced use of these pollutants may aid in the global control of diseases caused by environmental pollutants (Yoneda and Mokhtar, 2018). Persistent pollution from chemicals in the environment, on the other hand, is a major problem for the environment around the world. The degradation of these harmful chemicals has been studied using various chemical, physical, and biological methods. When degrading persistent environmental pollutants, the t1/2, physiochemical environment, and chemical degradation kinetics all play a role (Brunson and Sabatini, 2014). Various physiochemical techniques can be used, such as centrifugation and centrifugal settling, coagulation, hydrolysis, desorption, and photolysis. However, the efficacy and cost-effectiveness of these degradation methods are both low (Tsopela et al., 2014). When it comes to removing persistent chemicals from the environment, biological methods are both environmentally and economically friendly. Microorganisms produce enzymes used to degrade toxic chemicals through bond cleavage and the mineralization process. On the other hand, biotransformation can describe a process in which a chemical transformation takes place over some time. To transform toxic chemicals into less or nontoxic substances, microorganisms are effective (Calabi-Floody et al., 2012). As a result of this, microbial engineering was introduced, in which different genetic tools are used to modify microbial strains and meet modern bioremediation requirements. This includes *Dehalococcoides*, *Burkholderia*, *Pseudomonas*, and *Alcaligenes* strains and bacterial combinations with *Achromobacter*. *Alcaligenes* and *Rhodococcus* were developed to enhance the bioremediation of persistent environmental pollutants. Catabolic pathways for hazardous chemicals have been developed using various genetic engineering techniques (Hadibarata and Kristanti, 2012; Jiang et al., 2014; Moro et al., 2018). Bioremediation is the practice of using microbial strains to degrade persistent pollutants in the environment. Bioremediation also includes using advanced technologies to design microbial strains for pollutant degradation. New cells and enzymes and degradation pathways are designed to introduce microbial engineering into bioremediation processes (Tahirbegi et al., 2017). Using this bioremediation method, persistent environmental pollutants can be mineralized in a more environmentally friendly and cost-effective manner. For example, using a combination of protein and metabolic engineering, it is possible to remediate 1,2,3-trichloropropane (Wang et al., 2016). The degradability of organic pollutants and dyes has frequently been demonstrated using genetic engineering methods. In the same way, *Pseudomonas* sp. can be used to degrade various plastic polymers, such as polyethylene, polyvinyl chloride, polystyrene, and others, at various degradation rates. The engineering of microbial consortia is a good indicator of recent advances in synthetic biology (Yang et al., 2013b). A microbial consortium performs many times better than a microbial monoculture in efficiency and performance. The degradation of persistent environmental pollutants is accelerated through microbial consortia (Munir et al., 2022a). Numerous microorganisms work to degrade the pollutants in the environment, but each microbe works for degradation for a limited amount of time. A more efficient pollutant degrader than any wild-type is a microbially engineered strain. Engineered microorganisms that can degrade persistent pollutants are the main focus of this chapter (Singh et al., 2018b). It also discusses traditional and cutting-edge approaches to microbial strain engineering and optimization to degrade persistent environmental pollutants using these microbial strains (Ma et al., 2013).

11.2 PHYSICOCHEMICAL TREATMENT FOR DEGRADATION

For environmental and public health reasons, the destruction of emerging organic micropollutants or their transformation into harmless chemicals is widely desired during wastewater treatment (WWT) (Zhong et al., 2019). Physical, chemical, and biological methods are often used to remove emerging organic micropollutants from wastewater treatment plants (WWTPs). Adsorption onto surfaces, penetration into porous materials, and selective permeation are physicochemical technologies (Antonacci et al., 2018). Emerging micropollutants (EOMs) can be destroyed chemically or biologically during the tertiary treatment stage after the adsorption process. Some chemical therapies for emerging organic micropollutants' elimination have attracted growing interest, including advanced oxidation processes (AOPs), which generate radicals by producing hydroxyl radicals (H_2O_2) (Ali et al., 2022; Samal and Trivedi, 2020; Tabish et al., 2022). Rarely are emerging organic micropollutants destroyed entirely by AOP therapies, which use many resources. Emerging organic micropollutants can be completely degraded by biological means, making this an exciting area of research for WWT (Rafeeq et al., 2022b; Rasheed et al., 2022; Sher et al., 2022). For several emerging organic micropollutants, such as plasticizers, endocrine disrupters, and other plasticizers and surfactants, biodegradation is the most common removal method from municipal wastewater, particularly in the treated effluent. Adsorption of pollutants onto biomass or biofilm by microorganisms is one method of microbial biodegradation. Others, such as mineralization, result in inorganic compounds (such as CO_2 and H_2O) and biomass as the end products of the degradation. A biotransformation or detoxification process in which an emerging organic micropollutant undergoes a first transformation that results in more minor hazardous metabolites can also occur (through the metabolism of a microorganism or a group of them). Biotransformation can also produce toxic or persistent daughter products of the original molecule (Khan et al., 2022). As a carbon supply and energy source, microorganisms can use co-metabolism processes to biodegrade emerging organic micropollutants materials. We are interested in this process because emerging organic micropollutants may not be available in sufficient quantities to function as growth substrates (de Jesus et al., 2022b). Microorganisms which are used to degrade the emerging organic micropollutant during the treatment of wastewater through WWT, were studies including metabolic activities. Biological WWT operations benefit greatly from an in-depth knowledge of emerging organic micropollutant degrading routes and processes. On the other hand, biodegradation routes may comprise numerous events occurring simultaneously or sequentially, such as complicated redox reactions (Rasheed et al., 2022).

11.3 PHOTOCATALYSIS

It was seen that *Aminobacter* sp. MSH1 was capable of degrading the 2,6 dichlorphenamide (BAM), a metabolite of dichlobenil, in biologically active sand filters (Ali et al., 2022; Rasheed et al., 2022). Up to 96% of BAM (beginning concentration: 2.7 lg/L BAM) was removed from the infected filters with just 1.1 hours of residence time (Rafeeq et al., 2022b; Sher et al., 2022). Due to its high water solubility, this chemical has a rapid treatment time compared to the other pollutants on this list. pBAM1 and pBAM2 are the two plasmids that *Aminobacter* MSH1 uses to break down BAM into 2,6-dichlorobenzene acid (2,6-DCBA) and Krebs cycle intermediates, respectively. This investigation showed that the capacity to catabolize (i.e., the loss of catabolic genes) was unstable if the EOM was removed. Inorganic xenobiotic catabolic genes in organic-xenobiotic-degrading bacteria are dangerous, perhaps due to intramolecular gene rearrangements or complete plasmid loss, as previously described (de Jesus et al., 2022b; Khan et al., 2022).

The biodegradation of herbicide mecoprop, or 2-(4-chloro-2-methylphenoxy) propanoic acid (a combination of two enantiomers, where the (R)-enantiomer contains the herbicidal action), was examined. Carrier-based biofilms of immobilized microorganisms were produced in mecoprop-spiked low-nutrient wastewater (Qamar et al., 2022; Rafeeq et al., 2022a). A substantial preference for (S)-mecoprop, the enantiomer lacking herbicidal action, was seen after 200 hours of biofilm formation, but there was no biodegradation if the bacteria did not form biofilms. *Sphingobium, Rhodospirillales, Parvibaculum, Kaistia, Bradyrhizobium, Roseomonas,* and *Variovorax* were the most common bacterial taxa detected in this consortium (Dhiman et al., 2022; Ul-Abdin et al., 2022). The sequenced pooled DNA was filtered for known phenoxy herbicide catabolism genes before being assembled into a metagenomic assembly. According to newly discovered metabolite research, the fungus may use 4-chloro-2-methylphenol sulfate to link the sulfation and glucuronidation pathways in aromatic hydrocarbon metabolism. However, 3% of the metagenome was dedicated to fungal biomass and its genetic potential for phenoxypropionic acid pesticide biodegradation—chemicals used in industry (Shilova et al., 2022b; Singh, 2022).

An endocrine disruptor, nonylphenol, is a persistent breakdown product of nonionic surfactants. A range of treatment periods from 1.5 to 2,880 hours has yielded biodegradation efficiencies of more than 89% in pure cultures and recognized consortia, suggesting that the benzene ring is oxidized first, followed by a stepwise breakdown of the nonylphenol side chains (Lavanya et al., 2022; Shilova et al., 2022a). Nonylphenol biodegradation can also benefit from the presence of surfactants secreted by *Sphingomonas* sp. Research on biodegradation by strain TTNP3 of *Sphingomonas* sp., including nonylphenol isomers and bisphenol A (BPA), was also surveyed. To digest alkylphenols and their downstream metabolites (i.e., hydroquinone [HQ]), this bacterium uses a type II ipso-substitution mechanism and a patchwork of genes, including the opdA genes and the hqd gene cluster (genes for HQ degradation). According to the location of these genes and other *Sphingomonas* strains' catabolic capabilities, numerous research addressed in this article has hypothesized that these genes were acquired via transposition events (Rizwan et al., 2022; Uçar et al., 2022). A plastic intermediate called BPA has also been widely examined for degradation. An immobilized microbe consortium removed 87%–93% of the total nitrogen with significant nitrification activity (Bakhtiari et al., 2022a). A high association between removal and nitrification efficiency was discovered, indicating a probable competitive inhibition between the target chemical and ammonia. Because of this, the capacity of these bacteria to degrade BPA or nitrate ammonia appears to diminish. Ammonia-oxidizing bacteria and heterotrophic microorganisms from nitrifying activated sludge have previously been shown to break down BPA. *Nitrosomonas europaea* was used to degrade BPA in the absence of allylthiourea, an inhibitor of the ammonia monooxygenase enzyme. The researchers hypothesized that ammonia monooxygenase was responsible for the breakdown of BPA since *N. europaea* could not digest BPA when allylthiourea was present. An anaerobic treatment with nitrate- or sulfate-reducing conditions can also remove BPA (de Jesus et al., 2022a; Javed et al., 2022; Munir et al., 2022b; Qamar et al., 2022). Treatment periods ranged from 960 to 2,880 hours, despite the eradication rate being above 89% *Proteobacteria*, *Bacteroides*, *Chloroflexi*, *Firmicutes*, *Gemmatimonades*, and *Actinobacterium* were shown to be the dominant bacterial groups in anaerobic sediments digesting BPA (Bakhtiari et al., 2022b; Bilal et al., 2022; Uçar et al., 2022).

11.4 SONOCHEMICAL METHODS AND NANOREMEDIATION

Sonochemical methods have recently emerged as viable for removing persistent contaminants from the environment. The study of chemical reactions induced in solution by sound is known as sonochemistry (Ali Zomorodian et al., 2017; Perullini et al., 2014; Wang et al., 2014). Due to the high vapor temperature and vapor pressure caused by sound waves, environmental pollutants are pyrolytically degraded at the bubble–water interface. Ultrasonography is commonly used when it comes to destroying or simulating the elimination of liquid-phase pollutants. The conservative method of environmental pollution treatment used decomposition or oxidation processes to remediate contaminants. A series of compression and rarefaction creates positive pressure, whereas rarefaction generates negative pressure. Cavitation bubbles are formed due to this increased tensile strength of the liquid. While these bubbles obtain their energy from sound waves, when they reach an almost-burstable stage (5,000 K, 1,000 atm), the environment becomes oxidative as hydroxyl and hydroperoxyl radicals develop, which are used in the pyrolytic destruction of persistent contaminants. Ninety-six antipyrines were degraded by Leong and Chang (2022) using a sono-photocatalytic oxidative method in an aqueous solution (Yang et al., 2013a). Similarly, sonolysis and photolysis were utilized to remediate pharmaceutical industry effluent polluted with environmental contaminants. H_2O_2 and ultrasound were used to remove tinidazole from the aqueous solution efficiently. Goethite ZVI (natural/synthetic), carbon nanotubes, and TiO_2 nanomembranes have been used for environmental cleanup (Lambropoulou et al., 2017; Wang et al., 2009). Bimetallic nanoparticles have performed a significant role in removing contaminants from water systems. Carbon nanotubes have also been used to remediation hazardous heavy metals, such as Pb_{12}, Cr_{13}, and Zn_{12}, and metalloids, such as volatile organic and biological contaminants, dioxins, and arsenic compounds. Carbon nanotubes have been used to absorb synthetic pollutants from water bodies and have an excellent binding capability for many functional entities. Compared to traditional cleanup methods, nanoremediation takes significantly less time to complete the process. NZVI decreased TCE concentrations by 99% and shortened cleanup time from years to only a few days, according to Guo et al. (2021).

11.5 BIOTECHNOLOGICAL APPROACHES FOR MICROPOLLUTANT DEGRADATION

Different biotechnological strategies have been described for the bioremediation of environmental contaminants, and highly effective omics methodologies form the backbone of these methods. In bioremediation, a lot of information has been gathered throughout the years (Das et al., 2021; Paumo et al., 2021; Rtimi et al., 2016). Different bioinformatics methods might investigate proteins and genetic material to discover microorganisms' degrading potential (Deblonde and Hartemann, 2013;

Hartmann et al., 1999; Zenker et al., 2014). To improve the quality of the environment, microbes have been used. Several techniques can be used to clean up polluted areas, including system biology, genetic engineering (CRISPR/Cas), synthetic biology, enzyme immobilization, and microbiological electrochemistry (Akter et al., 2012; Verlicchi et al., 2010; Yuan et al., 2013). Since their conception and development, improved performance has been a constant goal of these biotechnological technologies. By creating distinct strains of microorganisms that may manufacture appealing and valuable products/molecules from environmental contaminants, these technologies have played a vital role in environmental sustainability. Environmental pollutants cannot be recycled in the primary biodegradation systems, but modified bacteria can quickly enhance the working capacity (Li et al., 2015; Trinh et al., 2016; Wu et al., 2015). Traditional cleanup procedures are more intrusive and expensive than biotechnological alternatives. Environmental contaminants can be degraded via biotechnology in contaminated environments. Technology that promotes and improves bioremediation's efficiency for cleaning up a polluted environment includes nanoscience, omics, gene discoveries and improvements, and novel material creation (Gothwal and Shashidhar, 2015; Khalfbadam et al., 2017; Mubedi et al., 2013; Thompson et al., 2013).

11.6 MICROBIAL ELECTROCHEMICAL SYSTEM

Microbiological electrochemical systems (MES) were seems to be promising technology when we talk about treatment aspect of micropollutants from wastewater. Figure 11.1 depicts a fundamental principle of an MES. To simultaneously release and convert liberated electrons from synthesized goods into electricity, MES is utilized

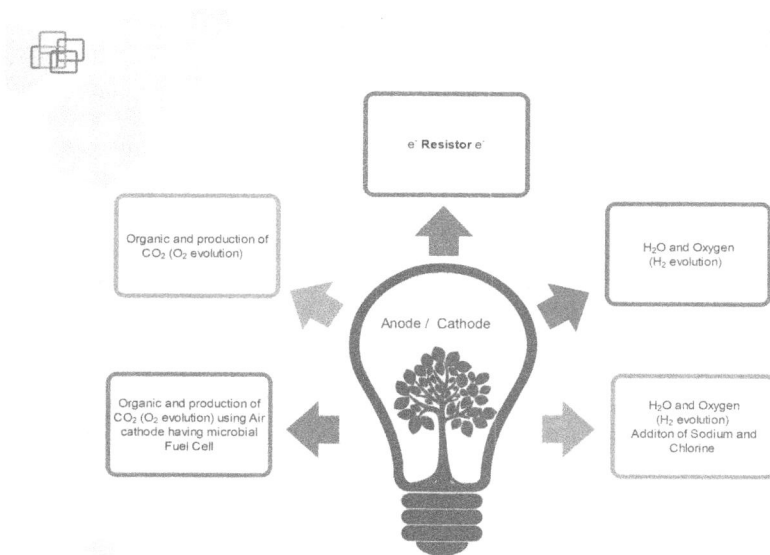

FIGURE 11.1   Basic principles in four typical microbial electrosynthesis.

(Khalfbadam et al., 2017; Li et al., 2015). Bacterial metabolism plays a vital role in the reduction or oxidation of pollutants to produce harmless products or intermediates. An MES has several advantages over other bioremediation methods, including the following: (1) the energy consumption is lower, (2) this results in energy or electricity production that may supply power to environmental systems, (3) the production of nonharmful, nontoxic products, and (4) it can cause redox reactions and be applied to many toxic synthetic products. There are two sections to the MES: the cathode and the anode. The ion-exchange membrane (IEM) separates them. The presence of bacteria in a biofilm or planktonic condition may be determined by placing an anode chamber there. Protons, electrons, and a variety of secondary metabolites are produced due to the microbial system's oxidation of substrates (Sathe et al., 2022). This external circuit transferred the generated electrons from one end of the device's output to the opposite end. However, in the IEM, the electrons at the cathode decreased the protons at the anode chamber to complete the circuit (Sathe et al., 2021).

The water oxidation produces electrons and protons at the anode of the MES in the presence of external electrical potential to create hydrogen (H_2), alcohol (R-OH), methane (CH_4), and acids (RCOOH). As a consequence, either hydrogen or electrical energy can be generated. Extra power may be needed in some circumstances for the redox reaction (Cantor et al., 1987; Tandukar et al., 2007; Yuan et al., 2013). The MES has been put to good use in wastewater treatment. There have been several research studies on the use of MES in soil bioremediation due to the soil's electrochemical properties. Anodic oxidation and anodic reduction are two methods for using MESs in soil remediation. Products like petroleum act as electron donors, and degradation occurs during oxidation–reduction reactions; anodic reduction, where synthetic products act as electron acceptors in a reducing atmosphere, also occurs. In an electric field, soil particles undergo hydrological and physicochemical changes; sorption of products occurs on electrodes or the biofilm that already exists on the electrodes; Anode-produced electrons can be re-acquired by using protons at the cathode when pH rises during O_2 reduction (cathode alkalization) and (cathode reduction). A bioelectrochemical system has been developed and named the "Bioelectric well" for usage in contaminated wells or groundwater sources (Emmanuel et al., 2013). Bioelectroventing, a new method for removing atrazine (a pesticide) from the soil, has been developed for this purpose. Analytic oxidation has rapidly destroyed numerous hydrocarbons in a short amount of time. It was shown that for total petroleum, the bioremediation effectiveness was between 82.1% and 89.7% after only 120 days of testing. According to reports, toxic metals including copper, cadmium, and lead may allegedly be transmitted and accumulated in cathode regions of electric field influence, which soil MES pushes. As of day 108, 44% of the anode's anode lead and cadmium were removed, respectively. An amount of copper equivalent to 200 mg/kg migrated to the cathodic area in 56 days in one-chambered MES (Alizadeh Kordkandi et al., 2018; Organization, 2012; U S Environmental Protection Agency, 2018).

11.7 IMMOBILIZED ENZYMES FOR MICROPOLLUTANT DEGRADATION

It is possible to fix enzymes on the surface of solids and employ this process to obtain reusable, active, and perhaps stable enzymes. Depending on the application, it is possible to use enzymes in either an immobilized or free state as catalysts. Because enzymes in the free form are unstable, readily denatured, rapidly attacked by protease, and difficult to repurpose, their function is restricted to the immediate vicinity of the enzyme's structure (Brodin et al., 2013; Cao et al., 2020). A novel alternative to free enzymes for bioremediation is the immobilization of degradative ones. "Emulsion immobilization" refers to any process in which several supports are used to attach an enzyme (either in an accessible form or soluble state). Using immobilized enzymes in environmental pollutant bioremediation, large-scale remediation of contaminants may be accomplished at a reasonable cost. Immobilization has several advantages, including long-term stability, high efficiency, reduced costs, and the potential to reuse enzymes for industrial applications. The strength of immobilized enzyme to heat is improved by immobilization which expands the enzyme pH and temperature range of action (Carlsson et al., 2006; von Sperling and Mascarenhas, 2005).

Additionally, the immobilization of intracellular enzymes in cell-free media might enhance their stability. Enzyme immobilization improves enzyme properties significantly without requiring any changes to the enzyme's structure or potential activity. Enzyme immobilization can be accomplished using various techniques, including entrapment crosslinking, affinity tag binding, encapsulation, and adsorption binding. Large surface area, affordability, and lack of restriction are all requirements for immobilizing the enzyme on a supporting material used for immobilization.

This technique's importance was enhanced by using an immobilized enzyme to degrade potentially dangerous xenobiotic compounds. There have been a variety of research studies on the impact of immobilization on synthetic product biodegradation. Because of the electrostatic contact, manganese peroxidase has been adsorbed onto a unique nanoclay and nanomaterial (Kundu et al., 2013; Vidal-Limon et al., 2018; Wang and Ren, 2013). The stability of nanoclay immobilized manganese peroxidase was superior to that of free manganese peroxidase across a wide range of pH and temperature conditions and in long-term storage. When tested for catalytic activity against polycyclic aromatic hydrocarbons (PAH), immobilization in solution had no harmful effects, indicating that the enzyme's structure had not altered during immobilization. Immobilized *Momordica charantia* (bitter gourd) peroxidase decolorized textile dyes better than free enzymes. Table 11.1 shows the degradation of EMP in different microbial cultures. Immobilized enzymes were compared. To eliminate atrazine pesticides from polluted soil, the efficacy of immobilized and free enzymes from *Arthrobacter* species HB-5 has been investigated. Both enzymes had the same catalytic potential for degrading atrazine, whereas immobilized enzymes showed a wide pH range. It has been demonstrated that horse-radish peroxidase can be immobilized in Ca-alginate beads and that this method can discolor acid orange-7 (75% of the time) and acid blue-25 (84% of the time) (Kelly and He, 2014; Wang et al., 2015; Xin et al., 2019). Acid Violet-109, an anthraquinone dye, was injured by the immobilized horse-radish peroxidase enzyme in 35 minutes. The detailed operation parameters and efficiency are illustrated in Table 11.2 at lab/pilot scales. Fungal strain-derived lignin peroxidase was used to immobilize carbon nanotubes. Remazol brilliant blue-R dye was effectively discolored by lignin peroxidase mounted on carbon nanotubes, which also demonstrated higher specific potential, catalytic activity, and lower Km values when compared to free enzymes. On the mesoporous

TABLE 11.1

Degradation of EMP in Different Microbial Cultures

S. No.	Pollutant	Microorganism	Efficiency	Operation Parameters	References
1.	2,6 Dichlorobenzamide	*Aminobacter* sp. strain	97% after 1.1 hours	Insignificant byproducts formed	Rizwan et al. (2022)
2.	Mecoprop 2-(4-chloro-2-methylphenoxy) propanoic acid	Mixed consortium	Up to 83% for 200 hours	Methanol improves efficiency	Zhong et al. (2019)
3.	Nonylphenol	*Sphingomonas* sp. strain	92% using membrane and 49% using activated sludge method	Bioaugmentation was a need, and structurally similar metabolites formed	Yang et al. (2013b)
4.	Bisphenol A	*Chloroflexi, Firmicutes, Gemmatimonadetes*	88% from 959–2,882 hours of operation	Bacterial diversity increased with nitrate and decreased due to Sulfur	Bakhtiari et al. (2022a)
5.	Triclosan	*Nitrosomonas europaea*	69% after 10 hours of operation	Work well in the absence of allylthiourea and with ammonia-oxidizing bacteria	Kolpin et al. (2002)
6.	Naproxen	*Stenotrophomonas maltophilia*	77% after 480 hours	Metabolized with glucose or phenol	Paper et al. (2016)
7.	Ibuprofen	Heterotrophic nitrifying microorganisms	99% after 383 hours	Not linked to microbial strain	Tahirbegi et al. (2017)

TABLE 11.2

Biological Removal of EMP in Lab/Full-Scale Treatment Systems

S.No.	Treatment	Pollutant Removed	Operation Condition	Efficiency	Remark	References
1.	Lab scale-bio-film activated sludge system	Benzotriazoles and hydroxybenzothiazole	pH = 7.1–8.1, MLSS = 3 g/L, Temp = 22°C	Up to 88%	Best performed with aerobic biodegradation	Li et al. (2019)
2.	Anoxic-MBR aerobic-MBR	Carbamazepine and sulfamethoxazole	pH = 7.8, HRT = 24 hours, MLSS = 10.3 g/L, Temp = 22°C	Maximum up to 39%	Near-anoxic conditions favored	Venkata Mohan et al. (2014)
3.	Anaerobic-MBR	Galaxolide, tonalide, phantolide, traseolide, cashmeran	pH = 6.9, HRT = 96 hours, MLSS = 10 g/L	Up to 97%	Biotransformation was the main mechanism	Kumar et al. (2017)
4.	MBR and activated carbon	Sulfamethoxazole carbamazepine	pH = 7.5, HRT = 24 hours, MLSS = 100 g/L	Up to 92%	Adsorption process	Wang and Ren (2013)
5.	Anaerobic granular sludge sequential batch reactor	Fluoxetine	pH = 7, MLSS = 10.3–18.3 g/L, Temp = 20°C	Max. up to 24%	The adsorption process favored granules formation	de Jesus et al. (2022a)

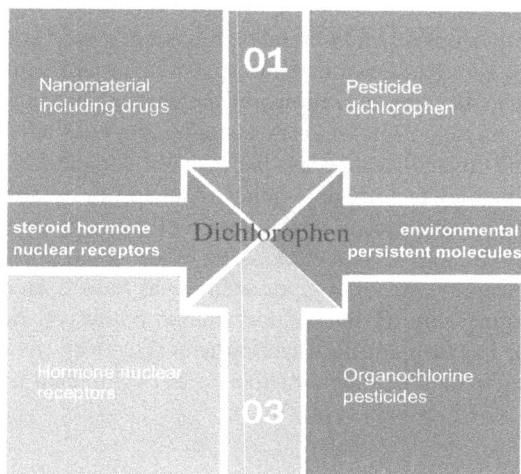

FIGURE 11.2   Degradation of pesticide dichlorophen.

nanoframework, laccase-immobilized *Coriolopsis gallica*-laccase was used to break down dichlorophen pesticide and potentially reduce its hazardous effects after oxidation as shown in Figure 11.2.

11.8   METABOLIC ENGINEERING APPROACHES FOR POLLUTANT DEGRADATION

Biodegradation processes can be improved by selecting and boosting the expression of relevant genes in indigenous bacteria, as well as other methods of metabolic engineering (Hadibarata and Kristanti, 2012; Jiang et al., 2014). Metabolic engineering methods are used to optimize regulatory and genetic systems to break down environmental contaminants. Metabolic systems are studied by metabolic engineers using various biobased systems and "omics" methods (Li et al., 2019; Yoneda and Mokhtar, 2018). The combination of metabolic engineering and sophisticated systems biology makes bioremediation of environmental contaminants desirable. Bioinformatics and systems biology may be used to understand better how our complex biosystems interact with the broader ecology. Metabolic engineering alters the genomes of the modified bacteria, causing them to change their degradation processes. Environmental contaminants may also be degraded as a consequence of modified biodegradation. The chemical palette of fungi and bacteria cells is broadened by using these strategies during degradation mechanisms. The bioremediation of environmental toxins can also be performed using metabolomics, an area of research (Haigh-Flórez et al., 2014).

Various methods have been used to get all of the metabolites investigated in the prior investigations. The metabolites produced during the bioremediation of environmental pollutants were studied using multiple techniques, including nuclear magnetic resonance, direct-injection mass spectrometry, high-performance liquid chromatography, gas chromatography-mass spectroscopy, and Fourier transform infrared spectroscopy. Phenanthrene bioremediation was studied using comparative metabolomics with the Sinorhizobium sp. C4 (Li et al., 2013; Perullini et al., 2014). Using metabolomics, the breakdown of metal ions and biphenyl linked with succinate was examined in *Pseudomonas pseudoalcaligenes* KF707. It is now possible to co-express and create several genes because of the rapid advancements in molecular and systems biology techniques—genes regulated by synthetic and molecular biology technologies breakdown a wide range of contaminants from the surrounding environment (Venkata Mohan et al., 2014; Wang et al., 2015; Wang and Ren, 2014). Since their inception, genomic-scale techniques have been extensively used for the metabolic engineering of diverse degradation pathways. Access to entire genome sequences for bacteria and information acquired from the genome acts as a catalyst for metabolic engineering activities in environmental contaminant degradation (Kumar et al., 2017; Kundu et al., 2013).

11.9 INVENTION OF NOVEL GENES INVOLVED IN BIOREMEDIATION

Biodegradation processes can be improved by selecting and boosting the expression of relevant genes in indigenous bacteria, as well as other methods of metabolic engineering. Through metabolic engineering techniques, regulatory and genetic systems may be optimized for the breakdown of environmental contaminants. The metabolic engineers understand the physiologies and metabolisms of microbial strains via a variety of biobased systems and "omics" methodologies. Advanced systems biology and metabolic engineering make the bioremediation of environmental toxins more appealing. It is possible to study complex biological processes and their relationship to the border environment using systems biology. Metabolic engineering alters the genomes of the modified bacteria, causing them to change their degradation processes (Calabi-Floody et al., 2012).

Additionally, the biodegradation of contaminants in the surrounding environment may be influenced by environmental changes. These strategies improve the chemical repertoire of fungal and bacterial cells throughout the degradation process. Metabolomics can also investigate the bioremediation potential of contaminants in the environment. Various methods have collected all of the metabolites investigated in the prior research during degradation. Nuclear magnetic resonance, direct-injection mass spectrometry, high-performance liquid chromatography, gas chromatography-mass spectrometry, and Fourier transform infrared analysis were used to study the bioremediation of environmental contaminants (Jiang et al., 2014).

Through comparing metabolomics, the bioremediation of phenanthrene was studied using the Sinorhizobium sp. C4 (Tsopela et al., 2014). Metabolomic processes were used to examine *Pseudomonas pseudoalcaligenes* KF707's degradation of succinate-associated metal ions and biphenyl. Because molecular and systems biology technologies are improving quickly, it is now possible to co-express and create many genes. Chemical and molecular biology techniques govern a gene cascade that breaks down many environmental contaminants. Metabolic engineering of different degradation pathways has been researched intensively since the emergence of genome-scale technologies. As a result of the availability of entire genome sequences for bacteria and the information acquired from the genome, each stage in the degradation of environmental contaminants is aided (Yoneda and Mokhtar, 2018).

11.10 ENHANCED BIOREMEDIATION VIA METABOLIC ENGINEERING PROCESSES

These materials were used to synthesize 4-amino-2, 6-dinitrotoluene (4-ADNT) and TNT derivatives successfully. This bioremediation method relied on polymerization and oligomerization to succeed. The catechol-laccase coupling process removed azocolours like methyl orange from aromatic amines. As seen in Figure 11.2, the food industry effluent was also treated with laccase in the coupling processes (Javed et al., 2022). During the microbial breakdown of environmental contaminants, bacteria, fungi, and yeast all produce enzymes. In the early stages of biodegradation, the polymeric chain cleavage occurs to break down polymers into smaller pieces. Mineralization takes place on the smaller mono/oligomers as the following step after being delivered to the cells. It also produces several other compounds, including adenosine triphosphate (ATP), salts, CO_2, CH_4, N_2, H_2, and minerals during this process (Bakhtiari et al., 2022a). Despite their chemical nature and dependence on the environment, enzymes play a crucial role in biodegradation (de Jesus et al., 2022a). Their capacity to speed up chemical reactions makes enzymes ideal for use in the natural world. Polypeptides and other proteins with molecular weights ranging from a few thousand to several million g/mol have been discovered.

Additionally, enzymes and their substrates interact because of an active site. Each form of the substrate has a different effect on the enzymes. Hence, it is essential to know which substrates each type of enzyme is most successful with. Vitamins, Adenosine 5′-triphosphate (ATP), and metal ions all contribute to the enzyme's ability to do its task. Due to the vast diversity of enzymes, numerous catalysis methods have been studied. In the degradation process, enzyme hydrolysis and oxidation are the two distinct processes (Li et al., 2019). The hydrolysis of ester bonds facilitates polymer breakdown, and several enzymes are involved in this process. There are a variety of enzymes with particular active sites under consideration. By the esterase enzyme activity, it is possible to break the ester bond (Hou et al., 2019). An intriguing field of research is now focused on isolating functional genes and improving their pollutant breakdown capacity. A synthetic genetic circuit is needed for the implantation. As a host–vector biosafety (HVB) strain, *Pseudomonas putida* was recommended by the recombinant DNA advisory committee. As a result of the widespread consensus that it is safe to release it into the environment (GRAS) (Ali Zomorodian et al., 2017). This makes *P. putida* an intriguing candidate for use in the next generation of synthetic biology frameworks since it can withstand a wide range of harsh environmental conditions, including pH, temperature, toxins, solvents, oxidative stress, and osmotic stress. Low nutritional needs, as well as an adaptable metabolism, distinguish *P. putida* from other algae. For environmental bioremediation purposes, the microbial biosorbent model might be used to attain this goal. Due to the design that includes promoter genes and gene expression, *P. putida* synthetic genetic circuit may also degrade persistent substances (Shekoohiyan et al., 2016).

11.11 CONCLUSIONS

Organic and inorganic pollutants are being released into the environment as synthetic chemicals are increasingly used in agriculture, industry, and medicine. Conventional remediation techniques are ineffective since they use just a few microorganisms to clean up polluted areas. Therefore, it is possible to enhance synthetic chemical cleanup *in situ* and *ex situ* by using contemporary strategies based on microbial engineering. Organic and environmental contaminants can be effectively remedied through microbial engineering. More research is being conducted to get a scientific knowledge of environmental pollutants and remove them utilizing effective technologies. This research is currently being processed. The use of metabolic processes can reduce soil and water pollution. A thorough understanding of microbial technology and ongoing research is essential for the development of novel approaches to the management and bioremediation of persistent contaminants in the environment. Biodegradation appears to be a potential process for eliminating organic EOMs during WWT. The total removal of EOMs is still highly unusual, and additional study is needed to achieve this. Knowledge of biodegradation mechanisms in complex systems, such as heterogeneous consortia, complex water matrices (e.g., EOM mixes and metabolites), and associated technologies might lead to a greater understanding of EOM biodegradation. Systems biology's top-down and bottom-up methodologies can be used in this study to identify the active metabolic pathways during the breakdown of EOMs. Improved genomic, proteomic, and metabolomic techniques for EOM biodegradation in sewage/WWTP samples are being developed. They are enabling the use of non-bacterial for EOM removal to overcome the limits of bacteria, which results in inadequate EOM removal. The fifth step is to take advantage of the biodegradation capabilities of certain EOM combinations and combine them with other biological WWTP. A good awareness of each treatment's limits is required when combining biological, physical, and pharmacological therapies. Non-conventional biodegradation systems, such as modified Conventional activate sludge (CAS) and Membrane biological reactor (MBR) systems, are being studied for their ability to be scaled up and used in the biodegradation of complex EOM combinations. They are incorporating bioremediation techniques into the secondary treatment of WWTPs. Through joint development, the progress and enrichment of environmental bioinformatics, big data analysis, and WWTs. According to the evidence presented here, future research should use a more sophisticated approach. Multifactorial experiments, omics analysis, and various in silico technologies may be used to evaluate and better interpret experimental data critically. By doing so, it will be possible to establish with greater precision the biodegradation processes, catabolic kinetics, and operational conditions required to construct WWTs with improved EOM biodegradation capacities.

REFERENCES

Akter, F., Amin, M.R., Osman, K.T., Anwar, M.N., Karim, M.M., Hossain, M.A., 2012. Ciprofloxacin-resistant Escherichia coli in hospital wastewater of Bangladesh and prediction of its mechanism of resistance. *World J. Microbiol. Biotechnol.* 28, 827–834. doi: 10.1007/s11274-011-0875-3.

Ali, A., Ahmed, Z., Maqbool, R., Shahzad, K., Shah, Z.H., Ali, M.Z., Alsamadany, H., Bilal, M., 2022. Nanobioremediation of insecticides and herbicides. In: H.M.N. Iqbal, M. Bilal, T.A. Nguyen, G. Yasin (Eds.), *Biodegradation and Biodeterioration at the Nanoscale.* Elsevier: Amsterdam, Netherlands. doi: 10.1016/b978-0-12-823970-4.00023-3.

Ali Zomorodian, S.M., Shabnam, M., Armina, S., O'Kelly, B.C., 2017. Strength enhancement of clean and kerosene-contaminated sandy lean clay using nanoclay and nanosilica as additives. *Appl. Clay Sci.* 140, 140–147. doi: 10.1016/j.clay.2017.02.004.

Alizadeh Kordkandi, S., Khoshfetrat, A.B., Faramarzi, A., 2018. Performance modelling of a partially-aerated submerged fixed-film bioreactor: Mechanistic analysis versus semi data-driven method. *J. Ind. Eng. Chem.* 61, 398–406. doi: 10.1016/j.jiec.2017.12.039.

Antonacci, A., Lambreva, M.D., Arduini, F., Moscone, D., Palleschi, G., Scognamiglio, V., 2018. A whole cell optical bioassay for the detection of chemical warfare mustard agent simulants. *Sensors Actuators, B Chem.* 257, 658–665. doi: 10.1016/j.snb.2017.11.020.

Bakhtiari, S., Doustkhah, E., Pedram, M.Z., Yarmohammadi, M., Seydibeyoğlu, M.Ö., 2022a. Effects of nanoparticles on the biodegradation of organic materials. In: H.M.N. Iqbal, M. Bilal, T.A. Nguyen, G. Yasin (Eds.), *Biodegradation and Biodeterioration at the Nanoscale.* Elsevier: Amsterdam, Netherlands, pp. 153–174. doi: 10.1016/b978-0-12-823970-4.00008-7.

Bakhtiari, S.S.E., Bakhsheshi-Rad, H.R., Razzaghi, M., Ismail, A.F., Sharif, S., Ramakrishna, S., Berto, F., 2022b. Effects of nanomaterials on biodegradation of biomaterials. In: H.M.N. Iqbal, M. Bilal, T.A. Nguyen, G. Yasin (Eds.), *Biodegradation and Biodeterioration at the Nanoscale.* Elsevier: Amsterdam, Netherlands. doi: 10.1016/b978-0-12-823970-4.00006-3.

Bilal, M., Bhatt, P., Nguyen, T.A., Iqbal, H.M.N., 2022. *Biodegradation and biodeterioration at the nanoscale*: An introduction. In: H.M.N. Iqbal, M. Bilal, T.A. Nguyen, G. Yasin (Eds.), *Biodegradation and Biodeterioration at the Nanoscale.* Elsevier: Amsterdam, Netherlands doi: 10.1016/b978-0-12-823970-4.00001-4.

Brodin, T., Fick, J., Jonsson, M., Klaminder, J., 2013. Dilute concentrations of a psychiatric drug alter behavior of fish from natural populations. *Science* 80, 814–815. doi: 10.1126/science.1226850.

Brunson, L.R., Sabatini, D.A., 2014. Practical considerations, column studies and natural organic material competition for fluoride removal with bone char and aluminum amended materials in the Main Ethiopian Rift Valley. *Sci. Total Environ.* 488–489, 580–587. doi: 10.1016/j.scitotenv.2013.12.048.

Calabi-Floody, M., Velásquez, G., Gianfreda, L., Saggar, S., Bolan, N., Rumpel, C., Mora, M.L., 2012. Improving bioavailability of phosphorous from cattle dung by using phosphatase immobilized on natural clay and nanoclay. *Chemosphere* 89, 648–655. doi: 10.1016/j.chemosphere.2012.05.107.

Cantor, K.P., Hoover, R., Hartge, P., Mason, T.J., Silverman, D.T., Altman, R., Austin, D.F., Child, M.A., Key, C.R., Marrett, L.D., 1987. Bladder cancer, drinking water source, and tap water consumption: A case-control study. *J. Natl. Cancer Inst.* 79, 1269–1279.

Cao, S.S., Duan, Y.P., Tu, Y.J., Tang, Y., Liu, J., Zhi, W. Di, Dai, C., 2020. Pharmaceuticals and personal care products in a drinking water resource of Yangtze River Delta Ecology and Greenery Integration Development Demonstration Zone in China: Occurrence and human health risk assessment. *Sci. Total Environ.* 721, 137624. doi: 10.1016/j.scitotenv.2020.137624.

Carlsson, C., Johansson, A.K., Alvan, G., Bergman, K., Kühler, T., 2006. Are pharmaceuticals potent environmental pollutants?. Part I: Environmental risk assessments of selected active pharmaceutical ingredients. *Sci. Total Environ.* 364, 67–87. doi: 10.1016/j. scitotenv.2005.06.035.

Das, A., Adak, M.K., Mahata, N., Biswas, B., 2021. Wastewater treatment with the advent of TiO$_2$ endowed photocatalysts and their reaction kinetics with scavenger effect. *J. Mol. Liq.* 338, 116479. doi: doi: 10.1016/j.molliq.2021.116479.

de Jesus, R.A., Costa, J.A.S., Paranhos, C.M., Bilal, M., Bharagava, R.N., Iqbal, H.M.N., Ferreira, L.F.R., Figueiredo, R.T., 2022a. Process of biodegradation controlled by nanoparticle-based materials: mechanisms, significance, and applications. In: H.M.N. Iqbal, M. Bilal, T.A. Nguyen, G. Yasin (Eds.), *Biodegradation and Biodeterioration at the Nanoscale.* Elsevier: Amsterdam, Netherlands. doi: 10.1016/b978-0-12-823970-4.00004-x.

de Jesus, R.A., de Assis, G.C., de Oliveira, R.J., Bilal, M., Bharagava, R.N., Iqbal, H.M.N., Ferreira, L.F.R., Figueiredo, R.T., 2022b. Metal oxide nanoparticles for environmental remediation. In: H.M.N. Iqbal, M. Bilal, T.A. Nguyen, G. Yasin (Eds.), *Biodegradation and Biodeterioration at the Nanoscale.* Elsevier: Amsterdam, Netherlands. doi: 10.1016/ b978-0-12-823970-4.00025-7.

Deblonde, T., Hartemann, P., 2013. Environmental impact of medical prescriptions: Assessing the risks and hazards of persistence, bioaccumulation and toxicity of pharmaceuticals. *Public Health.* doi: 10.1016/j.puhe.2013.01.026.

Dhiman, N.K., Sidhu, N., Agnihotri, S., Mukherjee, A., Reddy, M.S., 2022. Role of nanomaterials in protecting building materials from degradation and deterioration. In: H.M.N. Iqbal, M. Bilal, T.A. Nguyen, G. Yasin (Eds.), *Biodegradation and Biodeterioration at the Nanoscale.* Elsevier: Amsterdam, Netherlands doi: 10.1016/b978-0-12-823970-4.00024-5.

Emmanuel, J., Pieper, U., Rushbrook, P., Stringer, R., Townend, W., Wilburn, S., Zghondi, R., 2013. *Safe Management of Wastes from Health-Care Activities.* World Health Organization: Geneva, Switzerland.

European Environmental Agency (2007). *The European Environment – State and Outlook 2010 Synthesis.* Copenhagen: European Environment Agency, p. 228

Gothwal, R., Shashidhar, T., 2015. Antibiotic pollution in the environment: A review. *Clean Soil, Air, Water* 43, 479–489. doi: 10.1002/clen.201300989.

Guo, H., Li, Z., Xiang, L., Jiang, N., Zhang, Y., Wang, H., Li, J., 2021. Efficient removal of antibiotic thiamphenicol by pulsed discharge plasma coupled with complex catalysis using graphene-WO$_3$-Fe$_3$O$_4$ nanocomposites. *J. Hazard. Mater.* 403, 123673. doi: 10.1016/j.jhazmat.2020.123673.

Hadibarata, T., Kristanti, R.A., 2012. Identification of metabolites from benzo[a]pyrene oxidation by ligninolytic enzymes of Polyporus sp. S133. *J. Environ. Manage.* 111, 115–119. doi: 10.1016/j.jenvman.2012.06.044.

Haigh-Flórez, D., De la Hera, C., Costas, E., Orellana, G., 2014. Microalgae dual-head biosensors for selective detection of herbicides with fiber-optic luminescent O2 transduction. *Biosens. Bioelectron.* 54, 484–491. doi: 10.1016/j.bios.2013.10.062.

Hartmann, A., Golet, E.M., Gartiser, S., Alder, A.C., Koller, T., Widmer, R.M., 1999. Primary DNA damage but not mutagenicity correlates with ciprofloxacin concentrations in german hospital wastewaters. *Arch. Environ. Contam. Toxicol.* 36, 115–119. doi: 10.1007/ s002449900449.

Hou, J., Tang, J., Chen, J., Zhang, Q., 2019. Quantitative structure-toxicity relationship analysis of combined toxic effects of lignocellulose-derived inhibitors on bioethanol production. *Bioresour. Technol.* 289, 121724. doi: 10.1016/j.biortech.2019.121724.

Javed, M.R., Rashid, M.H., Tariq, A., Seemab, R., Ijaz, A., Abbas, S., 2022. Interaction of nanomaterials with microbes. In: H.M.N. Iqbal, M. Bilal, T.A. Nguyen, G. Yasin (Eds.), *Biodegradation and Biodeterioration at the Nanoscale.* Elsevier: Amsterdam, Netherlands. doi: 10.1016/b978-0-12-823970-4.00003-8.

Jiang, Y., Tang, W., Gao, J., Zhou, L., He, Y., 2014. Immobilization of horseradish peroxidase in phospholipid-templated titania and its applications in phenolic compounds and dye removal. *Enzyme Microb. Technol.* 55, 1–6. doi: 10.1016/j.enzmictec.2013.11.005.

Kelly, P.T., He, Z., 2014. Nutrients removal and recovery in bioelectrochemical systems: A review. *Bioresour. Technol.* 153, 351–360. doi: 10.1016/j.biortech.2013.12.046.

Khalfbadam, H.M., Ginige, M.P., Sarukkalige, R., Kayaalp, A.S., Cheng, K.Y., 2017. Sequential solid entrapment and in situ electrolytic alkaline hydrolysis facilitated reagent-free bio-electrochemical treatment of particulate-rich municipal wastewater. *Water Res.* 117, 18–26. doi: 10.1016/j.watres.2017.03.045.

Khan, A., Malik, S., Ali, N., Gao, X., Yang, Y., Bilal, M., 2022. Metal-organic framework for removal of environmental contaminants. In: H.M.N. Iqbal, M. Bilal, T.A. Nguyen, G. Yasin (Eds.), *Biodegradation and Biodeterioration at the Nanoscale.* Elsevier: Amsterdam, Netherlands. doi: 10.1016/b978-0-12-823970-4.00020-8.

Kolpin, D.W., Furlong, E.T., Meyer, M.T., Thurman, E.M., Zaugg, S.D., Barber, L.B., Buxton, H.T., 2002. Pharmaceuticals, hormones, and other organic wastewater contaminants in U.S. streams, 1999–2000: A national reconnaissance. *Environ. Sci. Technol.* 36, 1202–1211. doi: 10.1021/es011055j.

Kumar, G., Saratale, R.G., Kadier, A., Sivagurunathan, P., Zhen, G., Kim, S.H., Saratale, G.D., 2017. A review on bio-electrochemical systems (BESs) for the syngas and value added bio-chemicals production. *Chemosphere* 177, 84–92. doi: 10.1016/j.chemosphere.2017.02.135.

Kundu, A., Sahu, J.N., Redzwan, G., Hashim, M.A., 2013. An overview of cathode material and catalysts suitable for generating hydrogen in microbial electrolysis cell. *Int. J. Hydrogen Energy* 38, 1745–1757. doi: 10.1016/j.ijhydene.2012.11.031.

Lambropoulou, D., Evgenidou, E., Saliverou, V., Kosma, C., Konstantinou, I., 2017. Degradation of venlafaxine using TiO$_2$/UV process: Kinetic studies, RSM optimization, identification of transformation products and toxicity evaluation. *J. Hazard. Mater.* 323, 513–526. doi: 10.1016/j.jhazmat.2016.04.074.

Lavanya, N., Vallinayagam, S., Rajendran, K., 2022. Biodegradation of timber industry-based waste materials. In: H.M.N. Iqbal, M. Bilal, T.A. Nguyen, G. Yasin (Eds.), *Biodegradation and Biodeterioration at the Nanoscale.* Elsevier: Amsterdam, Netherlands. doi: 10.1016/b978-0-12-823970-4.00012-9.

Leong, Y. K., and Chang, J.-S., 2022. Valorization of fruit wastes for circular bioeconomy: Current advances, challenges, and opportunities. *Bioresour. Technol.* 359, 127459. doi: 10.1016/j.biortech.2022.127459

Li, J., Du, Y., Bao, T., Dong, J., Lin, M., Shim, H., Yang, S.T., 2019. n-Butanol production from lignocellulosic biomass hydrolysates without detoxification by Clostridium tyrobutyricum Δack-adhE2 in a fibrous-bed bioreactor. *Bioresour. Technol.* 289, 121749. doi: 10.1016/j.biortech.2019.121749.

Li, W., Niu, Q., Zhang, H., Tian, Z., Zhang, Y., Gao, Y., Li, Y.Y., Nishimura, O., Yang, M., 2015. UASB treatment of chemical synthesis-based pharmaceutical wastewater containing rich organic sulfur compounds and sulfate and associated microbial characteristics. *Chem. Eng. J.* 260, 55–63. doi: 10.1016/j.cej.2014.08.085.

Li, X., Xing, M., Yang, J., Lu, Y., 2013. Properties of biofilm in a vermifiltration system for domestic wastewater sludge stabilization. *Chem. Eng. J.* 223, 932–943. doi: 10.1016/j.cej.2013.01.092.

Ma, J., Xu, L., Jia, L., 2013. Characterization of pyrene degradation by Pseudomonas sp. strain Jpyr-1 isolated from active sewage sludge. *Bioresour. Technol.* 140, 15–21. doi: 10.1016/j.biortech.2013.03.184.

Moro, L., Pezzotti, G., Turemis, M., Sanchís, J., Farré, M., Denaro, R., Giacobbe, M.G., Crisafi, F., Giardi, M.T., 2018. Fast pesticide pre-screening in marine environment using a green microalgae-based optical bioassay. *Mar. Pollut. Bull.* 129, 212–221. doi: 10.1016/j.marpolbul.2018.02.036.

Mubedi, J.I., Devarajan, N., Faucheur, S. Le, Mputu, J.K., Atibu, E.K., Sivalingam, P., Prabakar, K., Mpiana, P.T., Wildi, W., Poté, J., 2013. Effects of untreated hospital effluents on the accumulation of toxic metals in sediments of receiving system under tropical conditions: Case of south india and democratic republic of congo. *Chemosphere* 93, 1070–1076. doi: 10.1016/j.chemosphere.2013.05.080.

Munir, H., Tahira, K., Bagheri, A.R., Bilal, M., 2022a. Biodegradation of materials in presence of nanoparticles. Biodegrad. Biodeterior. *Nanoscale* 9–30. doi: 10.1016/b978-0-12-823970-4.00002-6.

Munir, H., Tahira, K., Bagheri, A.R., Bilal, M., 2022b. Biodegradation of materials in presence of nanoparticles. In: H.M.N. Iqbal, M. Bilal, T.A. Nguyen, G. Yasin (Eds.), *Biodegradation and Biodeterioration at the Nanoscale*. Elsevier: Amsterdam, Netherlands. doi: 10.1016/b978-0-12-823970-4.00002-6.

Paper, C., Ministry, T.A., View, U.M., Changes, M., River, T., View, B.C., Abbas, T., 2016. Applicability of MBR technology for decentralized municipal wastewater treatment in Iraq applicability of MBR technology for decentralized.

Paumo, H.K., Dalhatou, S., Katata-Seru, L.M., Kamdem, B.P., Tijani, J.O., Vishwanathan, V., Kane, A., Bahadur, I., 2021. TiO2 assisted photocatalysts for degradation of emerging organic pollutants in water and wastewater. *J. Mol. Liq.* 331, 115458. doi: doi: 10.1016/j.molliq.2021.115458.

Perullini, M., Ferro, Y., Durrieu, C., Jobbágy, M., Bilmes, S.A., 2014. Sol-gel silica platforms for microalgae-based optical biosensors. *J. Biotechnol.* 179, 65–70. doi: 10.1016/j.jbiotec.2014.02.007.

Qamar, S.A., Hassan, A.A., Rizwan, K., Rasheed, T., Bilal, M., Nguyen, T.A., Iqbal, H.M.N., 2022. Biodegradation of micropollutants. In: H.M.N. Iqbal, M. Bilal, T.A. Nguyen, G. Yasin (Eds.), *Biodegradation and Biodeterioration at the Nanoscale*. Elsevier: Amsterdam, Netherlands. doi: 10.1016/b978-0-12-823970-4.00018-x.

Rafeeq, H., Qamar, S.A., Nguyen, T.A., Bilal, M., Iqbal, H.M.N., 2022a. Microbial degradation of environmental pollutants. In: H.M.N. Iqbal, M. Bilal, T.A. Nguyen, G. Yasin (Eds.), *Biodegradation and Biodeterioration at the Nanoscale*. Elsevier: Amsterdam, Netherlands. doi: 10.1016/b978-0-12-823970-4.00019-1.

Rafeeq, H., Qamar, S.A., Shah, S.Z.H., Ashraf, S.S., Bilal, M., Nguyen, T.A., Iqbal, H.M.N., 2022b. Biodegradation of environmental pollutants using horseradish peroxidase. In: H.M.N. Iqbal, M. Bilal, T.A. Nguyen, G. Yasin (Eds.), *Biodegradation and Biodeterioration at the Nanoscale*. Elsevier: Amsterdam, Netherlands doi: 10.1016/b978-0-12-823970-4.00022-1.

Rasheed, T., Rizwan, K., Shafi, S., Bilal, M., 2022. Nanobiodegradation of pharmaceutical pollutants. In: H.M.N. Iqbal, M. Bilal, T.A. Nguyen, G. Yasin (Eds.), *Biodegradation and Biodeterioration at the Nanoscale*. Elsevier: Amsterdam, Netherlands. doi: 10.1016/b978-0-12-823970-4.00026-9.

Rizwan, K., Rasheed, T., Bilal, M., 2022. Nano-biodegradation of polymers. In: H.M.N. Iqbal, M. Bilal, T.A. Nguyen, G. Yasin (Eds.), *Biodegradation and Biodeterioration at the Nanoscale*. Elsevier: Amsterdam, Netherlands. doi: 10.1016/b978-0-12-823970-4.00010-5.

Rtimi, S., Giannakis, S., Bensimon, M., Pulgarin, C., Sanjines, R., Kiwi, J., 2016. Supported TiO_2 films deposited at different energies: Implications of the surface compactness on the catalytic kinetics. *Appl. Catal. B Environ.* 191, 42–52. doi: 10.1016/j.apcatb.2016.03.019.

Samal, K., Trivedi, S., 2020. A statistical and kinetic approach to develop a Floating Bed for the treatment of wastewater. *J. Environ. Chem. Eng.* 8, 104102. doi: 10.1016/j.jece.2020.104102.

Sathe, S.M., Chakraborty, I., Sankar Cheela, V.R., Chowdhury, S., Dubey, B.K., Ghangrekar, M.M., 2021. A novel bio-electro-Fenton process for eliminating sodium dodecyl sulphate from wastewater using dual chamber microbial fuel cell. *Bioresour. Technol.* 341, 125850. doi: doi: 10.1016/j.biortech.2021.125850.

Sathe, S.M., Chakraborty, I., Dubey, B.K., Ghangrekar, M.M., 2022. Microbial fuel cell coupled Fenton oxidation for the cathodic degradation of emerging contaminants from wastewater: Applications and challenges. *Environ. Res.* 204, 112135. doi: 10.1016/j. envres.2021.112135.

Shekoohiyan, S., Moussavi, G., Naddafi, K., 2016. The peroxidase-mediated biodegradation of petroleum hydrocarbons in a H_2O_2-induced SBR using in-situ production of peroxidase: Biodegradation experiments and bacterial identification. *J. Hazard. Mater.* 313, 170–178. doi: 10.1016/j.jhazmat.2016.03.081.

Sher, F., Hazafa, A., Rashid, T., Bilal, M., Zafar, F., Mushtaq, Z., Nisa, Z.U., 2022. Effects of zeolite-based nanoparticles on the biodegradation of organic materials. In: H.M.N. Iqbal, M. Bilal, T.A. Nguyen, G. Yasin (Eds.), *Biodegradation and Biodeterioration at the Nanoscale*. Elsevier: Amsterdam, Netherlands. doi: 10.1016/b978-0-12-823970-4.00021-x

Shilova, O.A., Tsvetkova, I.N., Vlasov, D.Y., Ryabusheva, Y. V., Sokolov, G.S., Kychkin, A.K., Văn Nguyên, C., Khoroshavina, Y. V., 2022a. Microbiologically induced deterioration and environmentally friendly protection of wood products. In: H.M.N. Iqbal, M. Bilal, T.A. Nguyen, G. Yasin (Eds.), *Biodegradation and Biodeterioration at the Nanoscale*. Elsevier: Amsterdam, Netherlands. doi: 10.1016/b978-0-12-823970-4.00013-0.

Shilova, O.A., Vlasov, D.Y., Khamova, T.V., Zelenskaya, M.S., Frank-Kamenetskaya, O.V., 2022b. Microbiologically induced deterioration and protection of outdoor stone monuments. In: H.M.N. Iqbal, M. Bilal, T.A. Nguyen, G. Yasin (Eds.), *Biodegradation and Biodeterioration at the Nanoscale*. Elsevier: Amsterdam, Netherlands. doi: 10.1016/ b978-0-12-823970-4.00015-4.

Singh, N.B., 2022. Microbiologically induced deterioration of cement-based materials. In: H.M.N. Iqbal, M. Bilal, T.A. Nguyen, G. Yasin (Eds.), *Biodegradation and Biodeterioration at the Nanoscale*. Elsevier: Amsterdam, Netherlands. doi: 10.1016/ b978-0-12-823970-4.00016-6.

Singh, R., Bhunia, P., Dash, R.R., 2018a. Understanding intricacies of clogging and its alleviation by introducing earthworms in soil biofilters. *Sci. Total Environ.* 633, 145–156. doi: 10.1016/j.scitotenv.2018.03.156.

Singh, R., Bhunia, P., Dash, R.R., 2018b. COD removal index: A mechanistic tool for predicting organics removal performance of vermifilters. *Sci. Total Environ.* 643, 1652–1659. doi: 10.1016/j.scitotenv.2018.07.272.

Tabish, M., Zarin, A., Malik, M.U., Khan, M.A., Zhao, J., Yasin, G., 2022. Microbial-induced corrosion of metals with presence of nanoparticles. In: H.M.N. Iqbal, M. Bilal, T.A. Nguyen, G. Yasin (Eds.), *Biodegradation and Biodeterioration at the Nanoscale*. Elsevier: Amsterdam, Netherlands. doi: 10.1016/b978-0-12-823970-4.00027-0.

Tahirbegi, I.B., Ehgartner, J., Sulzer, P., Zieger, S., Kasjanow, A., Paradiso, M., Strobl, M., Bouwes, D., Mayr, T., 2017. Fast pesticide detection inside microfluidic device with integrated optical pH, oxygen sensors and algal fluorescence. *Biosens. Bioelectron.* 88, 188–195. doi: 10.1016/j.bios.2016.08.014.

Tandukar, M., Ohashi, A., Harada, H., 2007. Performance comparison of a pilot-scale UASB and DHS system and activated sludge process for the treatment of municipal wastewater. *Water Res.* 41, 2697–2705. doi: 10.1016/j.watres.2007.02.027.

Thompson, J.M., Gündoğdu, A., Stratton, H.M., Katouli, M., 2013. Antibiotic resistant Staphylococcus aureus in hospital wastewaters and sewage treatment plants with special reference to methicillin-resistant Staphylococcus aureus (MRSA). *J. Appl. Microbiol.* 114, 44–54. doi: 10.1111/jam.12037.

Trinh, T., van den Akker, B., Coleman, H.M., Stuetz, R.M., Drewes, J.E., Le-Clech, P., Khan, S.J., 2016. Seasonal variations in fate and removal of trace organic chemical contaminants while operating a full-scale membrane bioreactor. *Sci. Total Environ.* 550, 176–183. doi: 10.1016/j.scitotenv.2015.12.083.

Tsopela, A., Lale, A., Vanhove, E., Reynes, O., Séguy, I., Temple-Boyer, P., Juneau, P., Izquierdo, R., Launay, J., 2014. Integrated electrochemical biosensor based on algal metabolism for water toxicity analysis. *Biosens. Bioelectron.* 61, 290–297. doi: 10.1016/j.bios.2014.05.004.

US Environmental Protection Agency, 2018. *2018 Edition of the Drinking Water Standards and Health Advisories Tables.*

Uçar, N., Bakhtiari, S., Doustkhah, E., Yarmohammadi, M., Pedram, M.Z., Alyamaç, E., Seydibeyoğlu, M.Ö., 2022. Biodegradation of plastic-based waste materials. In: H.M.N. Iqbal, M. Bilal, T.A. Nguyen, G. Yasin (Eds.), *Biodegradation and Biodeterioration at the Nanoscale.* Elsevier: Amsterdam, Netherlands. doi: 10.1016/b978-0-12-823970-4.00009-9.

Ul-Abdin, Z., Anwar, W., Khitab, A., 2022. Microbiologically induced deterioration of concrete. In: H.M.N. Iqbal, M. Bilal, T.A. Nguyen, G. Yasin (Eds.), *Biodegradation and Biodeterioration at the Nanoscale.* Elsevier: Amsterdam, Netherlands, pp. 389–403. doi: 10.1016/b978-0-12-823970-4.00017-8.

Venkata Mohan, S., Velvizhi, G., Vamshi Krishna, K., Lenin Babu, M., 2014. Microbial catalyzed electrochemical systems: A bio-factory with multi-facet applications. *Bioresour. Technol.* 165, 355–364. doi: 10.1016/j.biortech.2014.03.048.

Verlicchi, P., Galletti, A., Masotti, L., 2010. Management of hospital wastewaters: The case of the effluent of a large hospital situated in a small town. *Water Sci. Technol.* 61, 2507–2519. doi: 10.2166/wst.2010.138.

Vidal-Limon, A., García Suárez, P.C., Arellano-García, E., Contreras, O.E., Aguila, S.A., 2018. Enhanced degradation of pesticide dichlorophen by laccase immobilized on nanoporous materials: A cytotoxic and molecular simulation investigation. *Bioconjug. Chem.* 29, 1073–1080. doi: 10.1021/acs.bioconjchem.7b00739.

von Sperling, M., Mascarenhas, L.C.A.M., 2005. Performance of very shallow ponds treating effluents from UASB reactors. *Water Sci. Technol.* 51, 83–90. doi: 10.2166/wst.2005.0432.

Wang, H., Luo, H., Fallgren, P.H., Jin, S., Ren, Z.J., 2015. Bioelectrochemical system platform for sustainable environmental remediation and energy generation. *Biotechnol. Adv.* 33, 317–334. doi: 10.1016/j.biotechadv.2015.04.003.

Wang, H., Ren, Z.J., 2013. A comprehensive review of microbial electrochemical systems as a platform technology. *Biotechnol. Adv.* 31, 1796–1807. doi: 10.1016/j.biotechadv.2013.10.001.

Wang, H., Ren, Z.J., 2014. Bioelectrochemical metal recovery from wastewater: A review. *Water Res.* 66, 219–232. doi: 10.1016/j.watres.2014.08.013.

Wang, J., Lv, Y., Zhang, Z., Deng, Y., Zhang, L., Liu, B., Xu, R., Zhang, X., 2009. Sonocatalytic degradation of azo fuchsine in the presence of the Co-doped and Cr-doped mixed crystal TiO_2 powders and comparison of their sonocatalytic activities. *J. Hazard. Mater.* 170, 398–404. doi: 10.1016/j.jhazmat.2009.04.083.

Wang, L., Guo, Z., Che, Y., Yang, F., Chao, J., Gao, Y., Zhang, Y., 2014. The effect of vermifiltration height and wet: DRy time ratio on nutrient removal performance and biological features, and their influence on nutrient removal efficiencies. *Ecol. Eng.* 71, 165–172. doi: 10.1016/j.ecoleng.2014.07.018.

Wang, L., Zhang, Y., Luo, X., Zhang, J., Zheng, Z., 2016. Effects of earthworms and substrate on diversity and abundance of denitrifying genes (nirS and nirK) and denitrifying rate during rural domestic wastewater treatment. *Bioresour. Technol.* 212, 174–181. doi: 10.1016/j.biortech.2016.04.044.

World Health Organization, 2012. Pharmaceuticals in drinking-water.

Wu, M., Xiang, J., Que, C., Chen, F., Xu, G., 2015. Occurrence and fate of psychiatric pharmaceuticals in the urban water system of Shanghai, China. *Chemosphere* 138, 486–493. doi: 10.1016/j.chemosphere.2015.07.002.

Xin, X., Chen, B.Y., Hong, J., 2019. Unraveling interactive characteristics of microbial community associated with bioelectric energy production in sludge fermentation fluid-fed microbial fuel cells. *Bioresour. Technol.* 289, 121652. doi: 10.1016/j.biortech.2019.121652.

Yang, H.Y., Han, Z.J., Yu, S.F., Pey, K.L., Ostrikov, K., Karnik, R., 2013a. Carbon nanotube membranes with ultrahigh specific adsorption capacity for water desalination and purification. *Nat. Commun.* 4. doi: 10.1038/ncomms3220.

Yang, J., Liu, J., Xing, M., Lu, Z., Yan, Q., 2013b. Effect of earthworms on the biochemical characterization of biofilms in vermifiltration treatment of excess sludge. *Bioresour. Technol.* 143, 10–17. doi: 10.1016/j.biortech.2013.05.099.

Yoneda, M., Mokhtar, M., 2018. *Environmental Risk Analysis for Asian-Oriented, Risk-Based Watershed Management: Japan and Malaysia.* Springer: Berlin, Germany. doi: 10.1007/978-981-10-8090-6.

Yuan, S., Jiang, X., Xia, X., Zhang, H., Zheng, S., 2013. Detection, occurrence and fate of 22 psychiatric pharmaceuticals in psychiatric hospital and municipal wastewater treatment plants in Beijing, China. *Chemosphere* 90, 2520–2525. doi: 10.1016/j.chemosphere.2012.10.089.

Zenker, A., Cicero, M.R., Prestinaci, F., Bottoni, P., Carere, M., 2014. Bioaccumulation and biomagnification potential of pharmaceuticals with a focus to the aquatic environment. *J. Environ. Manage.* doi: 10.1016/j.jenvman.2013.12.017.

Zhong, H., Liu, X., Zhu, L., Yang, Y., Yan, S., Zhang, X., 2019. Bioelectrochemically-assisted vermibiofilter process enhancing stabilization of sewage sludge with synchronous electricity generation. *Bioresour. Technol.* 289, 121740. doi: 10.1016/j.biortech.2019.121740.

12 The Wholistic Approach for Sewage Sludge Management

Ashish Dehal, K. Rathika,
Bholu Ram Yadav, and A. Ramesh Kumar
CSIR-National Environmental Engineering
Research Institute (CSIR-NEERI)
Academy of Scientific and Innovative Research (AcSIR)

CONTENTS

12.1 Introduction 204
12.2 Current Status of Sewage Management..205
12.3 Source of Sludge from Different Unit Operations or Processes of STP.......206
12.4 Characteristics of Sludge during Different Stages of Treatment.................207
12.5 Major Contaminants in Sludge: A Brief Overview 209
    12.5.1 Metallic Contaminants 209
    12.5.2 Organic Contaminants..209
    12.5.3 Pathogenic Organisms 211
12.6 Sludge Stabilization 211
    12.6.1 Biological Stabilization.. 212
        12.6.1.1 Anaerobic Digestion (AAD) 212
        12.6.1.2 Aerobic Digestion (AD) 213
12.7 Sludge Thickening and Dewatering.. 214
    12.7.1 Sludge Thickening 214
        12.7.1.1 Gravity Thickening.. 214
        12.7.1.2 Dissolved Air Flotation (DAF) Thickening....................... 215
        12.7.1.3 Gravity Belt (GB) Thickening... 215
        12.7.1.4 Rotary Drum (RD) Thickening 216
        12.7.1.5 Centrifugal Thickening.. 216
    12.7.2 Dewatering Process 216
        12.7.2.1 Belt Filter Press (BFP) 217
        12.7.2.2 Screw Press 218
        12.7.2.3 Rotary Press 218
        12.7.2.4 Centrifugal Sludge Dewatering 218
        12.7.2.5 Sludge Drying Beds 219
        12.7.2.6 Lagoons.. 219
        12.7.2.7 Electro-Dewatering (EDW) Process.................................. 220

DOI: 10.1201/9781003202431-12

12.7.3 Sludge Conditioning...220
        12.7.3.1 Inorganic Chemical Conditioning (ICC).........................221
        12.7.3.2 Organic Polymers..221
        12.7.2.3 Thermal Conditioning.......................................221
12.8 Pathogen Removal from Sludge...222
12.9 Assessment of Sludge Treatment and Disposal Options......................223
        12.9.1 Current Sludge Management Practices............................223
        12.9.2 Disposal Options..225
                12.9.2.1 Land Application....................................225
                12.9.2.2 Incineration...226
                12.9.2.3 Reutilization for Production of Building Materials..........227
                12.9.2.4 Landfilling..227
12.10 Conclusion...228
Acknowledgements...229
References...229

12.1 INTRODUCTION

The management and handling of sludge from sewage treatment plants is an expensive and complex process. With the increase in population, the quantity of sewage generation is also increasing day by day. The load on the existing sewage treatment plants is also increasing, due to which the efficiency of the treatment plants is affected. The number of sewage treatment plants is projected to rise in the near future to avoid the illegal disposal of sewage in water bodies. In developing countries, most of the sewage is treated using conventional biological treatment systems which generate a huge quantity of sludge. As a result, it is extrapolated that the quantity of sludge generated from sewage treatment plants will also increase.

In biological treatment methods, a part of the organic material is absorbed and turned into microbial biomass, referred to as secondary or biological sludge. Since it is mainly made up of biological solids, it is also known as a biosolid. An appropriate final destination for biosolids is fundamental to a successful sanitation system. Despite the fact that sludge accounts for just 1%–2% of the entire sewage volume, its management is extremely difficult, with expenses often ranging from 20% to 60% of the operational cost of the sewage treatment plant. Aside from its economic significance, the ultimate sludge destination is complicated since it is typically carried out outside the treatment plant's boundaries.

It is usual that in the design of wastewater treatment plants, the topic concerning sludge management is disregarded, causing this complex activity to be undertaken without previous planning by plant operators and frequently under emergency conditions. Because of this, inadequate alternatives for final disposal have been adopted, largely reducing the benefits accomplished by the sewerage systems. These factors indicate that sludge management is becoming a major issue in several countries. The issue will likely worsen in the coming years as more sewage treatment facilities are installed.

This chapter aims to provide a comprehensive overview of sewage sludge generation rates, types of sludge generation from sewage treatment plants, and their characteristics. After giving insights into common contaminants present in sludge, this chapter also enunciates various approaches to sludge stabilization, thickening, conditioning, and dewatering techniques. Eventually, widely adopted safe sludge handling and disposal trends have been detailed after a brief description of the sludge treatment of pathogenic removal techniques.

12.2 CURRENT STATUS OF SEWAGE MANAGEMENT

The global sewage generation rate is 359 billion liters per year. Only 63% of the generated sewage, equal to 225 billion liters, is collected by the existing sewer lines. Of the collected sewage, only 52% equal to 188 billion liters are properly treated every year. About 48% of the collected sewage and 37% of the total are left untreated (Jones et al. 2021).

The Central Pollution Control Board (CPCB) reports that 38,524 MLD of sewage is generated in India by class I cities and class II towns for 2008, India has a sewage handling capacity of 11,787 MLD. This indicates that 69.55% of the generated sewage (i.e., 26,797 MLD) is left untreated. Only 1724.6 MLD of sewage (equal to 6%) has been proposed to be addressed by sewage treatment plants (STPs) under the planning and construction stage. The sewage generation increased from 7,067 MLD in 1978–1979 to 62,000 MLD in 2014–2015, whereas the treatment capacity increased from 2,758 to 23,277 MLD only. According to the CPCB report on "National Inventory of sewage treatment plants, 2021", there are 1,631 STPs located across 35 states and union territories with a treatment capacity of 36,668 MLD. Out of the existing STPs, only 1,093 STPs are in operable condition. The 102 STPs are in non-operable conditions, 274 STPs are in the construction stage, and 162 STPs are in the planning stage (CPCB 2021). The current status of STPs in India, with details on the actual capacity of sewage sludge undergoing compiled treatment, is shown in Figure 12.1.

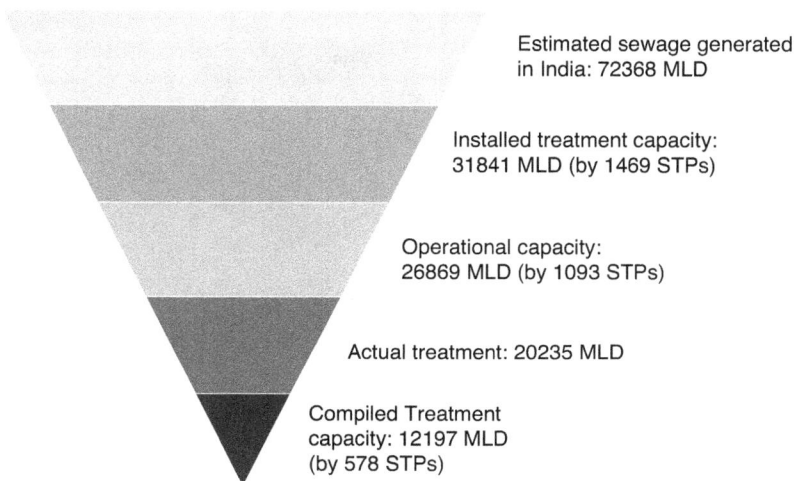

Estimated sewage generated in India: 72368 MLD

Installed treatment capacity: 31841 MLD (by 1469 STPs)

Operational capacity: 26869 MLD (by 1093 STPs)

Actual treatment: 20235 MLD

Compiled Treatment capacity: 12197 MLD (by 578 STPs)

FIGURE 12.1 Current status of STPs in India (CPCB 2021).

In 2014, 522 STPs were under operable conditions in India. It increased to a total of 1,093 STPs with a sewage handling capacity of 26,869 MLD in the year 2020 (CPCB, 2021). Although the number of operable STPs increased by 50%, for developing countries such as India, more infrastructure facilities are required to bridge the gap amid sewage treatment and synchronize in meeting future demands.

12.3 SOURCE OF SLUDGE FROM DIFFERENT UNIT OPERATIONS OR PROCESSES OF STP

The solid byproduct obtained from the wastewater treatment is termed "sludge". Although sludge constitutes only 1%–2% by volume of the treated wastewater, the treatment, and handling of sludge encapsulate 20%–60% of the total budget of sewage treatment facilities (EPA 2004). This economic dominance of sludge management proves that sludge management is a nexus procedure.

The sludge is obtained from various stages of unit operations or processes in a sewage treatment facility. Figure 12.2 elucidates the various unit processes in STPs. The first stage is screening which has a bar screen to remove large-size debris to avoid damage to valves and pumps of the treatment facility. The grit which cannot be trapped through screening is removed in the grit chamber. Primary clarifiers allow larger size particles to settle down, which can be removed from the bottom of the tank as primary sludge. The breakdown of organic matter from wastewater begins in the aeration tank. The continuous aeration aids in bacterial growth to degrade organic content and enhance the growth of biomass in the tank. The effluent from biological treatment is then fed to a secondary clarifier for organic sedimentation of treated wastewater. The effluent from the secondary clarifier is coined "secondary sludge". The sludge obtained at various stages of treatment is then stabilized and disposed of by STPs. Table 12.1 details the nature of sludge obtained from various sources and its percentage.

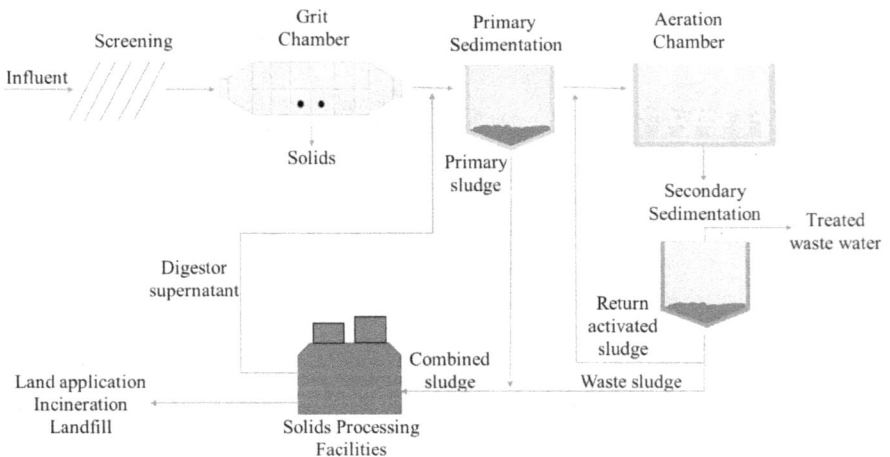

FIGURE 12.2   Unit operations in the CAS process.

TABLE 12.1
Sources of Sludge

Source of Sludge (Unit Operation/ Process)	Nature of Sludge	Remarks
Screening	Coarse solids	For the treatment of municipal wastewater, a sieve opening of 250–500 µm is adopted. When 20% of total suspended solids in sewage exceeds 350 µm, sieving is adopted as the foremost primary treatment process (Rusten and Ødegaard 2006)
Grit removal	Grit and scum	Particles with a size >0.21 mm and a specific gravity of 2.65 are termed "grit". Modern grit chambers have an efficiency of 75% for particles of size 0.15 mm (EPA, 2002). Most of the grit removal facilities neglect to integrate scum-removal facilities
Pre-aeration	Grit and scum	A pre-aeration tank should be preceded by a grit and scum removal facility to avert grit deposition in aerators. Sludge from aeration tanks is not usually collected as the biosolids contain beneficial microbes that help degrade organic fractions
Primary sedimentation	Primary solids and scum	It removes 90%–95% of settleable solids; 40%–60% of suspended solids; 10%–15% of total solids; 20%–50% of biological oxygen demand (BOD) and 25%–75% of bacteria (EPA 1998)
Biological treatment	Suspended solids	BOD conversion produces suspended solids. Thickening agents are required to remove biological sludge
Secondary sedimentation	Secondary biosolids and scum	Biosolids produced in secondary clarifiers are removed. The total suspended solids concentration in effluent from the secondary clarifier should be below 10 mg/L (Voutchkov 2005). Gravity thickening of secondary sludge captures 75%–80% of solids. Also, 80%–95% of solids are captured by flotation and centrifuge techniques (EPA 1987).
Solids processing facilities	Solids, compost, and ashes	Sludge obtained from various sources can have two fates. They can be further utilized as manure/compost or discarded after proper stabilization and treatment

12.4 CHARACTERISTICS OF SLUDGE DURING DIFFERENT STAGES OF TREATMENT

Sludge constitutes both water and solids. Solids present in the sludge are both suspended and dissolved. Dissolved solids and suspended solids constitute both organic and inorganic fractions. All organic solids are volatile in nature. Digested sludge has volatile solids to total solids ratio (VS/TS) of 0.6–0.65 and 0.75–0.8 for undigested sludge. The sludge density is a function of VS, fixed solids (FS), and water. As the distribution of the above-mentioned components varies at different stages of treatment, the sludge density varies correspondingly. The density of sludge can be computed from known specific gravities (SGs) of VS, FS, and water equal to 1, 2.5, and 1, respectively (Andreoli et al. 2007). Table 12.2 shows the characteristics of sludge from different sources.

TABLE 12.2
Characteristics of Sludge

Source of Sludge	Phase	The Ratio of VS to TS		% of Dry Solids		SG of Solids		SG of Sludge		Sludge Density[a]	
		LL	UL	LL	UL	LL	UL	LL	UL	LL	UL
Primary sludge	Liquid	0.75	0.8	2	6	1.14	1.18	1.003	1.01	1003	1,010
Secondary sludge (aerobic)	Liquid	0.55	0.6	3	6	1.32	1.37	1.01	1.02	1010	1,020
Secondary sludge from AS (aerobic)	Liquid	0.75	0.8	0.6	1	1.14	1.18	1.001	-	1001	-
Secondary sludge from EA (aerobic)	Liquid	0.65	0.7	0.8	1.2	1.22	1.27	1.002	-	1002	-
Stabilization pond	Liquid	0.35	0.55	5	20	1.37	1.64	1.02	1.07	1020	1,070
Primary sludge	Thickened	0.75	0.8	4	8	1.14	1.18	1.006	1.01	1006	1,010
Secondary sludge from AS (aerobic)	Thickened	0.75	0.8	2	7	1.14	1.18	1.003	1.01	1003	1,010
Secondary sludge from EA (aerobic)	Thickened	0.65	0.7	2	6	1.22	1.27	1.004	1.01	1004	1,010
Mixed sludge	Thickened	0.75	0.8	3	8	1.14	1.18	1.004	1.01	1004	1,010
Mixed sludge	Digested	0.6	0.65	3	6	1.27	1.32	1.007	1.02	1007	1,020
Mixed sludge	Dewatered	0.6	0.65	20	40	1.27	1.32	1.05	1.1	1050	1,100

[a] Density of sludge is expressed in kg/m^3.

AS, Activated sludge; EA, Extended aeration activated sludge; SG, Specific gravity; UL, Upper limit; LL, Lower limit

12.5 MAJOR CONTAMINANTS IN SLUDGE: A BRIEF OVERVIEW

Throughout the treatment process, the concentration of sludge constituents is dynamic. While several organic constituents have useful characteristics, such as enhancing the fertility of the soil, others may have undesirable characteristics as it imposes environmental and sanitary risks. The undesirable constituents are the contaminants present in sludge, and their presence and concentration are strong functions of wastewater origin and wastewater characteristics. The sludge contaminants can be widely classified into metallic contaminants, organic contaminants, and pathogenic contaminants.

It is crucial for the wastewater and sanitation management sector to adopt a well-established sludge management practice to handle the contaminants. As domestic and non-domestic wastewater have different contaminant concentrations, it is efficient and economical to adopt sludge management practices accordingly. When efficient and economical methods are in place, achieving sustainability is serene. When harmful contaminants such as heavy metals and pathogens can be effectively removed from sludge, agricultural application of treated sludge will ensure a circular economy. Various state-of-the-art disinfection techniques can achieve the reduction of pathogenic concentration to an acceptable level. Removing heavy metallic and organic contaminants from sludge can be economically and environmentally challenging, especially for developing countries, as it is energy-intensive and emits toxic chemicals.

12.5.1 METALLIC CONTAMINANTS

Chronic toxicity of metallic contaminants from domestic wastewater is not usually reported due to their low concentration. Effluent from chemical industries, electroplating industries, pharmaceutical industries, oil industries, tanneries, foundries, dye, and pigment manufacturing units are the primary source of acute concentration of heavy metals in sludge. Heavy metal contaminants which are predominantly present in sludge are Cd, Cu, Zn, Ni, Hg, Cr, Pb, As, and Se. Table 12.3 details the metallic contaminants present in the sludge in the order of descending from the most occurring metallic form to the least occurring form.

12.5.2 ORGANIC CONTAMINANTS

Since treated sewage sludge has excellent fertilizing properties, its usefulness in land application is a widely sought-after area of interest. Hence, a wide range of organic substances is receiving more and more attention as possible contaminants of land, vegetation, and water. Chlorinated hydrocarbons, insecticides, and polychlorinated biphenyls were the most investigated chemicals in the beginning. Later studies concentrated on other organic chemicals in sewage treatment systems. Several phthalates esters (dibutyl, diethyl) were found in 13%–25% of the sludges at amounts above 50 mg/kg, according to studies conducted in 25 cities across the US. Naphthalene, phenol, and toluene were also detected at concentrations >50 mg/kg in 11%–25% of the sludges. Several other contaminants (chlorinated benzene, ethane, and methane) were also detected in sludge in a concentration >1 mg/kg. Trace organics such as chlorinated hydrocarbons and phenols, acrylonitrile, hydroquinone, styrene, etc.,

TABLE 12.3
Metallic Contaminants of Anaerobically Digested Sludge
(Babel and Dacera 2006)

Metal	Metallic Form[a]	Form of Existence in Sludge[a]
Cadmium (Cd)	Residual	Bound to organic matter
	Carbonates	Present as inorganic precipitate
	Sulfides	
	Organically bound	
	Adsorbed cadmium	
	Exchangeable form	
Chromium (Cr)	Residual	Bound to organic matter
	Organic	Bound to inorganic matter
	Carbonates	
	Adsorbed chromium	
	Exchangeable form	
Copper (Cu)	Sulfides	Incorporated to organic matter
	Residual	Incorporated to organic mineral
	Carbonates	aggregates
	Organically bound	Bound to organic matter
	Adsorbed copper	
	Exchangeable form	
Manganese (Mn)	Organic	Bound to organic matter
	Carbonates	
	Residual	
	Exchangeable form	
	Adsorbed manganese	
Nickel (Ni)	Residual	Bound to iron and manganese
	Organic	Bound to organic matter
	Carbonates	Bound to inorganic matter
	Exchangeable form	
	Adsorbed nickel	
Lead (Pb)	Carbonates	Bound to organic matter
	Organic bound	Bound to inorganic matter
	Residual	Present as inorganic precipitates
	Sulfides	
	Exchangeable form	
Zinc (Zn)	Organic	Bound to oxides of iron and
	Carbonates	manganese
	Residual	Bound to organic matter
	Adsorbed zinc	Bound to inorganic matter
	Exchangeable form	Present as inorganic precipitates

[a] Metallic forms of contaminants are written in the order of descending from most occurring to least occurring.

were also detected in sludge. The literature implies that sewage sludge has the most trace organics present at a concentration of <10 mg/kg (Andreoli et al. 2007).

12.5.3 PATHOGENIC ORGANISMS

Microbes found in sludge may be parasites, symbionts, saprophytes, commensals, etc. Out of these, only parasites are pathogenic, meaning they can infect animals and humans. Bacteria, helminths, viruses, protozoa, and fungi are the five types of pathogenic microbes found in sludge. The pathogens may have arisen from anthropogenic excreta. The presence and concentrations of pathogens are indicators of the public's sanitation and health conditions in that area. These might even arise from animal sources, such as cat and dog feces. Epidemiological studies have revealed that the presence of pathogenic organisms leads to various health concerns for humans and animals. The number of pathogenic organisms found in sewage varies considerably and is based on:

- socio-economic status
- the state of sanitation
- geographical location
- agro-industries' presence
- sludge treatment type

The storage and multi-purpose application of sewage sludge without prior stabilization and sanitization treatment may lead to various infections in animals and humans due to the spread of pathogens. Infection can spread through direct or indirect pathways (via the mouth or aspiration). Some of the pathways of pathogenic contraction are:

- humans may directly inhale pathogen-containing particulates during the spreading of untreated sludge in soil;
- by handling or consuming raw vegetables cultivated in untreated sludge-fertilized soil;
- animals can also become directly infected, hence resulting in clinical difficulties or serving as living reservoirs for pathogens;
- drinking water contaminated with sludge containing infectious pathogens;
- eating animal meat formerly contaminated with pathogens or helminth eggs (Andreoli et al. 2007).

Removal of all the contaminants is necessary for the safe utilization of sludge, particularly for land application.

12.6 SLUDGE STABILIZATION

Raw sewage sludge is rich in pathogens, highly putrescible, and emits foul odors quickly. Stabilization methods were created to stabilize the biodegradable percentage of organic substance in sludge, lower the danger of putrefaction, and lower pathogen concentrations (Bux et al. 2002). The stabilization processes can be majorly divided into three types, as demonstrated in Figure 12.3.

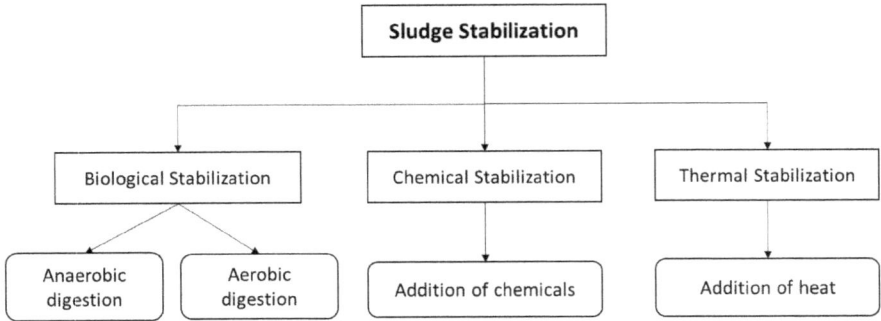

FIGURE 12.3   Various sludge stabilization approaches.

The present chapter focuses on the most widely used method of biological stabilization.

12.6.1   Biological Stabilization

In sewage treatment, the term digestion refers to the stabilization of organic materials by microbes in contact with sludge in a favorable environment for their growth, development, and reproduction. Sludge digestion processes may be divided into aerobic, anaerobic, or a mix of both.

12.6.1.1   Anaerobic Digestion (AAD)

The most often used sludge stabilization method is AAD. AAD biologically stabilizes the sludge in the absence of air. Converting volatile materials to biogas minimizes the quantity of volatile solids (CH_4, CO_2, and water). The biogas must then be processed further in order to collect and utilize the CH_4.

AAD can be operated either:

- At moderately thermophilic temperatures (50°C–60°C)
- At mesophilic temperatures (30°C–40°C)

The absolute ideal temperature within these ranges is determined by the intended biogas composition and yield, solid reduction, and energy efficiency.

Psychrophilic is defined as operation at temperatures below 20°C. It is more energy-efficient to operate at such low temperatures. However, the tank capacity requirement increases because biological reaction rates and corresponding gas production levels slow down at lower temperatures.

AAD is a multi-stage biochemical process that can stabilize a variety of organic materials. The process is divided into the following stages:

- The first step in AAD is hydrolysis. In this step, a reaction with water takes place where large organic molecules break down into smaller molecules.
- Enzymes convert complex molecules/components like lipids, cellulose, and proteins into soluble substances like alcohol, fatty acids, alcohol, CO_2, and NH_3. This stage of AAD is called acidogenesis.

- The earlier-stage products are converted by microbes into propionic acid, acetic acid, hydrogen, CO_2, and some other low-molecular-weight organic acids. This stage of AAD is called acetogenesis.
- In the last step of methanogenesis, two groups of CH_4-forming microbes take action: one group makes CH_4 from hydrogen and CO_2, while another transforms acetate into CH_4 and bicarbonates.

Similar to most of the sludge and sewage treatment processes, the design and effectiveness of AAD are also determined by feed characteristics. AAD is most commonly used in sewage treatment facilities to treat waste solids discharged from both primary and secondary sludge streams. The feed sludge quality influences the characteristics of the streams generated by the AAD process (i.e., biogas generated and stabilized solid product). Since AAD is a biological process, higher concentrations of toxic or potentially hazardous compounds, such as sulfates and heavy metals, can have an adverse effect on the process (Andreoli et al. 2007; Jenicek et al. 2010).

12.6.1.2   Aerobic Digestion (AD)

The biodegradation of organic sludge solids in the presence of oxygen is known as AD. The oxygen is delivered into the reactors as small air bubbles. The organic material present in the sludge is then converted to CO_2 and H_2O, while the NH_3 and amino species are converted to nitrates by the microbes in the sludge.

AD is similar to the conventional activated sludge (CAS) process, but it does not use wastewater as a feedstock and has prolonged solid retention times. However, because there is no feed water entering the system continuously, there is only an intermittent supply of organic substrate to nourish the microbes. Hence, they start dying and are used as food (substrate) by other bacteria. This stage is known as endogenous respiration and is responsible for lowering the sludge solids content. AD is generally less in capital cost than AAD for plants with less than 20,000 m^3/day wastewater treatment capacity.

Like the AAD process, the method generates a stable sludge end product while lowering pathogenic content. The main drawback of AD is that it is energy-intensive: the reaction uses air and produces CO_2 as the principal gaseous end product rather than CH_4. In addition, mechanical dewatering of digested sludge is more difficult. AD alike AAD can be configured as a single or multi-tank process, with the multi-tank having a higher efficiency by reducing the solids residence time. As with AAD, the process can be run at both mesophilic and thermophilic temperatures (Song et al. 2010).

Composting is also an aerobic digesting method. Prior to composting, sewage sludge can be blended with some other waste products such as straw, wood chips, green wastes, etc., to create a pasteurized product. Composting can convert nearly 20 to 30% of the volatile solids to CO_2. Composting involves natural mesophilic and thermophilic aerobic decomposition, mainly in a static system that is aerated by natural diffusion and hence requires little energy. It is a slow process that necessitates a vast amount of land. Windrow composting is the most well-known (employed) method of composting. If the metallic content of sludge is sufficiently low, the end product formed by the composting can be utilized for soil conditioning or other land practices.

Andreoli et al. (2007)

12.7 SLUDGE THICKENING AND DEWATERING

The key objective of sludge thickening and dewatering (both procedures) is to lower the water content in the sludge to minimize sludge volume. Conditioning of the sludge is also addressed in the present chapter as it encourages water removal. Sludge thickening is primarily employed for the sludge generated from primary sedimentation tanks, secondary settling tanks receiving wastewater from trickling filters, and activated sludge processes, with significant implications for sludge digester design and operation (Andreoli et al. 2007). The key difference between sludge thickening and dewatering is that the thickening process is adopted to remove the free water content of the sludge. Whereas the dewatering process is adopted to remove capillary water bound to flock particles of sludge. The total solid content of thickened sludge is 6%–10% and dewatered sludge is 40%–80%.

Various intermolecular interactions are accountable for binding water molecules to the sludge solids (Metcalf et al. 1991). Different classes of water bonding have been listed in Table 12.4, with corresponding techniques required for separation. Also, the water separation technique is enlisted in the decreasing order of ease of separation from top to bottom of the table.

12.7.1 SLUDGE THICKENING

There are several techniques available to reduce the water content from the sludge. The major types of sludge thickening processes have been explained in Figure 12.4.

12.7.1.1 Gravity Thickening

Gravity thickening improves solid concentrations by enabling particulates to sink to the vessel's bottom, resulting in a more concentrated (thickened) solid stream at the

TABLE 12.4
Type of Different Water Bonding and the Separation Techniques (Andreoli et al. 2007)

Sr. No.	Type of Class or Water Bonding	Techniques or Processes for Water Separation	Approximate Sludge Volume Reduction
1	Free water	Flotation or simple gravitational action (gravity thickeners and drying beds)	60% or more
2	Adsorbed water	Both require comparatively larger forces for	90%–95%
3	Capillary water	separation. The forces can be chemical (use of flocculants) or mechanical (use of centrifuges or filter presses)	
4	Cellular water	It can only be detached by thermal forces that cause alterations in the water's aggregation state. Freezing and evaporation (primarily) are two alternative ways to separate cellular water. Thermal drying can also be used for the separation	More than 95%

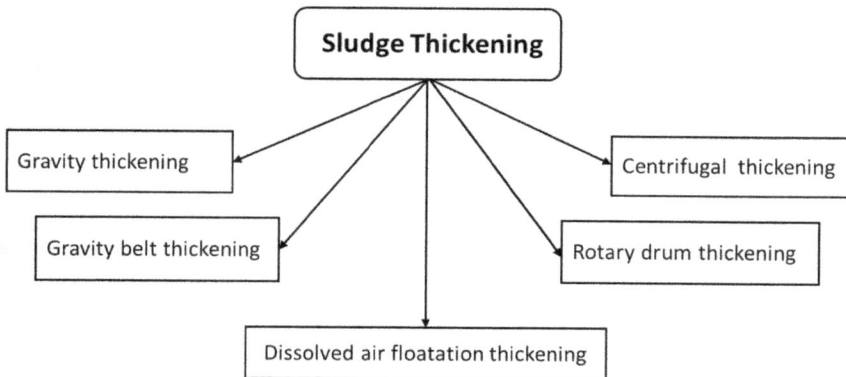

FIGURE 12.4 Types of sludge thickening.

bottom of the reactor and a less concentrated supernatant at the top. Sedimentation, often known as settling, is the most basic thickening method. The design elements of a gravity thickener are similar to those of a primary settling tank used for sewage treatment, and the tanks are usually circular in shape. The solids flux theory and zone settling principle govern the sludge dynamics within the thickeners. The size of a tank may be determined either by these principles or by analyzing the tank's solids and hydraulic loading rates (Hudson and Lowe 1996; EPA 1987). Eighty percent to 90% of the solids are retained in the sludge after gravity thickening. The efficiency of the gravity thickener increases with the depth of the sludge blanket. It should also be noted that the blanket depth also increases the retention time leading to the formation of biogas. Hence the optimal retention time for gravity thickeners for primary sludge is 24–48 hours. The retention time reduces to 18 and 30 hours when a combination of primary and secondary sludge is fed to gravity thickeners.

12.7.1.2   Dissolved Air Flotation (DAF) Thickening

DAF concentrates the sludge by enabling solids to float to the top instead of sinking (in contrast to that of gravity thickeners). DAF can effectively enhance sludge solids concentration in the supernatant when the solids are neutrally buoyant (i.e., they do not sink or float) and cannot be efficiently separated through the settling process. In DAF, thickening air is pumped into a solution held at high pressure. The air remains dissolved in such conditions. When the pressure is removed, dissolved air is expelled, producing tiny bubbles that bring sludge particulates to the top, which can be skimmed off. Excess activated sludge that does not thicken effectively in gravity thickening can be removed from DAF thickening (Hudson and Lowe 1996; Crossley and Valade 2006).

12.7.1.3   Gravity Belt (GB) Thickening

GB thickening allows filtrate/water to be removed from the sludge under the action of gravity by a moving belt (porous medium) over which the sludge lies, thus promoting the sludge thickening. GB thickening is a low-pressure method that enables water to drain naturally (under gravity) from the sludge. As water is drained through the porous-moving belt, the solid sludge concentrates over the belt length, forming a layer of solids on the surface of the belt. In a conventional conveyor belt, the belt is continually recirculated. The thickened solids are permitted to drop off the end of the belt into a collector tank, usually aided with scrapers (Weihua et al. 2010; Kozak et al. 2011).

12.7.1.4   Rotary Drum (RD) Thickening

In RD thickeners, the sludge is thickened by the agitation of the solids in a slowly rotating vessel having porous sides through which filtrate (water) drains out. The RD thickener works on the same principle as of GB thickener, in which filtrate (water) drains through a porous medium (Yang et al. 2017). Steel, polymers, and ceramics are some materials that can be used to manufacture porous walls. A water spray system is often used to clean the RD and keep the pores free from clogging.

12.7.1.5   Centrifugal Thickening

Centrifugal thickeners encourage solids to drift toward the walls of a swiftly spinning cylinder-shaped tank under the action of centrifugal force, hence increasing the sludge concentration. The water content of the sludge can be reduced using a variety of centrifuge systems. Centrifuges can be used for thickening as well as dewatering. The rotational speed and type of the concentrated solid product produced are the factors to be considered. Dewatering centrifuges require higher energy usage per unit of solid content (kg DS) to produce higher solids concentrations. The thickened product produced retains the fluid characteristics and can thus be pumped, but the dewatered product cannot be pumped, and a belt conveyor can only deliver it (Gable 2014).

12.7.2   Dewatering Process

The dewatering process eliminates a considerable amount of the sludge's water content, resulting in a more concentrated sludge with a dry solids content of 15%–45%. The product produced, termed a cake, is not free-flowing and forms lumps that can only be carried by a belt conveyor, mechanized earth-moving machinery, or a spade. Dewatering techniques use substantial mechanical effort to remove more water than thickening alone can (Chen et al. 2002; Mowla et al. 2013). Dewatering effectively reduces waste volume, transportation costs, and risk of spilling/leaking and increases wastewater recycling (To et al. 2016; Cao et al. 2021). The method of dewatering digested sludge affects the cost of transport and final disposal. Various sludge dewatering methods have been shown in Figure 12.5.

The following are the primary reasons for sludge dewatering (EPA 1987):

- dewatered sludge is easy to handle and hence, easy to transport
- cost savings associated with the transportation of a lower volume of sludge to the final disposal site
- the decrease in the water content of the sludge will increase its heating capacity, favoring incinerating operation of the sludge as the final disposal step
- reduction of the sludge volume destined for land disposal/landfills
- decrease in leachate volume generation if landfill disposal is practiced.

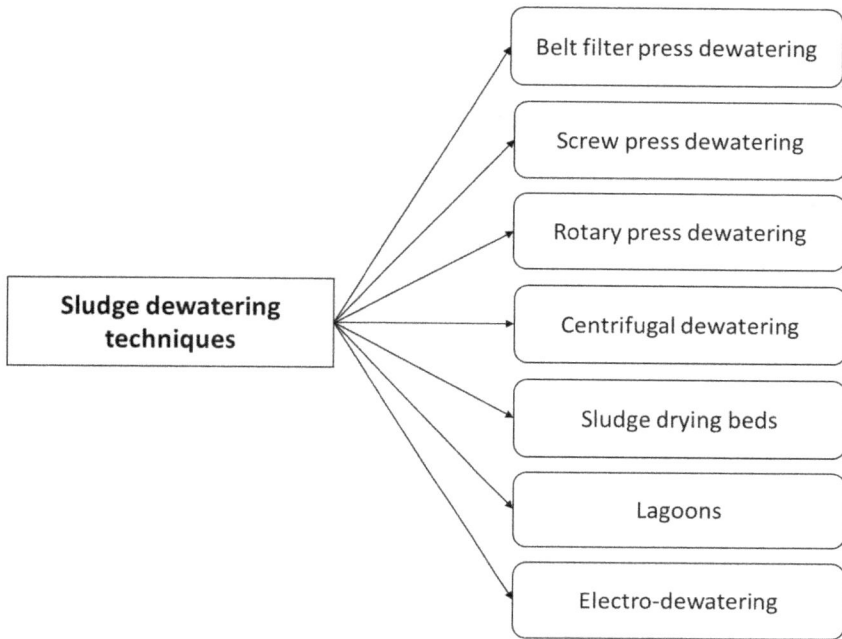

FIGURE 12.5 Various sludge dewatering techniques.

12.7.2.1 Belt Filter Press (BFP)

Sludge dewatering is accomplished with a BFP, which presses the sludge to expel filtrate (water) through a porous material. In the case of primary sludge, this method yields a cake (the dewatered product) with a dry solid concentration greater than or equal to 30%. Although the low-energy GB thickening technique has a similar working mechanism, it produces a maximum dry solid content of 10%. A BFP removes water by sequentially combining mechanical pressure and drainage. The BFP's overall performance can be determined by the feed sludge characteristics, sources, as well as dosing conditions (Hudson and Lowe 1996).

In most cases, the machinery consists of two to three recirculating belts. Two adjacent belts are assembled to overlap at one end to apply compression to the sludge and force water out of it. The sludge is initially conveyed through porous belts, as in the case of GB thickeners. It is then subjected to pressure as it passes through a wedge zone formed by two recirculating belts that run over a roller. A larger quantity of water is released by the pressing action of the two belts in this pressure region.

Many BFP methods include a second pressure area or zone, consisting of a set of rollers via which the two belts move with the sludge solids held between them. The rollers in the high-pressure region apply strain to the belts, applying both compressive and shearing force on the sludge, enabling it to lose even more water. Some BFP systems use a three-belt arrangement, with the GB operating independently of the two pressuring belts. This enables the recirculation rate of thickeners to be independently regulated from the dewatering operation (Vaxelaire and Olivier 2006).

12.7.2.2　Screw Press

A screw press dewaters the sludge by conveying it toward the inside of porous containers, usually in a cylindrical shape. The screw press is based on a gradually rotating Archimedean screw having RPM-5, encapsulated within a cylinder-shaped screen (also identified as a drum filter/screen). It is generally set at a 20° angle to the horizontal to aid water drainage into the sump. The screen is typically made of perforated metal or wedge wire and has less than a 0.5 mm aperture rating.

The screens used in this method are less prone to clogging than the filter medium used in other dewatering and thickening operations. As a result, spray cleaning is only used for around 24% of the operational period. Like other thickening and dewatering operations, the feed sludge quality also affects the dry solid content (cake). The cake's solid concentration was found inversely related to the feed sludge-volatile solids content (Loranger et al. 2019).

12.7.2.3　Rotary Press

A rotary press removes the water from sludge by forcing it through a narrow, rotating parallel-flow channel having porous sides. The rotary press dewaters the sludge using a cylindrical tank containing two sandwiched gently moving (<2 RPM) circular screens. At low applied pressure (0.1–0.5 bar), the sludge penetrates the channel formed by the two screens and continues a circular path around the channel. The pressure and shear force combination encourages water flow between the screens as the sludge passes through the channel (Chen et al. 2002).

The rotary press operation and performance are similar to the screw press, which runs at a low speed and employs a fine screen to keep sludge solids in the equipment. The rotary press screens cannot be used for filtering and spray cleaning simultaneously. Furthermore, the filtration area of a single rotary press is limited by the unit's diameter (up to 1.2 m). As a result, most RP installations are made up of several simultaneous units. The rotary presses have low energy consumption and are enclosed systems with excellent odor control and low noise levels.

12.7.2.4　Centrifugal Sludge Dewatering

Centrifugation is being used for thickening as well as dewatering sludge. The centrifuge technique used for both operations is almost identical. The following are the fundamental operational distinctions between the two methods: the rotational speed, the throughput, and the type of concentrated solid products produced (Dick et al. 1980).

Dewatering operation requires extra energy than thickening as more water must be drained to obtain greater solid concentration. After dewatering, the product with a dry solid concentration equal to or more than 50% resembles cake (semi-solid), which forms lumps instead of free-flowing fluid. As a result, it can only be transported by conveyor belts, whereas thickened material keeps the feed's fluid characteristics and may be pumped (Hudson and Lowe 1996).

The dewatering capacity and solid recovery of centrifugal dewatering are affected by the quality of the feed material and the dosing parameters. Most commonly, solid bowl centrifugation is used for dewatering, and this design may be modified to improve dewatering performance, resulting in a drier solid product by doing the following:

- increasing the length of the solid outflow section (also known as the beach).
- lowering the differential velocity.
- tweaking the screw to provide solid cake pressing in the beach area.

The first two adjustments increase the centrifuge's residence time. The third adjustment helps in additional dewatering (Chu and Lee 2001, 2002).

Solid bowl centrifugation is also used for the hybrid dewatering and drying process, in which the bowl is modified to allow for heating with a sweep gas. This process can produce dry solid concentrations of up to 90%.

12.7.2.5 Sludge Drying Beds

Sludge drying beds with lagoon systems are the most basic dewatering processes. They are often used effectively for un-thickened sludges, digested solids, and stabilized solids from extended aeration systems. Drying beds are simple to build and operate; they simply disseminate the sludge in thin layers (up to 30 cm) over a porous bed of gravel and sand. The water is then collected by the system of an embedded drainage network. Further evaporation helps in the exclusion of additional water from the sludge surface (Wang et al. 2010).

The amount of dewatering that can be achieved is determined by the drying time, change in weather conditions, and whether or not the bed is covered. The dry solids are mechanically removed when the sludge has been adequately dewatered, usually taking 10–15 days, and can be noticed by cracks, fissures, and dark brown surfaces. The front-end loader or specialized mechanized sludge removal device is usually used for the removal (Hudson and Lowe 1996; Manfio et al. 2018).

Since most drying beds are open/uncovered and are subject to local weather circumstances, water evaporation can be enhanced by enclosing the beds with a glass network or structure (a greenhouse). This adaptation is known as a solar-drying bed. The massive footprint imposed by the drying beds restricts their use in situations where land availability is not a constraint, despite their low energy and operational costs.

12.7.2.6 Lagoons

A lagoon can provide a dewatering and final disposal pathway for the digested sludge in places where the availability of land is not a constraint, potential ecological effects and danger to public health are minimum, and a low-cost alternative is desired. It is akin to a drying bed because it requires vast land areas and labor-intensive mechanized removal of the dry solids, but it is easier to construct as no filtrate drainage is needed (Wang et al. 2019).

A lagoon is a shallow pond with a sealed bottom that prevents sludge water from entering groundwater or other water bodies. Sludge is dumped into the lagoon, and the solids are left to settle. Water is removed via evaporation throughout the dewatering/dewatering operation cycle, which can last for several months. Meanwhile, the dense sediment layer at the bottom of the lagoon is stabilized by anaerobic biological activities.

The dried sludge is usually removed once the sludge has reached the target dry solid content of 25%–30%. The operational cycle of the lagoon is typically greater than one year, subject to the local climate conditions. A sludge drying lagoon typically takes a longer time than a sludge drying bed, where the drainage network removes a significant portion of the water.

12.7.2.7 Electro-Dewatering (EDW) Process

The EDW technique enhances removal in non-mechanized ways by employing an electric field. This method removes the water through an electro-osmosis mechanism. Electro-osmosis is an electrophoretic phenomenon in which the displacement of solids relative to the surrounding fluid, or fluid within a porous solid, can occur because of the charges at the liquid–solid interface when an electric field is applied (Mahmoud et al. 2010).

The sludge is pumped into an EDW system via a narrow horizontal tube. Under the effect of the electrophoretic force, water is removed from the sludge when an electric field is applied. There is no need for any other mechanical effort since the water removal is determined by the applied voltage and the charges at the particle surface.

This process can achieve dry solids up to 50% concentration. Although the method looks quite efficient, it consumes a massive amount of energy—about 300 kWh per ton of water drained. This restricts its usage to sludges with low water content, such as traditionally dewatered sludges. Due to high energy requirements, it competes with the thermal drying method rather than mechanical dewatering techniques for water removal, which uses less energy (20 kWh per ton of water drained) (Tuan et al. 2012; Mahmoud et al. 2010).

12.7.3 SLUDGE CONDITIONING

Conditioning is carried out prior to the dewatering process and positively impacts the effectiveness of the dewatering operation. Thermal treatment, organic polymers, and inorganic chemicals are usually used for accomplishing the conditioning process (Wu et al. 2020). The objective of conditioning is to alter the sludge particle size, distribution, surface charges, and particle interaction. In sewage sludge, the occurrence of colloidal particles with sizes ranging from 1 to 10 μm is pretty common. Biomass plays a vital role in trapping the particulates during biological treatment, reducing sludge dewaterability, and increasing the need for conditioning agents. The primary goal of the conditioning is to increase the particle's size by encapsulating the smaller particles into large flocs. Coagulation and flocculation are being used to achieve this. Coagulation causes the particles to become less stable, lowering the intensity of repulsive forces between them. It also facilitates the mutual attraction between particles. The agglomeration of thin solids and colloids takes place during flocculation (Mowla et al. 2013).

The number of conditioning agents being used depends on the characteristics of the sewage sludge and the equipment employed for dewatering. There can be variations in fine solids and water content levels depending on how the sludge is transported through pipes and the length of time the sludge is stored. The effects of various conditioning processes have been explained in Table 12.5.

TABLE 12.5
Effect of Conditioning Processes (EPA 1987)

	Inorganic Chemicals	Organic Polymers	Thermal Treatment (Heat)
Conditioning mechanism	Coagulation and flocculation	Coagulation and flocculation	Modifies surface properties, split cells, liberate chemicals, and initiates hydrolysis
Effect on admissible solid loading rate	Will increase	Will increase	Considerable increase
Effect on supernatant flow	Escalates solids capture	Escalates solids capture	A considerable amount of increase in color, suspended solids, N-NH$_3$, BOD, and COD
Effect on human resources	Little effect	Little effect	Needs trained workers and a reliable maintenance plan
Effect on sludge mass	Significantly increases	None	Decreases present mass but might increase the mass via recirculation

12.7.3.1   Inorganic Chemical Conditioning (ICC)

Pressure and vacuum filtration dewatering techniques are primarily linked with ICC. Ferric chloride and lime are majorly used chemicals used for the conditioning of sewage sludges. Ferrous sulfate, aluminum sulfate, and ferrous chloride are also sometimes used for sewage conditioning. The literature showed that power plant ash, pulverized coal, cement kiln dust, fly ash, sludge incinerator ash, etc., can also be used as conditioning agents for the sludge (Tuan et al. 2012; Wu et al. 2020).

12.7.3.2   Organic Polymers

Significant progress has been made in producing organic polymers to be used in sewage sludge treatment over the past few decades. Polymers are progressively more employed in sludge conditioning, as a wide range of choices is available (Mowla et al. 2013). It is vital to understand that these materials' composition, functional performance, and cost vary significantly. Polymers are water-soluble compounds with a long chain. They can be manufactured by adding functional monomers or groups to naturally existing polymers or entirely synthesized using individual monomers. Polymers were first employed to condition primary sludges and easy-to-dewater (combinations of primary and secondary) sludges. An increment in the efficiency of polymers has resulted in their growing use in all kinds of dewatering practices. Polymers also help in the reduction of operational and maintenance problems.

12.7.2.3   Thermal Conditioning

The thermal conditioning method improves the dewatering features of sludge by the combined effect of pressure and heat (EPA 1990). Thermal conditioning is a continuous flow procedure. In this process, the sludge is heated for 15–40 minutes to temperatures ranging from 350°F to 400°F and under a pressure of 1,720–2,750 kPa. There are basically two types of thermal conditioning operations: (1) low-pressure oxidation; (2) heat treatment. In low-pressure oxidation, air is added, whereas, in heat treatment, there is no addition of air during the process. Both these procedures result in producing biologically stable sludge with superb dewatering properties. Thermal conditioning enhances sludge dewaterability by introducing it to high pressure and temperature in a closed reactor vessel, coagulating solids, tearing down the sludge's gel-like structure, and enabling bound moisture to dissociate from solid particulates (Dohá6nyos et al. 2000). In addition, protein content in the sludge is hydrolyzed. Also, cells break down, and water seeps out, causing solid particles to agglomerate.

12.8 PATHOGEN REMOVAL FROM SLUDGE

Pathogen removal in sludge ensures a sufficiently poor amount of pathogenicity, decreases health hazards to the public and personnel who handle it, and prevents adverse effects on the environment if applied to the soils. The sludge application in gardens, agricultural fields, and parks has a greater degree of sanitary requirement. These needs can be addressed by pathogen reduction techniques, temporary limits on use, and public access (EPA 1992).

There are mainly four mechanisms to reduce pathogens from sewage sludge:

   a. **Thermal treatment**: The thermal method minimizes pathogenic organisms by combining two variables: temperature and sludge detention time. Since sludge has variable thermal diffusivities, which is a function of the total solids concentration, EPA recommends four distinct temporal temperature regimes which take into consideration how heat interacts with the sludge mass, mixing ease, sludge solids concentration, and heat exchange capacity.
   b. **Chemical treatment**: Alkaline chemicals elevate the pH of the sludge, negatively influencing the colloidal form of pathogenic microorganisms' cell protoplasm and providing an uncomfortable environment for the pathogens. Temperature rises can co-occur as pH rises, depending on the type of chemicals utilized. This increases the inactivation efficiency of pathogenic microorganisms and regulates the temperature–time relationship.
   c. **Biological treatment**: More research and data consistency are needed to ensure repeatability and scientific acceptability of the biological pathway for inactivating sludge pathogens. Vermiculture is among the most well-known solutions. Biodegradable matter is consumed by various detritivorous earthworms and subsequently excreted, resulting in humus (with a high agronomic value) that is easily taken up (assimilated) by the vegetation. When earthworms consume organic materials, they also consume pathogens found in the sludge, which they inactivate due to their stomach activity.
   d. **Radiation treatment**: Radiations can influence cell colloidal structures. Beta and gamma radiation can be employed to inactivate pathogenic microbes due to their detrimental effect on cell structures. Their efficiency in eliminating sludge pathogenic agents is dependent on the radiation dose used. Since radiation cannot penetrate deeply into the sludge mass, its

efficiency necessitates administering it through a shallow layer of sludge. Since the organic matter in the sludge is unaffected by radiation, pathogens may re-grow if the sludge becomes contaminated again (EPA 1992; Andreoli et al. 2007).

Solar radiation, notably ultraviolet rays, is well-known for its bactericidal properties. Many studies have reported that some pathogens are inactivated when sludge is subjected to sun radiation. Nonetheless, there are now minimal reliable available data about this topic and whether or not pathogen eradication may be accomplished to decrease the detection thresholds. A comparison of various sludge pathogen removal methods during the implementation and operation phases has been demonstrated in Tables 12.6 and 12.7.

12.9 ASSESSMENT OF SLUDGE TREATMENT AND DISPOSAL OPTIONS

Evaluating different options for sludge treatment and final disposal is highly complicated owing to the confluence of economic, technical, and ecological aspects. Ultimate sludge disposal is sometimes overlooked in the planning and designing of sewage treatment facilities on account of being expensive and complex. Operators are sometimes required to handle the ultimate disposal of sludge on an emergency basis, along with all the associated expenses, operational challenges, and unfavorable environmental implications that could undercut the benefits of sewage treatment facilities (Davis 1994; Ding et al. 2021).

12.9.1 CURRENT SLUDGE MANAGEMENT PRACTICES

Due to the sheer expansion of sewage treatment facilities, sludge output dramatically increases in developing countries. Along with the rise in sludge generation, stricter rules for higher sludge quality are increasingly being implemented to minimize the negative environmental repercussions. Mechanical dewatering devices have become more prevalent in recent years due to their increased water removal efficiency. Sludge palletization, thermal drying, and other innovative procedures aimed at improving sludge quality, such as alkaline stabilization, composting, and several patented methods, are gaining popularity.

Many states and nations have recognized that landfill disposal of sludge is not a viable/sustainable approach due to the unavailability of land, the extra cost of long-distance transport, and escalating environmental regulations. Along with the influence of recycling policies, these factors describe a clear drift toward the practice of landfills entirely for non-recyclable waste. Improvements in the energy efficiency of incineration operations and energy recovery from landfills and anaerobic processes have been reported. The incineration trend of waste is increasing in Europe while declining in the US (Kelessidis and Stasinakis 2012).

Recycling has the most excellent prospects globally since it is the most cost-effective and environmentally friendly option. This ultimate disposal practice should be considered a beneficial amendment for farming lands when implemented under

TABLE 12.6
Comparison of Different Sludge Pathogen Removal Technologies for Implementation (Andreoli et al. 2007)

Process	Area	Skilled Personnel	Chemicals	External Power	External Biomass	Construction Cost	Operation and Maintenance Cost
Composting (windrow)	SI	LNI	LNI	LNI/MI	SI	LNI	LNI
Composting (in-vessel)	MI	MI	LNI	MI	SI	MI	MI
Auto-thermal aerobic digestion	MI	MI	LNI	MI	LNI	MI	MI
Pasteurization	MI	MI	LNI	SI	LNI	MI	MI
Lime treatment	MI	LNI/MI	MI	LNI	LNI	LNI	MI
Thermal drying	LNI	SI	LNI	SI	LNI	SI	SI
Incineration	LNI	SI	LNI	SI	LNI	SI	SI

LNI, Little or Non-existent importance; MI, Moderate importance; SI, Significant importance.

TABLE 12.7
Comparison of Different Sludge Pathogen Removal Technologies for Operation (Andreoli et al. 2007)

Process	Effect against Pathogens			Product Stability	Odor Potential
	Bacteria	Virus	Eggs		
Composting (windrow)	SI/MI	MI/LNI	SI/MI	SI	SI
Composting (in-vessel)	SI	SI/MI	SI	SI	MI
Auto-thermal aerobic digestion	SI/MI	SI/MI	SI	MI	MI
Pasteurization	SI	SI	SI	MI	MI
Lime treatment	SI/MI	SI	SI/MI	MI/LNI	SI/MI
Thermal drying	SI	SI	SI	SI	LNI
Incineration	SI	SI	SI	SI	LNI

LNI, Little or Non-existent importance; MI, Moderate importance; SI, Significant importance.

strong technical guidance to ensure a safe ecological and sanitary solution that boosts farmers' income. Environmental regulations in Europe are stricter than those in the US, particularly concerning metals. The nitrogen level of sludge frequently limits its applicability. The permitted nitrogen application rate in Europe's sensitive zones has been decreased from 210 to 170 kgN/ha/year (Campbell 2000; Andreoli et al. 2007).

The most significant elements influencing the widespread acceptance of sludge are smell issues during storage and processing. Recently, stabilization, dewatering, and other sophisticated sludge processing techniques have substantially developed. Still, due to a lack of proper planning, many treatment facilities lack the basic infrastructure required for such activities and must be retrofitted to handle the generated sludge effectively.

In many developing countries, private–public water and sanitation corporations address the demands for enhanced global environmental quality. In these countries, sewage treatment facilities are progressively being installed, resulting in a rise in sludge production. Before funding and/or licensing these facilities, certain nations have recently issued land application standards that demand a practical sludge disposal strategy.

12.9.2 Disposal Options

In this study, sludge disposal represents sludge's final or ultimate destination after treatment. Primary sludge disposal methods are land application, incineration, building material production, landfills, etc. (EPA 1979).

12.9.2.1 Land Application
Land application of sewage sludge can be divided into two types (EPA 1983, 1993):

- **Beneficial use**: application of treated sludge (biosolids) when the soil conditioning and fertilizing qualities of sewage sludge are utilized.
- **Discard/landfarming**: final sludge disposal, wherein soil is used as a substratum for residue storage or decomposition, without beneficial reuse of sludge residuals.
- **Beneficial use**: Agronomically, biosolids contain nutrients needed for plant growth, and their availability in biosolids depends on the influent wastewater and the sewage and sewage treatment techniques utilized. Nitrogen and phosphorus are abundant, whereas Ca and Mg are significant, especially in biosolids treated with alkaline stabilization. Potassium exists in very small concentrations but in a form that is easily absorbed by plant roots.

  Micro-elements are present in varying amounts in sludges with Fe, Mn, Cu, and Zn often having higher amounts than B and Mo. When biosolids are used as the sole source of nitrogen for plants, the administered levels of micronutrients are usually sufficient to ensure the nutritional requirement of the plants. It is vital to note that micro-elements are required in modest amounts, and harmful consequences may occur if biosolids are administered in greater amounts than crop farming demands (Granato and Pietz 1992).

- **Landfarming**: In landfarming, no beneficial use of sludge organic matter and nutrients takes place during land disposal. The method's purpose is for sludge to be biodegraded by soil microbes present in the tillable profile, whereas metals are retained on the upper soil layers. Soil supports microbes and organic material's oxidation reactions. Since the land dedicated to the landfarming operation does not attempt to cultivate any crops, the sludge volume applied is substantially greater than that used for agricultural purposes. Nonetheless, many environmental problems are applicable to both agriculture and landfarming applications, although with varying restrictions since landfarming requires more scientific interventions to prevent environmental contamination.

Since the major aim is the biodegradation of organic material on topsoil, the parameters such as rainfall, soil pH, temperature, nutrient balance, aeration, sludge characteristics, etc., define the process efficiency. Because of the excessive sludge application rates that have been in place for many years, the principal environmental considerations are connected to the potential contamination of both surface and underground water bodies. As a result, an effective and well-maintained drainage system should be implemented from the start of the operation.

12.9.2.2   Incineration

The sludge stabilization procedure that gives the highest volume reduction is incineration (Hudson and Lowe 1996). The residual ashes are typically below 4% of the volume of dewatered sludge supplied to incineration. Incinerators may accept sludge from many treatment facilities and are typically designed to serve more than 5 lakh population equivalents, having capacities >1 ton/hour. By combustion of sludge in the presence of sufficient oxygen, incineration eliminates organic compounds and harmful organisms. To drastically limit air pollutant emissions, incinerators must utilize advanced filtration systems. To guarantee operating efficiency and safety, gases emitted into the environment should be checked regularly.

There are mainly two types of sludge incinerators that are used in the current scenario:

- Fluidized bed incinerator
- Multiple chamber incinerator

Fluidized bed incinerators are currently preferred over multiple chamber incinerators due to lower operating costs and better air quality emitted by the chimney. At temperatures exceeding 815°C, operations under the autogenous condition ensure the complete eradication of volatile organic molecules at a cost-effective rate. Incinerator emissions can be reduced by regulating combustion and employing air filters (Corella and Toledo 2000).

Some of the air pollutants released during the incineration process are NO_x, CO, furans, dioxins, volatile organic pollutants (chlorinated solvents, toluene, etc.), acidic gases (SO_2, HCl, HF), etc. Solids are also included in exhaust from incinerators, consisting of fine particulates composed of metals and suspended particles condensable at ambient temperature (Liang et al. 2021). The metal concentrations in suspended solids are directly related to incinerated sludge quality. Electrostatic precipitators are commonly used devices to remove particulates from incinerator exhaust (Shaaban 2007; Kelessidis and Stasinakis 2012).

Despite the significant decrease in sludge volume, incineration cannot be regarded as a final disposal option since residue ashes require proper disposal. The risks of inappropriate ash disposal are linked with metal leaching and subsequent absorption by flora. Ultimate dumping in landfills is the most appropriate ash disposal option.

The latest approaches use a cement and ash mixture to guarantee adequate metal retention. Co-incineration of sludge in thermoelectric power plants or cement kilns using mineral coal as fuel is also a possible solution (Kacprzak et al. 2017).

12.9.2.3 Reutilization for Production of Building Materials

The characteristics of dewatered sludge permit constructive utilization in the manufacturing of ceramics, bricks, cement, cementitious materials, lightweight aggregate, etc. According to Huang et al. (2005), up to 15% of sewage treatment sludge can be added to make first-degree bricks at temperatures normally attained in brick kilns. Teixeira et al. (2011) utilized sludge in ceramic materials and discovered that the sludge might partially replace the clays used for ceramic brick production. Solid bricks that meet Brazilian technical specifications are manufactured by incorporating up to 10% sludge and a firing temperature of <1,000°C. However, beyond this temperature, 20% incorporation of sludge is viable for manufacturing roof tiles and bricks (Ahmad et al. 2016a, b).

Furthermore, Wang et al. (1998) reported that heavy metals in the sludge are retained (get intact) in the final product because sintering at higher temperatures bonds the components together, generating sintered matrices. The sintering also results in substantial strength and exceptionally low leachability. Chen et al. (2010) discovered that when shale was substituted by sludge from 4% to 10% in cement manufacturing, the 3- and 7-day strength was greater than the corresponding specimen in all cases. However, the 28-day strength of cement increases considerably only up to 5.5%, and starts rapidly reducing after 7% replacement. Some studies reported that sludge could be substituted for sand and cementitious material in the preparation of concrete and cement mortar (Zamora et al. 2008; Rodríguez et al. 2011).

12.9.2.4 Landfilling

A landfill is a method for securely disposing of solid urban waste onto the soil with minimal health risks and ecological effects. Landfill employs engineered methods that confine the disposed waste to the minimum possible volume, enclosed with a soil stratum after every working day or at shorter intervals, if essential. There is no concern regarding nutrient uptake or sludge utilization for any practical purpose when sludge is disposed of in landfills (Wang et al. 2010; Kelessidis and Stasinakis 2012). Anaerobic biodegradation occurs in confined sludge within cells, producing a variety of byproducts, including CH_4. The disposal of sludge into landfills is determined by sludge qualities and landfill parameters. There are generally two basic forms of landfill disposal of sludge:

Dedicated sanitary landfills: Planned and built specifically to collect sewage sludge, with specific measures to deal with particular sludge qualities and to meet environmental limits. Generally, it requires thermally dried sludge with a solid content greater than 30%.

Co-disposal with municipal solid waste (**MSW**): sewage sludge is disposed of with MSW in the landfill. Combining sewage sludge with MSW tends to hasten the biodegradation mechanism as a function of the sludge inoculation potential and nitrogen concentration.

Landfills are a versatile option since they can handle a wide range of sludge quantities, absorb surplus demand from other types of ultimate destinations, and function independently of external circumstances (Griffin et al. 1992). When choosing a landfill as a final disposal option, sludge properties such as pathogen level or degree of stabilization are not significant. The availability of sufficient land near the sewage treatment facility is a substantial concern for implementing landfills. Site selection should be based on a comprehensive investigation using multi-disciplinary considerations to identify the optimum ecological and financial choice. Aside from environmental agency clearance and full compliance with stringent criteria, the surrounding community of a planned dumpsite should be listened to, and any concerns should be addressed during the planning and building phases (Kelessidis and Stasinakis 2012).

A detailed study of the environmental implications of various installation and operation phases of the landfill is one of the first tasks that must be completed during sanitary landfill construction. Landfill site selection is crucial, and if the chosen location has favorable qualities, various consequences can be avoided or reduced. When a landfill is not properly constructed or maintained, it can pollute:

- air via bad odors, poisonous gases, or particulates
- groundwater or surface water bodies, as a result of seepage or leaching due to run-off
- soil via intrusion of percolated liquids.

Non-porous liners, gas collection systems, leachate collection systems, leachate treatment facilities, landfill closure, landfill monitoring, etc., should be carefully considered when designing the landfill (Taylor and Allen 2006).

12.10 CONCLUSION

The "grave to cradle" approach is an inevitable agenda to ensure a circular economy. Where management of available freshwater is crucial for sustainability in the contemporary world, it is also equally vital to ensure proper management of sewage sludge, which is the ultimate end product of water usage. Rather than disposing of the treated sewage, the literature strongly recommends that the treated sewage be reused. When usable materials can be redeemed from sludge, the cost of pollution control and abatement methods can be reduced significantly. It also ensures the conservation of important resources. When treated sewage is used for land application, further pollution and environmental hazards caused by the fertilizer manufacturing sector can be minimized to a considerable extent.

The hurdle to achieving this environmentally friendly, sustainable, and economic management practice is that the advancement in old cities of any developing country is very high that the existing sewer lines and treatment facilities cannot handle the overload. Also, developing cities are struggling to synchronize the development rate and the rate of establishing the required infrastructure for waste management. Indigenous climate, geography, and economy are the primary factors affecting the existence of any sewage treatment technology. Hence, it would be an intricate initiative to establish appropriate sewage treatment technology whose effluent characteristics match the effluent standards.

Any environmental awareness program can be brought to light of success with proper education. Unitary efforts from administrators, politicians and the public are inevitable to achieve successful implementation of sewage treatment practices and harness their many usefulness. This statement particularly stands true in the case of developing countries. It is the responsibility of the governments to build the infrastructure for sewage collection, handling, and treatment that matches the need of the corresponding locality. Administrative personnel should be accountable for ensuring the implementation of proper standards. The public should be responsible for cooperating with the government and administrators and avoid involving neglectful activities that may pose a burden to the infrastructure.

ACKNOWLEDGEMENTS

All authors acknowledge Director, CSIR-NEERI, for granting permission to publish the book chapter. Ashish Dehal thanks the University Grants Commission (UGC-NET) for providing financial assistance in the form of fellowship (JRF-210510115261, Environmental Sciences). Rathika K thanks the CSIR-Human Resource Development Group (CSIR-HRDG) for providing financial assistance in the form of fellowship (File No: 31/GATE/19(26)/2020-EMR-I).

REFERENCES

Ahmad, T., Ahmad, K., & Alam, M. (2016a). Characterization of water treatment plant's sludge and its safe disposal options. *Procedia Environmental Sciences*, *35*, 950–955.
Ahmad, T., Ahmad, K., & Alam, M. (2016b). Sustainable management of water treatment sludge through 3'R'concept. *Journal of Cleaner Production*, *124*, 1–13.
Andreoli, C. V., Von Sperling, M., & Fernandes, F. (2007). *Sludge Treatment and Disposal*. IWA Publishing: London.
Babel, S., & del Mundo Dacera, D. (2006). Heavy metal removal from contaminated sludge for land application: A review. *Waste Management*, *26*(9), 988–1004.
Bux, M., Baumann, R., Quadt, S., Pinnekamp, J., & Mühlbauer, W. (2002). Volume reduction and biological stabilization of sludge in small sewage plants by solar drying. *Drying Technology*, *20*(4–5), 829–837.
Campbell, H. W. (2000). Sludge management–future issues and trends. *Water Science and Technology*, *41*(8), 1–8.
Cao, B., Zhang, T., Zhang, W., & Wang, D. (2021). Enhanced technology based for sewage sludge deep dewatering: A critical review. *Water Research*, *189*, 116650.

Chen, G., Lock Yue, P., & Mujumdar, A. S. (2002). Sludge dewatering and drying. *Drying Technology*, *20*(4–5), 883–916.

Chen, H., Ma, X., & Dai, H. (2010). Reuse of water purification sludge as raw material in cement production. *Cement and Concrete Composites*, *32*(6), 436–439.

Chu, C. P., & Lee, D. J. (2001). Experimental analysis of centrifugal dewatering process of polyelectrolyte flocculated waste activated sludge. *Water Research*, *35*(10), 2377–2384.

Chu, C. P., & Lee, D. J. (2002). Dewatering of waste activated sludge via centrifugal field. *Drying Technology*, *20*(4–5), 953–966.

Corella, J., & Toledo, J. M. (2000). Incineration of doped sludges in fluidized bed. Fate and partitioning of six targeted heavy metals. I. Pilot plant used and results. *Journal of Hazardous Materials*, *80*(1–3), 81–105.

CPCB (2021). National inventory of sewage treatment plants. Central pollution control board. https://cpcb.nic.in/status-of-stps/.

Crossley, I. A., & Valade, M. T. (2006). A review of the technological developments of dissolved air flotation. *Journal of Water Supply: Research and Technology—AQUA*, *55*(7–8), 479–491.

Davis, R. D. (1994). Planning the best strategy for sludge treatment and disposal operations. *Water Science and Technology*, *30*(8), 149.

Dick, R. I., Ball, R. O., & Novak, J. T. (1980). Sludge dewatering. *Critical Reviews in Environmental Science and Technology*, *10*(4), 269–337.

Ding, A., Zhang, R., Ngo, H. H., He, X., Ma, J., Nan, J., & Li, G. (2021). Life cycle assessment of sewage sludge treatment and disposal based on nutrient and energy recovery: A review. *Science of the Total Environment*, *769*, 144451.

Doháyos, M., Zabranska, J., Jeníček, P., Štěpová, J., Kutil, V., & Horejš, J. (2000). The intensification of sludge digestion by the disintegration of activated sludge and the thermal conditioning of digested sludge. *Water Science and Technology*, *42*(9), 57–64.

EPA (1979) *Process Design Manual: Sludge Treatment and Disposal*. United States Environmental Protection Agency: Washington, DC.

EPA (1983) *Land Application of Municipal Sludge: Process Design Manual*. EPA/625/1-83-016, United States Environmental Protection Agency: Cincinnati.

EPA (1987) *Design Manual: Dewatering Municipal Wastewater Sludges*. EPA/625/1-87/014, United States Environmental Protection Agency: Cincinnati.

EPA (1990) *Autothermal Thermophilic Aerobic Digestion of Municipal Wastewater Sludge*. EPA/625/10-90/007, United States Environmental Protection Agency: Cincinnati.

EPA (1992) *Control of Pathogens and Vector Attraction in Sewage Sludge*. EPA/625/R-92/013, United States Environmental Protection Agency: Cincinnati.

EPA (1993) *Land Application of Biosolids: Process Design Manual*. United States Environmental Protection Agency: Cincinnatti.

EPA (1998) *Operation of Wastewater Treatment Plants*, vol. 1, 4th edition. USEPA, California State University: Sacramento.

EPA (2002) *Wastewater Technology Fact Sheet Screening and Grit Removal*. United States Environmental Protection Agency: Cincinnati.

EPA (2004) *Estimating Sludge Management Costs: Handbook*. EPA/625/6-85/010, US Environmental Protection Agency: Washington, DC.

Gable, J. J. (2014). Technical, environmental, and economic assessment of sludge thickening processes: A comparison of conventional thickening and energy-efficient centrifugal thickening technologies (Doctoral dissertation). College of Engineering, University of Wisconsin-Madison.

Granato, T. C., & Pietz, R. I. (1992). Chapter 9: Sludge application to dedicated beneficial reuse sites. In: *Municipal Sewage Sludge Management: Processing, Utilization, and Disposal*, C. Lue-Hing, D. R. Zenz, & R. Kuchenrither (Eds.), Water Quality Management Library, vol. 4. Technomic Publishing: Lancaster, PA, pp. 417–454.

Griffin, R. A., Lue-Hing, C., Sieger, R. B., Uhte, W. R., & Zenz, D. (1992). Municipal sewage sludge management at dedicated land disposal sites and landfills. In: *Municipal Sewage Sludge Management: Processing, Utilization, and Disposal*, C. Lue-Hing, D. R. Zenz, & R. Kuchenrither (Eds.), Water Quality Management Library, vol. 4. Technomic Publishing: Lancaster, PA, pp. 409–486.

Huang, C., Pan, J. R., & Liu, Y. (2005). Mixing water treatment residual with excavation waste soil in brick and artificial aggregate making. *Journal of Environmental Engineering*, *131*(2), 272–277.

Hudson, J. A., & Lowe, P. (1996). Current technologies for sludge treatment and disposal. *Water and Environment Journal*, *10*(6), 436–441.

Jenicek, P., Koubova, J., Bindzar, J., & Zabranska, J. (2010). Advantages of anaerobic digestion of sludge in microaerobic conditions. *Water Science and Technology*, *62*(2), 427–434.

Jones, E. R., van Vliet, M. T., Qadir, M., & Bierkens, M. F. (2021). Country-level and gridded estimates of wastewater production, collection, treatment and reuse. *Earth System Science Data*, 13(2), 237–254.

Kacprzak, M., Neczaj, E., Fijałkowski, K., Grobelak, A., Grosser, A., Worwag, M., ... & Singh, B. R. (2017). Sewage sludge disposal strategies for sustainable development. *Environmental Research*, *156*, 39–46.

Kelessidis, A., & Stasinakis, A. S. (2012). Comparative study of the methods used for treatment and final disposal of sewage sludge in European countries. *Waste Management*, *32*(6), 1186–1195.

Kozak, J., Patel, K., Abedin, Z., Lordi, D., O'Connor, C., Granato, T., & Kollias, L. (2011). Effect of ferric chloride addition and holding time on gravity belt thickening of waste activated sludge. *Water Environment Research*, *83*(2), 140–146.

Liang, Y., Xu, D., Feng, P., Hao, B., Guo, Y., & Wang, S. (2021). Municipal sewage sludge incineration and its air pollution control. *Journal of Cleaner Production*, *295*, 126456.

Loranger, É., Lanouette, R., Bousquet, J. P., & Martinez, M. (2019). Dewatering parameters in a screw press and their influence on the screw press outputs. *Chemical Engineering Research and Design*, *152*, 300–308.

Mahmoud, A., Olivier, J., Vaxelaire, J., & Hoadley, A. F. (2010). Electrical field: A historical review of its application and contributions in wastewater sludge dewatering. *Water Research*, *44*(8), 2381–2407.

Manfio, D. V., Tonetti, A. L., & Matta, D. (2018). Dewatering of septic tank sludge in alternative sludge drying bed. *Journal of Water, Sanitation and Hygiene for Development*, *8*(4), 792–798.

Metcalf, L., Eddy, H. P., & Tchobanoglous, G. (1991). *Wastewater Engineering: Treatment, Disposal, and Reuse*, vol. 4. McGraw-Hill: New York.

Mowla, D., Tran, H. N., & Allen, D. G. (2013). A review of the properties of biosludge and its relevance to enhanced dewatering processes. *Biomass and Bioenergy*, *58*, 365–378.

Rodríguez, N. H., Martínez-Ramírez, S., Blanco-Varela, M. T., Guillem, M., Puig, J., Larrotcha, E., & Flores, J. (2011). Evaluation of spray-dried sludge from drinking water treatment plants as a prime material for clinker manufacture. *Cement and Concrete Composites*, *33*(2), 267–275.

Rusten, B., & Ødegaard, H. (2006). Evaluation and testing of fine mesh sieve technologies for primary treatment of municipal wastewater. *Water Science and Technology*, *54*(10), 31–38.

Shaaban, A. F. (2007). Process engineering design of pathological waste incinerator with an integrated combustion gases treatment unit. *Journal of Hazardous Materials*, *145*(1–2), 195–202.

Song, L. J., Zhu, N. W., Yuan, H. P., Hong, Y., & Ding, J. (2010). Enhancement of waste activated sludge aerobic digestion by electrochemical pre-treatment. *Water Research*, *44*(15), 4371–4378.

Taylor, R., & Allen, A. (2006). Waste disposal and landfill: Potential hazards and information needs. In: *Protecting Groundwater for Health: Managing the Quality of Drinking-Water Sources*, O. Schmoll, G. Howard, J. Chilton, & I. Chorus (Eds). World Health Organization: Geneva, Switzerland, pp. 339–362.

Teixeira, S. R., Santos, G. T. A., Souza, A. E., Alessio, P., Souza, S. A., & Souza, N. R. (2011). The effect of incorporation of a Brazilian water treatment plant sludge on the properties of ceramic materials. *Applied Clay Science, 53*(4), 561–565.

To, V. H. P., Nguyen, T. V., Vigneswaran, S., & Ngo, H. H. (2016). A review on sludge dewatering indices. *Water Science and Technology, 74*(1), 1–16.

Tuan, P. A., Mika, S., & Pirjo, I. (2012). Sewage sludge electro-dewatering treatment: A review. *Drying Technology, 30*(7), 691–706.

Vaxelaire, J., & Olivier, J. (2006). Conditioning for municipal sludge dewatering. From filtration compression cell tests to belt press. *Drying Technology, 24*(10), 1225–1233.

Voutchkov, N. (2005). Clarifier performance monitoring and control. *Proceedings of the Water Environment Federation, 2005*(9), 6375–6399.

Wang, K. S., Chiang, K. Y., Perng, J. K., & Sun, C. J. (1998). The characteristics study on sintering of municipal solid waste incinerator ashes. *Journal of Hazardous Materials, 59*(2–3), 201–210.

Wang, W., Luo, Y., & Qiao, W. (2010). Possible solutions for sludge dewatering in China. *Frontiers of Environmental Science & Engineering in China, 4*(1), 102–107.

Wang, Y., Zhou, Y., Feng, D., Wu, C., Wang, X., & Min, F. (2019). Effects of chemical conditioners on deep dewatering of urban dewatered sewage sludge in the temporary sludge lagoon. *Journal of Environmental Engineering, 145*(10), 04019063.

Weihua, P., Pawlowski, J., Eads, S., & Cammack, B. (2010). In the thick of things: A new guideline helps operators optimize gravity belt thickener performance. *Water Environment & Technology, 10*, 30–33.

Wu, B., Dai, X., & Chai, X. (2020). Critical review on dewatering of sewage sludge: Influential mechanism, conditioning technologies and implications to sludge re-utilizations. *Water Research, 180*, 115912.

Yang, S., Phan, H. V., Bustamante, H., Guo, W., Ngo, H. H., & Nghiem, L. D. (2017). Effects of shearing on biogas production and microbial community structure during anaerobic digestion with recuperative thickening. *Bioresource Technology, 234*, 439–447.

Zamora, R. R., Alfaro, O. C., Cabirol, N., Ayala, F. E., & Moreno, A. D. (2008). Valorization of drinking water treatment sludges as raw materials to produce concrete and mortar. *American Journal of Environmental Sciences, 4*(3), 223.

13 Enhanced Biogas Production from Treatment Plant Sludges

*Mohd Imran Siddiqui, Izharul Haq Farooqi,
Hasan Rameez, and Farrukh Basheer*
Aligarh Muslim University

CONTENTS

13.1 Introduction 234
13.2 Anaerobic Digestion 236
    13.2.1 Microbiology of Anaerobic Digestion 237
13.3 Pre-Treatment...240
    13.3.1 Biological Pre-Treatment 241
        13.3.1.1 Aerobic Pre-Treatment......................................242
        13.3.1.2 Anaerobic Pre-Treatment..................................242
        13.3.1.3 Enzyme Pre-Treatment.....................................242
        13.3.1.4 Fungal Pre-Treatment.......................................242
    13.3.2 Chemical Pre-Treatment 243
        13.3.2.1 Alkaline and Acidic Pre-Treatment 243
        13.3.2.2 Fenton Pre-Treatment.......................................243
        13.3.2.3 Ionic Liquid Pre-Treatment.............................244
        13.3.2.4 Ozonation Pre-Treatment 244
    13.3.3 Physical Pre-Treatment 244
        13.3.3.1 Mechanical Pre-Treatment................................244
        13.3.3.2 Microwave Pre-Treatment.................................245
        13.3.3.3 High-Pressure Homogenization Pre-Treatment 245
        13.3.3.4 Pulse Electric Field Pre-Treatment 245
        13.3.3.5 Thermal Pre-Treatment.....................................246
        13.3.3.6 Ultrasonic Pre-Treatment..................................246
    13.3.4 Combined Pre-Treatment...246
13.4 Pre-Treatment Challenges and Its Scope 247
13.5 Conclusion 249
References..249

DOI: 10.1201/9781003202431-13

13.1 INTRODUCTION

With the fast expansion of the global population in the past few decades, energy demand is also proliferating, resulting in the depletion of fossil fuels at alarming rates. To meet this massive energy demand, there has been a rapid increase in the consumption of fossil fuels. For years, the extensive use of fossil fuels to meet the ever-increasing energy demands has raised worries about global warming, national energy security, and a continuous decline in fossil reserves. Since these fossil fuel reserves are limited and non-replenishable, consuming them injudiciously will soon deprive future generations. To meet the present energy needs while also safeguarding the sustainability of the resources for future generations, there is a dire need to minimize the dependence on non-renewable energy sources. This dilemma has prompted researchers to shift their focus toward eco-friendly and renewable energy sources such as hydel energy, wind energy, biomass energy, solar energy, etc. The researchers are making a conscious effort to develop techniques to harness these energies economically. There is worldwide support for developing such technologies to reduce the burden on non-renewable energy sources while switching to more sustainable alternatives with minimum environmental implications (Charters, 2001; Chynoweth et al., 2001).

The last few decades have witnessed tremendous growth in urbanization and industrialization, leading to a multi-fold increase in water consumption. This increased water consumption for various domestic, commercial, or industrial activities has ultimately led to increased wastewater production. Wastewater treatment plants deploy various physical, chemical, and biological processes to treat this wastewater before releasing them into local waterways. These treatment processes, particularly the conventional biological processes such as activated sludge, produce a large volume of biomass or sludge. This sludge is primarily composed of organic detritus, water, and active microorganisms, which are generally protein (30%), carbs (40%), and lipids (30%), all of which are present in particle form (Lin et al., 1999). As shown in Table 13.1, typical sewage sludge is composed of high-concentration

TABLE 13.1
Sewage Sludge Characterization as per EPA 1995

Parameters	Unit	Value
Solids	%	02–12 wet sludge
		12–40 dry sludge
Organic matter	% (dry solid basis)	75.0–80.0
Fecal coliforms	/100 mL	10^9
Virus	/100 mL	2,500–70,000
Helminth	/100 mL	200–1,000
N-total	% (dry solid basis)	0.10–17.6
P	% (dry solid basis)	0.10–14.3
K	% (dry solid basis)	0.02–2.64
Cr	mg/kg (dry solid basis)	119 (mean concentration)
Cu	mg/kg (dry solid basis)	741 (mean concentration)
Hg	mg/kg (dry solid basis)	5.2 (mean concentration)
Pb	mg/kg (dry solid basis)	134.4 (mean concentration)

organic matter, a plethora of pathogens, and even some toxic metals. For this reason, the efficient management of sludge remains a considerable challenge for academicians and experts. Many new techniques have lately been developed; however, they are challenging to put into practice on a broad scale. Landfill, incineration, and land application have all been used for decades as sludge disposal methods. These practices are neither environmentally friendly nor sustainable due to specific issues such as emissions of greenhouse gas (CH_4 and CO_2), soil contamination, water pollution, etc., resulting in various environmental and health hazards. Before disposing, the sludge must be treated to bring down its organic content to an extent where it does not create any harm or nuisance. Sludge treatment and disposal rules are getting more rigorous day by day, necessitating more efficient solutions (Christensen and Dick, 1985). Sludge treatment alone accounts for approximately 50%–60% of overall treatment costs (Appels et al., 2008; Pilli et al., 2011).

When this sludge is allowed to degrade under controlled conditions, it becomes a source of energy. Anaerobic digestion is one of the most extensively used methods for treating biodegradable organic waste such as sewage, municipal trash, industrial waste, manure, plant residues, kitchen wastes, and agricultural wastes. This process reduces the volume of the substrate and produces biogas and nutrient-rich fertilizer as the end products. Other value-added products such as bio-alcohol (Moya et al., 2016), bio-hydrogen (Asadi and Zilouei, 2017), bio-diesel (Duman et al., 2019; Uğuz et al., 2019), and biogas (Sztancs et al., 2020) are also made. Table 13.2 shows the biogas yields of various organic wastes. Because of its high organic content, activated sludge is considered a rich substrate for the anaerobic digestion process (Ebenezer et al., 2015).

TABLE 13.2
Biogas Yield of Different Organic Waste Types (Ariunbaatar et al., 2014)

Organic Waste	% Solid (Total)	% Solid (Volatile)	Biogas Yield (m³/kg VS)
Brewery spent grain	20–26	80–95	0.5–1.1
Corn Silage	20–40	94–97	0.6–0.7
Cattle manure (liquid)	6–11	68–85	0.1–0.8
Cattle excreta	25–30	75–85	0.6–0.8
Chicken excreta	10–29	67–77	0.3–0.8
Fermentation Residues	4–8	90–98	0.4–0.7
Grass Cutting (from lawns)	20–37	86–93	0.7–0.8
Grass Silage	21–40	87–93	0.6–0.8
Horse excreta	25–30	70–80	0.4–0.6
Municipal organic waste	15–30	80–95	0.5–0.8
Municipal wastewater sludge	3–5	75–85	0.3–0.5
Pig manure (liquid)	2–13	77–85	0.3–0.8
Pig excreta	20–25	75–80	0.2–0.5
Rumen content (untreated)	12–16	85–88	0.3–0.6
Straw from cereals	~86	89–94	0.2–0.5
Sheep excreta	18–25	80–85	0.3–0.4
Vegetable Wastes	5–20	76–90	0.3–0.4
Whey	4–6	80–92	0.5–0.9
Yeast	10–18	90–95	0.5–0.7

In anaerobic digestion, the rate of generation of CH_4 from carbs, proteins, and lipids is 370, 740, and 1,014 NL per kg of volatile solids (VS) (Harris and McCabe, 2015). Biogas is a flammable gas, mainly made up of CH_4, CO_2, and trace amounts of H_2, NH_3, and H_2S. CH_4 has a 35.8 kJ/L heat value, while H_2 has a 10.8 kJ/L heat value. Considering the potential of biogas as a fuel and the gigantic quantities in which the sludge (substrate for biogas production) is produced, this energy is considered to be one of the cleaner alternatives to replace the conventional non-renewable ones. Because of its minimal carbon footprint and sustainable production, it can fulfill energy requirements to some extent.

However, there are certain difficulties associated with the anaerobic digestion of sludge. Due to its complicated structure and sluggish biodegradability, poor CH_4 yield, long retention times, and partly digested substrates are often faced. Certain pre-treatment techniques are applied to the sludge prior to anaerobic digestion to address these issues. To break sludge flocks, pre-treatment procedures (biological, chemical, physical, etc.) may be applied alone or in combination, followed by hydrolysis of complex organics to simple organics that are readily accessible to microorganisms. Consequently, enhanced biogas yield and higher solid reduction is achieved (Carrère et al., 2010). Microbial cells abound in secondary sludge, and their cell walls prevent exoenzyme decomposition (Ebenezer et al., 2015). The predominant organic part of the floc structure is an exocellular polymeric material. The digestibility of activated sludge is limited by the binding mechanism of exocellular polymeric molecules (Şahinkaya and Sevimli, 2013). The chosen pre-treatment procedure is designed to break down cell walls and help in the release of intracellular components into the digestive environment (Karuppiah and Azariah, 2019), hence increasing biodegradability and biogas production. This chapter delves into the different sludge pre-treatment methods for improved bio-methanation, as well as provides a quick look into anaerobic digestion. This chapter's purpose is to give an in-depth look at contemporary pre-treatment technology development trends and earlier works done in the same area.

13.2 ANAEROBIC DIGESTION

Anaerobic digestion has been a continually growing method for organic waste treatment and stabilization since the 1970s (van Lier et al., 2001). Due to the oil crisis, waste disposal issues, renewable energy legislation, and other factors, development reached its pinnacle in the previous decade. It is a complicated biochemical process, including microbial activity and syntrophic consortium formation. It is chosen over aerobic digestion because of its cheap operating costs, low energy footprint, and moderate performance, particularly sludge digestion and stabilization (Appels et al., 2008). According to most academics and experts, anaerobic digestion is a sustainable and effective process for sludge treatment and disposal (Pilli et al., 2011). This method has much potential for many types of organic waste management and can be employed as a commercially viable renewable bio-energy source.

Anaerobic digestion is a microbial process in which complex organic molecules are transformed into CH_4 and CO_2 without free oxygen. Hydrolysis, acidogenesis, acetogenesis, and methanogenesis are the four basic steps of the process. Due to the presence of high-molecular-weight organics, macro-organics (extracellular polymeric substances), toxic byproduct formation (complex heterocyclic compounds), and undesirable volatile fatty acids formed during hydrolysis, many researchers have reported that solubilization or microbial hydrolysis is the rate-limiting step in the process of anaerobic digestion (Dohányos et al., 2004; Li and Noike, 1992; Tiehm et al., 2001; Wang et al., 1999). In contrast, methanogenesis is the rate-limiting step in the case of a readily biodegradable substrate (Gavala et al., 2003; Rozzi and Remigi, 2004). The hydrolysis of carbohydrates takes a few hours, proteins and lipids take a few days, while lignin and lignocellulose require several days (Atelge et al., 2020b). The digestive process will also be hindered if the substrate includes a lot of lignin and lignocellulose. Microorganisms produce a number of hydrolytic enzymes, but they are inadequate to break down the substrate's very complicated structure (Christy et al., 2014); therefore, the hydrolysis phase is the rate-determining step. As a result, identifying the composition of a substrate is critical in assessing its biochemical methane potential. The methane potential of a substrate may be determined by its macromolecular composition. Hydrolysis molecules like carbohydrates, proteins, and lipids have a high biodegradability, while hydrolysis compounds like lignocellulose and biopolymers have a low biodegradability (Edwiges et al., 2018). Seasonality, collection locations, human behavior, and storage conditions may all have an impact on the quality of the same waste. Furthermore, substrates may be broken down fast in the hydrolysis phase due to the acidogenesis phase, resulting in an uneven pH in the reactor, leading methanogenic bacteria to be harmed and the process to cease (Fisgativa et al., 2016). Pre-treatment and/or varied anaerobic digestion setups are necessary to counteract the substrate's negative effects.

The anaerobic degradation of organic waste may be done in batch or continuous-flow reactors. Fresh substrates are put into batch reactors. The digestate is totally removed after the completion of the digestion process (as determined by the stoppage of biogas generation). The procedure then begins all over again. A batch reactor has the benefit of a cheap initial investment cost. The operating expenses, on the other hand, are quite substantial. Continual-flow digester technology refers to the feeding of fresh substrates to digesters on a continuous basis. Biogas is produced continuously and predictably by the system. Continuous-flow systems may contain many tanks and reactors in various designs, such as horizontal and vertical flow (Al Seadi, 2008).

The anaerobic digestion process is further divided into dry and wet processes based on the moisture content of the biomass. In the wet process, the total solid content of the biomass is less than or equal to 10%. The mixed digester materials have less viscosity, so it makes it easy to mix digester material in the reactor. If the solid content of the biomass is between 15% and 35%, the process is categorized as a dry process (Weiland, 2010). While the wet anaerobic digestion process runs continuously, the dry anaerobic digestion process is operated in a continuous or batch-fed mode. The wet anaerobic digestion process is mostly used to digest agricultural waste (Murphy et al., 2011). Table 13.3 depicts the comparison between the wet and dry processes.

13.2.1 MICROBIOLOGY OF ANAEROBIC DIGESTION

The anaerobic digestion process occurs in four steps, i.e., hydrolysis, acidogenesis, acetogenesis, and methanogenesis. A consortium of microorganisms carries out

TABLE 13.3

Difference between Wet and Dry Anaerobic Digestion Process (Meegoda et al., 2012)

Properties	Process (Wet)	Process (Dry)
Water (%)	>90%	~65% to 85%
Total solid content	Up to 10%	~15% to 35%
Required reactor volume	Large	Small
Mixing conditions	Simple	Problematic
Process mode	Continuous	Batch or continuous
Solid–liquid separation	Problematic	Simple

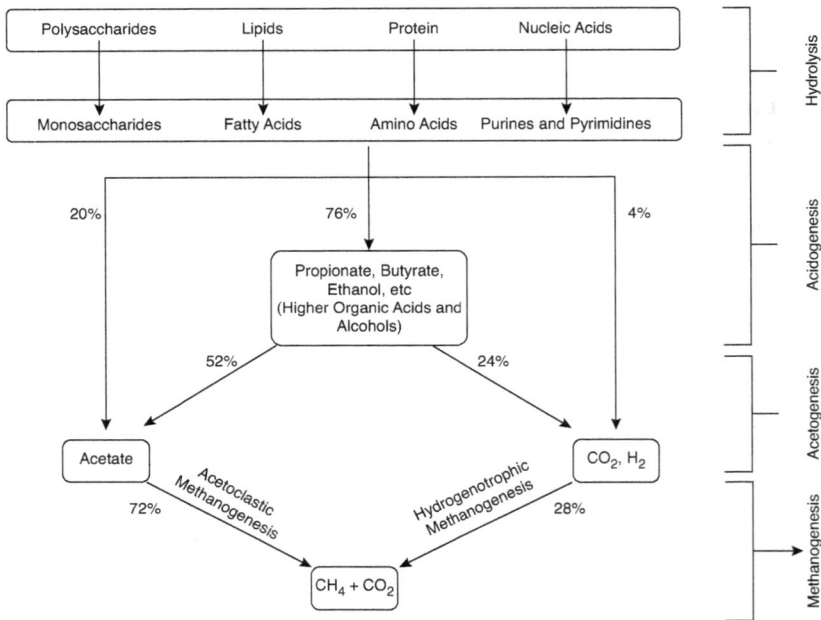

FIGURE 13.1   Schematic depiction of different stages of anaerobic digestion.

these digestion processes. Bacteria and Archaea are two prokaryotic kingdoms that act together to digest organic matter anaerobically. Figure 13.1 depicts the four stages of anaerobic degradation, along with the conversion percentages of organic matter through different pathways (Metcalf et al., 1991).

Organic matter primarily comprises complex insoluble polymers such as carbohydrates, lipids, and proteins. As these insoluble compounds cannot penetrate the microbial cell membrane, the microbes are unable to break down these compounds. The first step is hydrolysis in the anaerobic digestion of the sludge, in which the complex polymers are first converted into the simpler mono- and oligomers that the microbes can utilize. Polysaccharides are broken down into monosaccharides, lipids are broken down into long-chain fatty acids, and proteins are converted into amino acids (Gumisiriza et al., 2017). The hydrolysis of cellulose to glucose is shown in Eq. (13.1) (Ostrem et al., 2004; Zupančič and Grilc, 2012). This breakdown occurs under the action of extracellular enzymes secreted by the hydrolytic bacteria (Shah et al., 2014). The major bacterial species responsible for hydrolysis include *Clostridium, Bacteroides, Cellulomonas, Succinivibrio, Erwinia, Prevotella, Ruminococcus, Firmicutes, Microbispora, Acetovibrio, Fibrobacter,* etc. (Guo et al., 2013, 2015).

$$C_6H_{10}O_5(\text{Cellulose}) + H_2O \rightarrow C_6H_{12}O_6(\text{Glu cos e}) \tag{13.1}$$

The second step in the anaerobic digestion process is acidogenesis, during which the products of hydrolysis are converted into short-chain volatile fatty acids (VFAs) (acetic acid), higher VFAs (propionic acid and butyric acid), and alcohols (C_2H_5OH) (Karuppiah and Azariah, 2019). Other byproducts such as NH_3, CO_2, and H_2 gas are also formed during acidogenesis. A few chemical reactions during acidogenesis are shown in Eqs. (13.2)–(13.4) (Barua and Dhar, 2017; Dang et al., 2016; Rotaru et al., 2014a, b). The bacterial groups responsible for carrying out the process of acidogenesis are called acidogens. Such bacterial species include *Clostridium, Lactobacillus, Peptoccus, Geobacter, Bacteroides, Desulfovibrio, Desulfobacter, Sarcina, Phodopseudomonas, Eubacterium,* etc. (Gonzalez-Fernandez et al., 2015; Sun et al., 2015).

$$C_6H_{12}O_6(\text{Cellulose}) \rightarrow 2CH_3CH_2OH + 2CO_2 \tag{13.2}$$

$$C_6H_{12}O_6(\text{Cellulose}) + 2H_2 \rightarrow 2CH_3CH_2COOH + 2H_2O \tag{13.3}$$

$$C_6H_{12}O_6(\text{Cellulose}) \rightarrow 3CH_3COOH \tag{13.4}$$

The third step of the anaerobic digestion process is acetogenesis. During acetogenesis, certain bacterial species known as acetogens oxidize the higher VFAs and alcohols to acetic acid, carbon dioxide, and hydrogen gas (Appels et al., 2008). Chemical reactions during acetogenesis are shown in Eqs. (13.5) and (13.6) (Ostrem et al., 2004; Zupančič and Grilc, 2012). This conversion is greatly favored by the low concentration levels of the end products of the reaction. The forward reaction is favored at acetate concentrations of 10^{-4} to 10^{-1} mol/L and a H_2 partial pressure of 10^{-3} to 7×10^{-3} atm for butyrate and 10^{-6} to 7×10^{-4} atm for propionate (Dhar et al., 2015; McCarty and Smith, 1986). The bacteria responsible for acetogenesis include *Syntrophus, Moorella, Syntrophobacter, Pelotomaculum, Syntrophomonas, Desulfovibrio, Syntrophothermus,* etc. (Cai et al., 2016; Guo et al., 2015).

$$CH_3CH_2COOH + 3H_2O \rightarrow CH_3COOH + H_2CO_3 + 3H_2 \tag{13.5}$$

$$CH_3CH_2OH + 2H_2O \rightarrow CH_3COOH + 3H_2O \tag{13.6}$$

The final step is methanogenesis in which various methanogenic archaea convert the end products of acetogenesis into methane and carbon dioxide (Zhen et al., 2017).

The chemical reactions taking place during methanogenesis are depicted in Eqs. (13.7)–(13.9) (Ostrem et al., 2004; Zupančič and Grilc, 2012). The conversion of acetate to methane, known as acetoclastic methanogenesis, is carried out by two major archaea, namely *Methanosarcina* and *Methanosaeta*. In contrast, the conversion of H_2 and CO_2 to CH_4 is carried out via hydrogenotrophic methanogenesis, by *Methanobacterium* and *Methanoculleus* (Lu et al., 2015; Okudoh et al., 2014).

$$CH_3COOH \rightarrow CH_4 + CO_2 \tag{13.7}$$

$$CO_2 + 4H_2 \rightarrow CH_4 + H_2O \tag{13.8}$$

$$2CH_3CH_2OH + CO_2 \rightarrow CH_4 + 2CH_3COOH \tag{13.9}$$

13.3 PRE-TREATMENT

Pre-treatment of sludge or other organics is a procedure that converts non-biodegradable or hardly biodegradable organics to readily biodegradable organics by cell lysis or rupturing and releasing intercellular materials in the aqueous phase. Pre-treatment is used to prepare substrates for the microorganism in the anaerobic digestion process. The pre-treated substrate's molecular size is reduced, and the simpler and smaller composition of the substrate becomes more accessible to the bacterial consortia in the reactor (Li et al., 2017). As a consequence, the surface area of the pre-treated substrate increases and enzymic activity increases; as a result, the solubility of the substrate, rate, and, in certain situations, the extent of biodegradability increases. Higher VS decomposition, quicker biogas, and methane generation rate and ultimate yields, and a reduced volume of residual solid for final disposal after the anaerobic digestion process are indicators of these improvements (Mills et al., 2014; Zhen et al., 2017). In 1979, researchers also developed biochemical methane potential (BMP) tests and anaerobic toxicity test procedures to measure the anaerobic treatability of organics (Lin et al., 1999). After preliminary screening research, the BMP test was judged to be the best technique for a quick evaluation of the anaerobic digestion process. As a result, a growing number of researchers are using the BMP test to evaluate the effectiveness of anaerobic digestion and system kinetics (Lin et al., 1999). Furthermore, pre-treatment increases organic loading capacity and reduces hydraulic retention time, resulting in a smaller reactor volume/footprint. As a result, the capital cost of a biogas plant is lowered. Several pre-treatment procedures have been developed as a result of these factors. Pre-treatment procedures are classified into four categories: biological, chemical, physical, and combination. Pre-treatment may be used on any form of sludge, including primary, secondary (aerobic/anaerobic), chemical, and mixed sludge. Most of the study on sludge pre-treatment was done on waste-activated sludge, followed by mixed and primary sludge (Neumann et al., 2016). Because of the lack of readily biodegradable organic matter in anaerobic digestion, sludge pre-treatment is required. For non-pre-treated sludge, the first-order digestion rate constant was determined as 0.15/day (Shimizu et al., 1993). Methods for sludge pre-treatment have been evolving since the late 1970s. The first

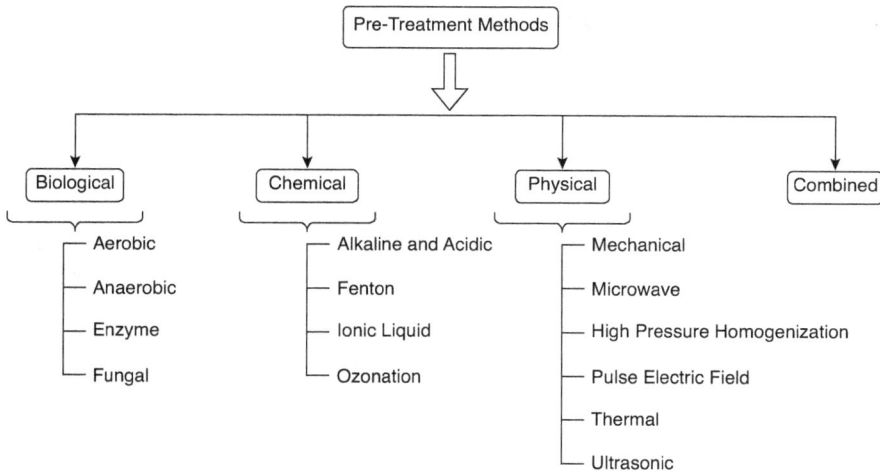

FIGURE 13.2 Broad classification of pre-treatment.

pre-treatment of sludge was thermal pre-treatment, resulting in significant improvements in anaerobic bio-degradability and dewaterability (Haug et al., 1978).

To be cost-effective, pre-treatment procedures must satisfy specific requirements, including decreasing the size of the substrate and increasing its porosity, increasing the solubility and degradability of a given substrate, eliminating inhibitory generation, and requiring low energy input (Singh et al., 2015). Additionally, certain hazardous compounds such as aliphatic acids and furans may be generated during the pre-treatment process. Eliminating them adds to the expense of pre-treatment procedures (Achinas and Euverink, 2016). Appropriate pre-treatment procedures must be chosen based on the physical and chemical features of the substrate in order to achieve these criteria. They may be classified into four broad categories based on their methods: biological, chemical, physical, and combined pre-treatments, as seen in Figure 13.2.

This chapter discusses several pre-treatment methods, including physical, chemical, biological, and combined. Almost all of the research indicates, pre-treatment prior to digestion results in a decrease in hydraulic retention time and an increase in biogas production.

13.3.1 BIOLOGICAL PRE-TREATMENT

Biological pre-treatment is an ecologically beneficial procedure that involves organisms directly or indirectly. To optimize the anaerobic digestion process, scientists have attempted to maximize the theoretical yield of a given substrate by using pre-digestion (aerobic or anaerobic), the addition of enzymes, and biosurfactant routes. The primary goal of biological pre-treatment is to degrade insoluble materials such as cellulose and protein using exoenzymes (Oh et al., 2000). Some different biological pre-treatments are discussed here.

13.3.1.1 Aerobic Pre-Treatment

Because of its self-heating capacity and quick breakdown rate, aerobic digestion has been offered as an alternative to physical and chemical pre-treatments (Jang et al., 2014). The presence of proteolytic bacteria in activated sludge, such as *Geobacillus stearothermophilus*, allows for aerobic thermophilic pre-treatment without the need for bio-augmentation, which has important implications for the whole stabilization process (Dumas et al., 2010). Furthermore, aerobic processes may break down materials that would not deteriorate in anaerobic circumstances, improving overall stability (Carrère et al., 2010). Aerobic pre-treatment is carried out in the presence of O_2 using aerobic and anaerobic mixed cultures. Organisms eat organic matter and generate CO_2, H_2O, and nitrate in the process (Fitzgerald, 2013). Following this process, organic materials, particularly lignocellulose, become more available to anaerobic digestion microbes. This pre-treatment enables the anaerobic digestion process to attain stable conditions for methane generation more quickly (Suchowska-Kisielewicz et al., 2013).

13.3.1.2 Anaerobic Pre-Treatment

Anaerobic pre-treatment is essentially a technique of phase separation. Pohland and Ghosh (1971) first proposed the concept of phase separation as a two-phase system. The concept was to run an anaerobic digestion process in stages to optimize it for many different cultures participating in biochemical reactions throughout the process. Each phase has its own set of optimal parameters for microbes, including hydraulic retention time requirements, temperature, and pH. In general, the first two steps (Hydrolysis and Acid production) are carried out concurrently. In contrast, the methanogenesis phase is carried out in a separate reactor, referred to as a di-phase digester (Abbasi et al., 2011). The hydrolysis phase is kept between 5.5 and 6.5 pH to optimize biodegradation, whereas the methanogenic phase is kept between 6.8 and 7.2 pH (Weiland, 2010). Due to the demand for more sophisticated control systems, this phase separation process is often utilized on a large scale for municipal and industrial wastes. This pre-treatment is known as a multi-phase fermentation or temperature-phased anaerobic digestion (TPAD) because digesters have variable temperatures and pH levels. Pre-treatment also boosts biogas methane content because of CO_2 availability in three distinct forms: at high pH (CO_3), at neutral pH (HCO_3), and at low pH (CO_2) (Bochmann and Montgomery, 2013).

13.3.1.3 Enzyme Pre-Treatment

By introducing enzymes, polymers in the feedstock, particularly those derived from lignocellulose, may be split down into minute pieces. Pre-treatment with enzyme improves sludge solubilization. Different enzymes such as a-amylase, cellulase, lipase, endo-xylanase, protease, and dextranase are introduced to the anaerobic digester according to the organic substrate combinations (Burgess and Pletschke, 2008; Wawrzynczyk et al., 2008).

13.3.1.4 Fungal Pre-Treatment

As a component of biological pre-treatment, fungal pre-treatment has many benefits, including fast growth and degradation of lignocellulose via the use of numerous white-, brown-, and soft-rot fungi (Shirkavand et al., 2017). Numerous process factors, including the size of the substrate, temperature, duration of pre-treatment, and moisture content, all affect the breakdown of lignin during this pre-treatment. While this pre-treatment does not require any additional chemicals and consumes little energy, some disadvantages include the requirement for specific environmental conditions, a large pre-treatment area, a relatively long treatment time, and the requirement for specific species to grow continuously (Sindhu et al., 2016).

13.3.2 CHEMICAL PRE-TREATMENT

Chemical pre-treatment is a frequently utilized technique for successfully breaking down complicated materials into smaller ones and decreasing the HRT of the digester (Zhou et al., 2012). Chemical pre-treatment using different acids and alkalis, Fenton, ozonation, and ionic liquids is used to promote biodegradability under various circumstances. As a result, it boosts the substrate's biodegradability and digestibility. Some types of chemical pre-treatment are discussed here.

13.3.2.1 Alkaline and Acidic Pre-Treatment

Chemical pre-treatment with varying concentrations of acids and bases is employed under various situations to increase waste biodegradability. Hydrolysis is resistant to lignocellulosic materials. The alkali pre-treatment removes the acetate structure from the feedstock, making it readily accessible to hydrolytic enzymes (Karp et al., 2015). The primary purpose of alkaline and acidic pre-treatment is to generate hydroxyl radicals that destroy or liquefy cell walls (Carlsson et al., 2012). Due to the increased surface area created by these processes, the substrate becomes more accessible to microorganisms (Banu and Kavitha, 2017). When an extreme pH is applied, proteins denature, and the substrate lipid saponifies (Carlsson et al., 2012). The extracellular polymeric framework of the substrate cells is damaged, and the substrate cells cannot maintain their turgor pressure (Banu and Kavitha, 2017). As a result, intracellular components are released into the surrounding media. It improves the substrate's biodegradability and digestibility. Additionally, the alkaline pre-treatment causes the substrate to expand, similar to thermal pre-treatment. The solution may be made using varying amounts of CaO, KOH, and NaOH for alkali pre-treatment (Bochmann and Montgomery, 2013) or acids such as H_2SO_4, HCl, H_3PO_4, and HNO_3 for acidic pre-treatment (Zhen et al., 2017). Historically, the pH ranges used for alkaline and acidic pre-treatments were pH 10–13 and pH 1–4, respectively (Pedersen et al., 2011).

13.3.2.2 Fenton Pre-Treatment

Another chemical pre-treatment method is the Fenton procedure, which involves the addition of hydrogen peroxide and ferrous iron. At a pH of between 2 and 3, ferrous iron interacts with H_2O_2 to form highly reactive hydroxyl (OH) radicals. The OH radicals attack and degrade different chemical compounds without producing any harmful and inhibiting byproducts (Atay and Akbal, 2016; Erden and Filibeli, 2011). The concentration of H_2O_2, the contact duration, the initial pH, and the temperature all affect this pre-treatment.

13.3.2.3 Ionic Liquid Pre-Treatment

Pre-treatment with ionic liquids is another promising technique that utilizes ionic solvents with low melting points (<100°C) and non-volatile properties (Holm and Lassi, 2011). Ionic liquid reacts with lignocellulosic-type biomass and converts into tiny components that encourage anaerobic digestion owing to an increased active surface area for the enzyme. Pre-treatment converts cellulose crystals to cellulose chains (Lara-Serrano et al., 2018).

13.3.2.4 Ozonation Pre-Treatment

Ozonation pre-treatment differs from other chemical pre-treatment in that it does not need any chemicals other than ozone. This pre-treatment approach has the advantages of no odor generation and no pathogen remaining in the substrate (Atelge et al., 2020a). The substrate undergoes partial hydrolysis and oxidation in the ozone zone, but not total oxidation. The ozonation pre-treatment mechanism is the direct oxidation process dependent on the combination of substrates. The second process uses hydroxyl radicals in an indirect reaction. Substrates are broken down into extremely small and diffused particles, which subsequently diffuse into the liquid media containing soluble organic compounds (Pérez-Elvira et al., 2006). Some particular components are split down into microscopic particles throughout the process (Pérez-Elvira et al., 2006). Researchers' recommendations for ozone dosage vary from 0.05 to 0.5 g O_3/g TS in the literature; however, the optimum suited dose is dependent on the kind of substrates (Atay and Akbal, 2016; Deublein and Steinhauser, 2011; Pérez-Elvira et al., 2006; Salihu and Alam, 2016). While it has a high energy requirement, this pre-treatment positively influences sludge dewaterability (Pérez-Elvira et al., 2006).

13.3.3 PHYSICAL PRE-TREATMENT

There are no extra components such as chemicals, enzymes, or fungus introduced during physical pre-treatment, and it is a well-known strategy for increasing biogas output. For instance, with the particular substrate, the temperature may raise by thermal pre-treatment, and the substrate's solubility at a specific temperature and process duration determined. Physical preparation enhances the substrate's surface area and facilitates interaction between the substrate and the microorganism. The tiny size particles of the substrate result in a reduction in viscosity, which facilitates mixing and prevents the formation of a floating layer in the reactor. The floating layer prevents biogas generated in the reactor from escaping, which may be troublesome. Physical pre-treatment has also been shown to increase the substrate's cellulose surface area, enhancing its accessibility to microbial activity (R. Singh et al., 2016). However, this method's primary disadvantage is its excessive energy consumption. Some different physical pre-treatments are discussed here.

13.3.3.1 Mechanical Pre-Treatment

Mechanical pre-treatment often enhances the substrate's bulk density, surface area, porosity, and accessibility of substrate to microbes. This pre-treatment process modifies the waste's structural features, such as the crystallinity of lignocellulosic material (O'Dwyer et al., 2008). The tiny particle size of substrates results in a drop in mixing viscosity and a decrease in the floating layer in the reactor. The floating layers might obstruct the outflow of generated biogas from the reactor, resulting in operational difficulties (Bochmann and Montgomery, 2013). Mechanical pre-treatment's primary purpose is to increase the surface area and reduce the crystallinity of the lignocellulose substrate, resulting in lowered polymerization and crystallinity (Merklein et al., 2016). Historically, mechanical pre-treatments such as knife mills (chopping) and hammer mills (crushing) have been utilized.

13.3.3.2 Microwave Pre-Treatment

Microwave pre-treatment is one technique for thermal pre-treatment. This pre-treatment technique is based on the application of electromagnetic radiations, which operate at a broad frequency range of 300 MHz to 300 GHz and wavelengths ranging from 1 mm to 1 m. Microwave pre-treatment has a dual effect on a particular substrate: thermal and non-thermal. With electrical field variation, the former alters the bi-polar component of the substrate such as H_2O, protein, lipid, and other complex molecules (Eskicioglu et al., 2007; Rani et al., 2013). The latter is a bi-polar pathway that occurs randomly inside the polarized region of the cell wall. The modification of the bi-polar pathway results in the dissolution of hydrogen bonds (Eskicioglu et al., 2007). By altering the sludge matrix with this pre-treatment, proteins in the sludge are released into the aqueous phase (Tang et al., 2010). Microwave heating promotes polar molecules by altering the dielectric characteristics of a particular substrate (Amin et al., 2017). As a result, the heating process is more efficient and consistent and requires lesser energy than standard thermal pre-treatment methods (Pellera and Gidarakos, 2017). Microwave pre-treatment reduces complicated substrate structures to tiny, homogeneous components, increasing their availability to the microbes and, hence, their digestibility. Additionally, the electromagnetic field facilitates biochemical and biological processes that include structural changes, such as the transition from a crystalline to a supermolecular structure (Quitain et al., 2013).

13.3.3.3 High-Pressure Homogenization Pre-Treatment

Pre-treatment using high-pressure homogenization exposes sludge to a pressure of up to 600 bars and then depressurizes it as it goes through the valve. As depressurized sludge exits the valve, it collides with a solid surface, and releases the cell content due to pressure gradient, turbulence, shear stress, and cavitation. The primary process disruptors are high pressure and result in a drop in pressure along the valve (Zhang et al., 2012).

13.3.3.4 Pulse Electric Field Pre-Treatment

The application of the electrical field is commonly used in modern times in the biochemistry domain to enhance the pace and output of reactions. Electrochemical pre-treatment is often used for sewage sludge as a pre-treatment procedure for anaerobic digestion. The sludge is passed by a conduit where high voltage is supplied, creating an electrical force. The complex structured biomass is split down into tiny fragments on the application of this force. Pre-treatment such as pulse electric field is used to lyse organic waste microbial cell membranes, releasing intra and extracellular chemicals into the liquid phase (Martinez et al., 2020). Two electrodes are inserted into the pipe, and a voltage of up to 10 kV is delivered with a pulse period of

0.01 seconds (Atay and Akbal, 2016; Tyagi and Lo, 2011). This application is simple to implement and is well-suited for biogas facilities. Before entering the anaerobic digestion reactor, it is often added to the sludge treatment line to boost biogas production (Chiavola et al., 2015). This pre-treatment increased sludge digestion by 9% and biogas generation by 20% (Kopplow et al., 2004). Thus, it had a beneficial effect on biogas production by altering the mineral particles, microorganisms, proteins, fatty acids, carbohydrates, nucleic acids, short-chain alkanes, and extracellular polymeric substances (Yu et al., 2014).

13.3.3.5 Thermal Pre-Treatment

Thermal pre-treatment is a full-established technique, often known as the liquid hot water pre-treatment. The process is pressure, temperature and time dependent. By adding heat to the substrate's structure, it is converted to soluble from insoluble, increasing the substrate's biodegradability (Zhou et al., 2015). Thermal pre-treatment breaks down the chemical connections that hold the cell walls together, allowing the cellular component to escape into the liquid phase. Additionally, this pre-treatment enhances the release of water from the sludge matrix, which increases the sludge's dewaterability (Banu and Kavitha, 2017). Additionally, heat sterilizes and decreases microorganisms in waste (Zhou et al., 2015). If the substrate's moisture content is insufficient, extra water is required for operation. Substrates expand due to the process's pressure, temperature, and water impacts. As a result, the hydrogen bonds that maintain the crystallinity of celluloses structure and other complex structures are destroyed. This enhances the surface area and time of contact between microbes and substrates in anaerobic digestion, allowing for a lower hydraulic retention time and a smaller digester.

13.3.3.6 Ultrasonic Pre-Treatment

Ultrasonic pre-treatment is often used in wastewater treatment facilities to treat sewage sludge. Ultrasonic low frequency is in the range of 20–40 kHz generally uses. Due to the medium's continual compression and rarefaction, ultrasonic waves generate liquid-free tiny bubbles. In a few microseconds, the microbubbles rupture. A cavitation is an extreme event in which the temperature and pressure surrounding the microbubbles exceed 5,000 K and 500 bars, respectively (Zhen et al., 2017). As a physical outcome of this state, a significant hydro-mechanical shear force is created. Additionally, cavitation results in the generation of H and OH radicals, which may aid in the degradation of substrates (Wang et al., 2018). Thus, when ultrasonic pre-treatment is used, sludge disintegration is predicted to proceed in two ways, primarily via hydro-mechanical shear force and oxidizing action of the OH radical (Banu and Kavitha, 2017). Flocculated sludge and co-substrate (if present) degrade.

13.3.4 COMBINED PRE-TREATMENT

As previously stated, there are several pre-treatment techniques available to increase biogas and methane production from complex organic wastes. However, each pre-treatment technique has inherent limitations that might have a detrimental effect on CH_4 yield improvement at some stages, such as excessive energy consumption and the generation of unexpected types of substrates when alone used. One of the most effective ways to increase biogas and methane output is to use several chosen pre-treatment procedures concurrently or sequentially. This allows for a reduction in the pre-treatment intensity of each approach and a reduction in the input energy or operational cost of pre-treatment without sacrificing pre-treatment efficiency in certain circumstances. By optimizing the availability of substrate components, an appropriate combination of pre-treatment increases the degradability of a selected substrate cost-effectively. According to studies, when biological and physical or chemical pre-treatments are combined, the combined pre-treatment is more efficient than a solo pre-treatment of a given substrate.

13.4 PRE-TREATMENT CHALLENGES AND ITS SCOPE

As per the literature, it is discovered that pre-treatment procedures had a number of beneficial effects on biogas generation, reaction rates, solubility and biodegradability of the substrate, and pathogen control. Each one of them, though, has its own flaws. Certain considerations for pre-treatment procedures include economic viability and environmental stewardship. The performance of pre-treatment procedures is largely determined by the properties of the substrates. However, comparing pre-treatment procedures is difficult due to their non-standardized operational parameters, including substrate type, pre-treatment conditions, scales, anaerobic digestion process types, environmental conditions, and units. The bulk of pre-treatment investigations used batch-size BMP assays, and the improvement may not apply to the continuous-feed anaerobic digestion systems. As a result, the data available in the literature, except for a few research studies on combined pre-treatment, do not allow for an appropriate comparison.

Attaining favorable energy and economic balances is vital for developing sustainable pre-treatment technologies. While most pre-treatment treatments increase anaerobic digestion performance, their high energy costs preclude their widespread use in the field. It is observed that the thickening of sludge is seen as one of the significant steps in lowering pre-treatment energy expenditures (Braguglia et al., 2011; Gianico et al., 2013; Pérez-Elvira et al., 2009). However, high solid concentrations of sludge may be challenging to handle because of its rheology. Sludge yield stresses develop drastically with the increase in solid contents (Mori et al., 2006), which may alter mixing and pumping energy needs and promote the presence of dead areas in the digester and other equipment (Cheng, 1986). Additionally, it has been noted that increasing the substrate concentration during anaerobic digestion reduces CH_4 yield and rates owing to a drop in the rate of hydrolysis, mass transfer limits, and VFA buildup (Abbassi-Guendouz et al., 2012). Additionally, the efficiency of ultrasonic pre-treatments might be impacted by large solid concentration that absorbs sound waves (Pilli et al., 2011).

Notably, laboratory and industrial pre-treatment equipment may provide significantly different energy consumption findings (Pérez-Elvira et al., 2009; Zielewicz and Sorys, 2008). Laboratory equipment is often inefficient compared to industrial devices, and as a result, energy balance based on laboratory testing may not always reflect expenditures on a broader scale. As a result, a comprehensive examination of the scaling implications is required to accurately evaluate the energy and capital requirements of laboratory-tested pre-treatment techniques.

Environmental performance is another important issue to consider when scaling up sludge pre-treatment. Sludge application to the soil caused considerable toxicological and eutrophic hazards, and energy consumption had a greater impact on the environment than chemical use (terrestrial eco-toxicity and human toxicity). Local factors such as the cost and availability of chemical additives, as well as the local energy matrix and legislation, may all have an impact on the process's technical and economic viability. As a result, choices should be weighed in light of these considerations. COD solubilization has been widely used to evaluate pre-treatment effectiveness, but it does not necessarily correspond with an increase in anaerobic digestion performance (Kim et al., 2013). As a consequence, additional characteristics impacting methane generation and solid reduction following pre-treatment, such as particle size reduction, improved rheology, hydrolytic enzyme activation, and improved mass transfer conditions, should be explored. Examining the solubilization product after pre-treatment may also provide information about the mechanisms that lead to an improvement in anaerobic digestion (Tian et al., 2015). Furthermore, processes such as lower temperature and ultrasonic pre-treatment have been shown to increase biological sludge hydrolytic activity via stimulation and enzyme solubilization (Chu et al., 2011; Guo et al., 2011; Yan et al., 2010), leading to the development of pre-treatment methods that incorporate physical effects and the inherent hydrolytic capacity of secondary sludge (Carvajal et al., 2013). When the viscosity of sludge decreases as a consequence of different pre-treatments (Bougrier et al., 2006), mixing and mass transfer may improve as a result of increased diffusivity coefficient and decreased resistance to flow during digestions (Abbassi-Guendouz et al., 2012; Terashima et al., 2009).

While the impact of pre-treatment on biogas production and solid reduction has been widely studied, there have been few investigations on the impact of pre-treatment on sludge quality (Braguglia et al., 2015; Carballa et al., 2009). Traditional mesophilic digestion results in moderate pathogen inactivation and organic matter reduction. As a result, digested sludge is used sparingly in agriculture to reduce disease transmission and contamination of water and land. Increased sludge quality by pre-treatment may allow for greater sludge application in soil, reducing sludge disposal costs and yielding more acceptable results. Furthermore, since pre-treatment encourages the removal of biodegradable organic matter, the relative amount of humic compounds may increase, boosting the sludge's agricultural potential. When biomolecules go through physical, chemical, and microbiological transformations, humic compounds are generated. Humic compounds are significant in agriculture because they have an impact on the soil's quality and productivity. Humic and fulvic acids may account for roughly 20% of the organic content in sewage sludge, with humic acid dominating. While research on sludge quality has been conducted, none of the papers reviewed provided a full assessment of digested sludge's agricultural quality. The presence of organic micro-contaminants and heavy metals is another aspect that determines the quality of digested sludge. Although sludge pre-treatment may help in the removal of organic pollutants (El-Hadj et al., 2007), the mechanisms behind this effect are unclear, and operational parameters during digestion might have a considerable impact on the reported results. Furthermore, there were no data on the impact of pre-treatment on the recovery or removal of heavy metals from digested sludge. The concentration of heavy metals in sludge may grow during pre-treatment and digesting due to the considerable effects of pre-treatment on sludge properties, such as higher organic matter removal rates. Furthermore, since metals tend to solubilize at low pH levels, various pre-treatment conditions may help in the recovery of heavy metals. It would be interesting to see whether pre-treatments used before anaerobic digestion have any influence on the presence of heavy metals in digested sludge and if current heavy metal removal techniques may be coupled to improve sludge digestion and quality.

13.5   CONCLUSION

Depending on local conditions, pre-treatment of sewage sludge may improve anaerobic digestion. Pre-treatment has recently become a major study subject due to its potential benefits. New technologies (such as electric pulses) and process combinations are gaining interest (such as thermal and physical–chemical combinations). Other effects studied include macromolecule behavior changes, particle size reduction, rheological changes, enzymatic or biological stimulation, organic pollutant elimination, and changes in microbial dynamics. While these investigations provide some insights into sludge pre-treatment methods, more are needed.

The high energy expenditures of sludge pre-treatment operations render them unsustainable economically and environmentally. Energy integration and a greater understanding of pre-treatment and digestive processes might help overcome this constraint. Our increased understanding of how these processes work is critical to enhancing current technologies and developing new ones. To overcome the limits of currently available pre-treatments, combining pre-treatments to achieve synergistic effects looks interesting. Low-temperature pre-treatments and other "gentle" activities may increase anaerobic digestion while using lesser energy and chemicals.

Finally, pre-treatment may be used to improve sludge quality in the future. Sludge control is critical in wastewater treatment. Decreased sludge volume and contamination potential is the key economic and environmental goal. The manufacture of a useful product with some economic value from stabilized sludge can improve the process's economic balance and improve agricultural operations' environmental performance by replacing commercially available fertilizers with greater carbon footprints.

REFERENCES

Abbasi, T., Tauseef, S. M., & Abbasi, S. A. (2011). *Biogas Energy* (Vol. 2). Springer Science & Business Media: Berlin, Germany.

Abbassi-Guendouz, A., Brockmann, D., Trably, E., Dumas, C., Delgenès, J.-P., Steyer, J.-P., & Escudié, R. (2012). Total solids content drives high solid anaerobic digestion via mass transfer limitation. *Bioresource Technology, 111*, 55–61.

Achinas, S., & Euverink, G. J. W. (2016). Consolidated briefing of biochemical ethanol production from lignocellulosic biomass. *Electronic Journal of Biotechnology, 23*, 44–53.

Al Seadi, T. (2008). Biogas Handbook. University of Southern Denmark: Esbjerg, Denmark.

Amin, F. R., Khalid, H., Zhang, H., Rahman, S., Zhang, R., Liu, G., & Chen, C. (2017). Pretreatment methods of lignocellulosic biomass for anaerobic digestion. *AMB Express*, 7(1), 1–12.

Appels, L., Baeyens, J., Degrève, J., & Dewil, R. (2008). Principles and potential of the anaerobic digestion of waste-activated sludge. Progress in Energy and Combustion Science, 34(6), 755–781. doi: 10.1016/j.pecs.2008.06.002.

Ariunbaatar, J., Panico, A., Esposito, G., Pirozzi, F., & Lens, P. N. L. (2014). Pretreatment methods to enhance anaerobic digestion of organic solid waste. Applied Energy, 123, 143–156. doi: 10.1016/j.apenergy.2014.02.035.

Asadi, N., & Zilouei, H. (2017). Optimization of organosolv pretreatment of rice straw for enhanced biohydrogen production using Enterobacter aerogenes. *Bioresource Technology*, 227, 335–344.

Atay, Ş., & Akbal, F. (2016). Classification and effects of sludge disintegration technologies integrated into sludge handling units: An overview. *Clean–Soil, Air, Water*, 44(9), 1198–1213.

Atelge, M. R., Atabani, A. E., Banu, J. R., Krisa, D., Kaya, M., Eskicioglu, C., Kumar, G., Lee, C., Yildiz, Y., Unalan, S., Mohanasundaram, R., & Duman, F. (2020a). A critical review of pretreatment technologies to enhance anaerobic digestion and energy recovery. *Fuel*, 270, 117494. doi: 10.1016/j.fuel.2020.117494.

Atelge, M. R., Krisa, D., Kumar, G., Eskicioglu, C., Nguyen, D. D., Chang, S. W., Atabani, A. E., Al-Muhtaseb, A. H., & Unalan, S. (2020b). Biogas production from organic waste: Recent progress and perspectives. *Waste and Biomass Valorization*, 11(3), 1019–1040.

Banu, J. R., & Kavitha, S. (2017). Various sludge pretreatments: Their impact on biogas generation. In: L. Singh, & V. C. Kalia (Eds.), *Waste Biomass Management: A Holistic Approach* (pp. 39–71). Springer: New York.

Barua, S., & Dhar, B. R. (2017). Advances towards understanding and engineering direct interspecies electron transfer in anaerobic digestion. *Bioresource Technology*, 244, 698–707.

Bochmann, G., & Montgomery, L. F. R. R. (2013). Storage and pre-treatment of substrates for biogas production. In: A. Wellinger, J. Patrick Murphy, & D. Baxter (Eds.), *The Biogas Handbook: Science, Production and Applications* (pp. 85–103). Elsevier Inc: Amsterdam, Netherlands. doi: 10.1533/9780857097415.1.85.

Bougrier, C., Albasi, C., Delgenès, J. P., & Carrère, H. (2006). Effect of ultrasonic, thermal and ozone pre-treatments on waste activated sludge solubilisation and anaerobic biodegradability. *Chemical Engineering and Processing: Process Intensification*, 45(8), 711–718. doi: 10.1016/j.cep.2006.02.005.

Braguglia, C. M., Coors, A., Gallipoli, A., Gianico, A., Guillon, E., Kunkel, U., Mascolo, G., Richter, E., Ternes, T. A., Tomei, M. C., et al. (2015). Quality assessment of digested sludges produced by advanced stabilization processes. *Environmental Science and Pollution Research*, 22(10), 7216–7235.

Braguglia, C. M., Gianico, A., & Mininni, G. (2011). Laboratory-scale ultrasound pre-treated digestion of sludge: Heat and energy balance. *Bioresource Technology*, 102(16), 7567–7573.

Burgess, J. E., & Pletschke, B. I. (2008). Hydrolytic enzymes in sewage sludge treatment: A mini-review. *Water SA*, 34(3), 343–350.

Cai, M., Wilkins, D., Chen, J., Ng, S.-K., Lu, H., Jia, Y., & Lee, P. K. H. (2016). Metagenomic reconstruction of key anaerobic digestion pathways in municipal sludge and industrial wastewater biogas-producing systems. *Frontiers in Microbiology*, 7, 778.

Carballa, M., Omil, F., & Lema, J. M. (2009). Influence of different pretreatments on anaerobically digested sludge characteristics: suitability for final disposal. *Water, Air, and Soil Pollution*, 199(1), 311–321.

Carlsson, M., Lagerkvist, A., & Morgan-Sagastume, F. (2012). The effects of substrate pre-treatment on anaerobic digestion systems: A review. *Waste Management*, 32(9), 1634–1650.

Carrère, H., Dumas, C., Battimelli, A., Batstone, D. J., Delgenès, J. P., Steyer, J. P., & Ferrer, I. (2010). Pretreatment methods to improve sludge anaerobic degradability: A review. *Journal of Hazardous Materials*, *183*(1–3), 1–15. doi: 10.1016/j.jhazmat.2010.06.129.

Carvajal, A., Peña, M., & Pérez-Elvira, S. (2013). Autohydrolysis pretreatment of secondary sludge for anaerobic digestion. *Biochemical Engineering Journal*, *75*, 21–31.

Charters, W. W. S. W. S. (2001). Developing markets for renewable energy technologies. *Renewable Energy*, *22*(1–3), 217–222. doi: 10.1016/S0960-1481(00)00018-5.

Cheng, D. C. H. (1986). Yield stress: A time-dependent property and how to measure it. *Rheologica Acta*, *25*(5), 542–554.

Chiavola, A., Ridolfi, A., D'Amato, E., Bongirolami, S., Cima, E., Sirini, P., & Gavasci, R. (2015). Sludge reduction in a small wastewater treatment plant by electro-kinetic disintegration. *Water Science and Technology*, *72*(3), 364–370.

Christensen, G. L., & Dick, R. I. (1985). Specific resistance measurements: Methods and procedures. *Journal of Environmental Engineering*, *111*(3), 258–271. doi: 10.1061/(asce)0733-9372(1985)111:3(258).

Christy, P. M., Gopinath, L. R., & Divya, D. (2014). A review on anaerobic decomposition and enhancement of biogas production through enzymes and microorganisms. *Renewable and Sustainable Energy Reviews*, *34*, 167–173.

Chu, L., Wang, J., & Wang, B. (2011). Effect of gamma irradiation on activities and physicochemical characteristics of sewage sludge. *Biochemical Engineering Journal*, *54*(1), 34–39.

Chynoweth, D. P., Owens, J. M., & Legrand, R. (2001). Renewable methane from anaerobic digestion of biomass. *Renewable Energy*, *22*(1–3), 1–8. doi: 10.1016/S0960-1481(00)00019-7.

Dang, Y., Holmes, D. E., Zhao, Z., Woodard, T. L., Zhang, Y., Sun, D., Wang, L.-Y., Nevin, K. P., & Lovley, D. R. (2016). Enhancing anaerobic digestion of complex organic waste with carbon-based conductive materials. *Bioresource Technology*, *220*, 516–522.

Deublein, D., & Steinhauser, A. (2011). *Biogas from Waste and Renewable Resources: An Introduction*, John Wiley & Sons: Hoboken, NJ.

Dhar, B. R., Elbeshbishy, E., Hafez, H., & Lee, H.-S. (2015). Hydrogen production from sugar beet juice using an integrated biohydrogen process of dark fermentation and microbial electrolysis cell. *Bioresource Technology*, *198*, 223–230.

Dohányos, M., Zábranská, J., Kutil, J., & Jeníček, P. (2004). Improvement of anaerobic digestion of sludge. *Water Science and Technology*, *49*(10), 89–96. doi: 10.2166/wst.2004.0616.

Duman, F., Sahin, U., & Atabani, A. E. (2019). Harvesting of blooming microalgae using green synthetized magnetic maghemite (γ-Fe_2O_3) nanoparticles for biofuel production. *Fuel*, *256*, 115935.

Dumas, C., Perez, S., Paul, E., & Lefebvre, X. (2010). Combined thermophilic aerobic process and conventional anaerobic digestion: Effect on sludge biodegradation and methane production. *Bioresource Technology*, *101*(8), 2629–2636.

Ebenezer, A. V., Kaliappan, S., Kumar, S. A., Yeom, I.-T., & Banu, J. R. (2015). Influence of deflocculation on microwave disintegration and anaerobic biodegradability of waste activated sludge. *Bioresource Technology*, *185*, 194–201.

Edwiges, T., Frare, L., Mayer, B., Lins, L., Triolo, J. M., Flotats, X., & de Mendonça Costa, M. S. S. (2018). Influence of chemical composition on biochemical methane potential of fruit and vegetable waste. *Waste Management*, *71*, 618–625.

El-Hadj, T. B., Dosta, J., Márquez-Serrano, R., & Mata-Alvarez, J. (2007). Effect of ultrasound pretreatment in mesophilic and thermophilic anaerobic digestion with emphasis on naphthalene and pyrene removal. *Water Research*, *41*(1), 87–94.

Erden, G., & Filibeli, A. (2011). Effects of Fenton pre-treatment on waste activated sludge properties. *Clean–Soil, Air, Water*, *39*(7), 626–632.

Eskicioglu, C., Kennedy, K. J., & Droste, R. L. (2007). Enhancement of batch waste activated sludge digestion by microwave pretreatment. *Water Environment Research*, *79*(11), 2304–2317.

Fisgativa, H., Tremier, A., & Dabert, P. (2016). Characterizing the variability of food waste quality: A need for efficient valorisation through anaerobic digestion. *Waste Management*, *50*, 264–274.

Fitzgerald, G. C. (2013). Pre-processing and treatment of municipal solid waste (MSW) prior to incineration. In: N. B. Klinghoffer, & M. J Castaldi (Eds.), *Waste to Energy Conversion Technology* (pp. 55–71). Elsevier: Amsterdam, Netherlands.

Gavala, H. N., Yenal, U., Skiadas, I. v., Westermann, P., & Ahring, B. K. (2003). Mesophilic and thermophilic anaerobic digestion of primary and secondary sludge. Effect of pretreatment at elevated temperature. *Water Research*, *37*(19), 4561–4572. doi: 10.1016/S0043-1354(03)00401-9.

Gianico, A., Braguglia, C. M., Cesarini, R., & Mininni, G. (2013). Reduced temperature hydrolysis at 134°C before thermophilic anaerobic digestion of waste activated sludge at increasing organic load. *Bioresource Technology*, *143*, 96–103. doi: 10.1016/j.biortech.2013.05.069.

Gonzalez-Fernandez, C., Sialve, B., & Molinuevo-Salces, B. (2015). Anaerobic digestion of microalgal biomass: Challenges, opportunities and research needs. *Bioresource Technology*, *198*, 896–906.

Gumisiriza, R., Hawumba, J. F., Okure, M., & Hensel, O. (2017). Biomass waste-to-energy valorisation technologies: A review case for banana processing in Uganda. *Biotechnology for Biofuels*, *10*(1), 1–29.

Guo, J., Peng, Y., Ni, B.-J., Han, X., Fan, L., & Yuan, Z. (2015). Dissecting microbial community structure and methane-producing pathways of a full-scale anaerobic reactor digesting activated sludge from wastewater treatment by metagenomic sequencing. *Microbial Cell Factories*, *14*(1), 1–11.

Guo, J., Xu, Y., et al. (2011). Review of enzymatic sludge hydrolysis. *Journal of Bioremediation and Biodegradation*, *2*(5), 130. doi: 10.4172/2155-6199.1000130.

Guo, W. Q., Yang, S. S., Pang, J. W., Ding, J., Zhou, X. J., Feng, X. C., Zheng, H. S., & Ren, N. Q. (2013). Application of low frequency ultrasound to stimulate the bio-activity of activated sludge for use as an inoculum in enhanced hydrogen production. *RSC Advances*, *3*(44), 21848–21855. doi: 10.1039/c3ra41723a.

Harris, P. W., & McCabe, B. K. (2015). Review of pre-treatments used in anaerobic digestion and their potential application in high-fat cattle slaughterhouse wastewater. *Applied Energy*, *155*, 560–575.

Haug, R. T., Stuckey, D. C., Gossett, J. M., & McCarty, P. L. (1978). Effect of thermal pretreatment on digestibility and dewaterability of organic sludges. *Journal of the Water Pollution Control Federation*, *50*(1), 73–85.

Holm, J., & Lassi, U. (2011). Ionic liquids in the pretreatment of lignocellulosic biomass, Chapter 24. In: A. Kokorin (Ed.), *Ionic Liquids: Applications and Perspectives* (pp. 545–560). InTech: Rijeka, Croatia.

Jang, H. M., Cho, H. U., Park, S. K., Ha, J. H., & Park, J. M. (2014). Influence of thermophilic aerobic digestion as a sludge pre-treatment and solids retention time of mesophilic anaerobic digestion on the methane production, sludge digestion and microbial communities in a sequential digestion process. *Water Research*, *48*, 1–14.

Karp, E. M., Resch, M. G., Donohoe, B. S., Ciesielski, P. N., O'Brien, M. H., Nill, J. E., Mittal, A., Biddy, M. J., & Beckham, G. T. (2015). Alkaline pretreatment of switchgrass. *ACS Sustainable Chemistry & Engineering*, *3*(7), 1479–1491.

Karuppiah, T., & Azariah, V. E. (2019). Biomass pretreatment for enhancement of biogas production. In: J. R. Banu (Ed.), Anaerobic Digestion. InTech: London. doi: 10.5772/intechopen.82088

Kim, D.-H., Cho, S.-K., Lee, M.-K., & Kim, M.-S. (2013). Increased solubilization of excess sludge does not always result in enhanced anaerobic digestion efficiency. *Bioresource Technology*, *143*, 660–664.

Kopplow, O., Barjenbruch, M., & Heinz, V. (2004). Sludge pre-treatment with pulsed electric fields. *Water Science and Technology*, 49(10), 123–129.

Lara-Serrano, M., Sáez Angulo, F., Negro, M. J., Morales-delaRosa, S., Campos-Martin, J. M., & Fierro, J. L. G. (2018). Second-generation bioethanol production combining simultaneous fermentation and saccharification of IL-pretreated barley straw. *ACS Sustainable Chemistry & Engineering*, 6(5), 7086–7095.

Li, Y., Jin, Y., Li, J., Li, H., Yu, Z., & Nie, Y. (2017). Effects of thermal pretreatment on degradation kinetics of organics during kitchen waste anaerobic digestion. *Energy*, 118, 377–386.

Li, Y. Y., & Noike, T. (1992). Upgrading of anaerobic digestion of waste activated sludge by thermal pretreatment. *Water Science and Technology*, 26(3–4), 857–866. doi: 10.2166/wst.1992.0466.

Lin, J.-G., Ma, Y.-S., Chao, A. C., & Huang, C.-L. (1999). BMP test on chemically pretreated sludge. *Bioresource Technology*, 68(2), 187–192. doi: 10.1016/S0960-8524(98)00126-6.

Lu, X., Zhen, G., Chen, M., Kubota, K., & Li, Y.-Y. (2015). Biocatalysis conversion of methanol to methane in an upflow anaerobic sludge blanket (UASB) reactor: Long-term performance and inherent deficiencies. *Bioresource Technology*, 198, 691–700.

Martinez, J. M., Delso, C., Álvarez, I., & Raso, J. (2020). Pulsed electric field-assisted extraction of valuable compounds from microorganisms. *Comprehensive Reviews in Food Science and Food Safety*, 19(2), 530–552.

McCarty, P. L., & Smith, D. P. (1986). Anaerobic wastewater treatment. *Environmental Science & Technology*, 20(12), 1200–1206.

Meegoda, J. N., Hsieh, H.-N., Rodriguez, P., & Jawidzik, J. (2012). Sustainable community sanitation for a rural hospital in Haiti. *Sustainability*, 4(12), 3362–3376.

Merklein, K., Fong, S. S., & Deng, Y. (2016). Biomass utilization. In: C. E. Eckert, & C. T. Trinh (Eds.), Biotechnology for Biofuel Production and Optimization (pp. 291–324). Elsevier Inc: Amsterdam, Netherlands. doi: 10.1016/B978-0-444-63475-7.00011-X.

Metcalf, L., Eddy, H. P., & Tchobanoglous, G. (1991). *Wastewater Engineering: Treatment, Disposal, and Reuse* (Vol. 4). McGraw-Hill: New York.

Mills, N., Pearce, P., Farrow, J., Thorpe, R. B., & Kirkby, N. F. (2014). Environmental & economic life cycle assessment of current & future sewage sludge to energy technologies. *Waste Management*, 34(1), 185–195.

Mori, M., Seyssiecq, I., & Roche, N. (2006). Rheological measurements of sewage sludge for various solids concentrations and geometry. *Process Biochemistry*, 41(7), 1656–1662.

Moya, A. J., Mateo, S., Puentes, J. G., Fonseca, B. G., Roberto, I. C., & Sánchez, S. (2016). Comparing bioalcohols production from olive pruning biomass by biotechnological pathway with Candida guilliermondii and Pichia stipitis. *Waste and Biomass Valorization*, 7(6), 1369–1375.

Murphy, J., Braun, R., Weiland, P., & Wellinger, A. (2011). Biogas from crop digestion. *IEA Bioenergy Task*, 37(37), 1–23.

Neumann, P., Pesante, S., Venegas, M., & Vidal, G. (2016). Developments in pre-treatment methods to improve anaerobic digestion of sewage sludge. *Reviews in Environmental Science and Biotechnology*, 15(2), 173–211. doi: 10.1007/s11157-016-9396-8.

O'Dwyer, J., Zhu, L., Granda, C. B., Chang, V. S., Holtzapple, M. T., O'Dwyer, J. P., Zhu, L., Granda, C. B., Chang, V. S., & Holtzapple, M. T. (2008). Neural network prediction of biomass digestibility based on structural features. *Biotechnology Progress*, 24(2), 283–292. doi: 10.1021/bp070193v.

Oh, Y.-S., Shih, L., Tzeng, Y.-M., & Wang, S.-L. (2000). Protease produced by Pseudomonas aeruginosa K-187 and its application in the deproteinization of shrimp and crab shell wastes. *Enzyme and Microbial Technology*, 27(1–2), 3–10.

Okudoh, V., Trois, C., Workneh, T., & Schmidt, S. (2014). The potential of cassava biomass and applicable technologies for sustainable biogas production in South Africa: A review. *Renewable and Sustainable Energy Reviews*, 39, 1035–1052.

Ostrem, K., Themelis, N. J., et al. (2004). Greening waste: Anaerobic digestion for treating the organic fraction of municipal solid wastes. Earth Engineering Center Columbia University, 6–9.

Pedersen, M., Johansen, K. S., & Meyer, A. S. (2011). Low temperature lignocellulose pretreatment: effects and interactions of pretreatment pH are critical for maximizing enzymatic monosaccharide yields from wheat straw. *Biotechnology for Biofuels*, *4*(1), 1–10.

Pellera, F.-M., & Gidarakos, E. (2017). Microwave pretreatment of lignocellulosic agroindustrial waste for methane production. *Journal of Environmental Chemical Engineering*, *5*(1), 352–365.

Pérez-Elvira, S., Fdz-Polanco, M., Plaza, F. I., Garralón, G., & Fdz-Polanco, F. (2009). Ultrasound pre-treatment for anaerobic digestion improvement, *Water Science and Technology*, *60*(6), 1525–1532.

Pérez-Elvira, S. I., Nieto Diez, P., & Fdz-Polanco, F. (2006). Sludge minimisation technologies. *Reviews in Environmental Science and Biotechnology*, *5*(4), 375–398. doi: 10.1007/s11157-005-5728-9.

Pilli, S., Bhunia, P., Yan, S., LeBlanc, R. J., Tyagi, R. D., & Surampalli, R. Y. (2011). Ultrasonic pretreatment of sludge: A review. *Ultrasonics Sonochemistry*, *18*(1), 1–18. doi: 10.1016/j.ultsonch.2010.02.014.

Pohland, F. G., & Ghosh, S. (1971). Developments in anaerobic stabilization of organic wastes-the two-phase concept. *Environmental Letters*, *1*(4), 255–266.

Quitain, A. T., Sasaki, M., & Goto, M. (2013). Microwave-based pretreatment for efficient biomass-to-biofuel conversion. In: Z. Fang (Ed.), *Pretreatment Techniques for Biofuels and Biorefineries* (pp. 117–130). Springer: Berlin, Germany.

Rani, R. U., Kumar, S. A., Kaliappan, S., Yeom, I., & Banu, J. R. (2013). Impacts of microwave pretreatments on the semi-continuous anaerobic digestion of dairy waste activated sludge. *Waste Management*, *33*(5), 1119–1127.

Rotaru, A.-E., Shrestha, P. M., Liu, F., Markovaite, B., Chen, S., Nevin, K. P., & Lovley, D. R. (2014a). Direct interspecies electron transfer between Geobacter metallireducens and Methanosarcina barkeri. *Applied and Environmental Microbiology*, *80*(15), 4599–4605.

Rotaru, A.-E., Shrestha, P. M., Liu, F., Shrestha, M., Shrestha, D., Embree, M., Zengler, K., Wardman, C., Nevin, K. P., & Lovley, D. R. (2014b). A new model for electron flow during anaerobic digestion: Direct interspecies electron transfer to Methanosaeta for the reduction of carbon dioxide to methane. *Energy & Environmental Science*, *7*(1), 408–415.

Rozzi, A., & Remigi, E. (2004). Methods of assessing microbial activity and inhibition under anaerobic conditions: A literature review. *Reviews in Environmental Science and Biotechnology*, 3(2), 93–115. doi: 10.1007/s11157-004-5762-z.

Şahinkaya, S., & Sevimli, M. F. (2013). Sono-thermal pre-treatment of waste activated sludge before anaerobic digestion. *Ultrasonics Sonochemistry*, *20*(1), 587–594.

Salihu, A., & Alam, M. Z. (2016). Pretreatment methods of organic wastes for biogas production. *Journal of Applied Sciences*, *16*(3), 124–137.

Shah, F. A., Mahmood, Q., Shah, M. M., Pervez, A., & Asad, S. A. (2014). Microbial ecology of anaerobic digesters: The key players of anaerobiosis. *The Scientific World Journal*, *2014*.

Shimizu, T., Kudo, K., & Nasu, Y. (1993). Anaerobic waste-activated sludge digestion: A bioconversion mechanism and kinetic model. *Biotechnology and Bioengineering*, *41*(11), 1082–1091. doi: 10.1002/bit.260411111.

Shirkavand, E., Baroutian, S., Gapes, D. J., & Young, B. R. (2017). Pretreatment of radiata pine using two white rot fungal strains Stereum hirsutum and Trametes versicolor. *Energy Conversion and Management*, *142*, 13–19.

Sindhu, R., Binod, P., & Pandey, A. (2016). Biological pretreatment of lignocellulosic biomass: An overview. *Bioresource Technology*, *199*, 76–82.

Singh, J., Suhag, M., & Dhaka, A. (2015). Augmented digestion of lignocellulose by steam explosion, acid and alkaline pretreatment methods: A review. *Carbohydrate Polymers*, *117*, 624–631.

Singh, R., Krishna, B. B., Kumar, J., & Bhaskar, T. (2016). Opportunities for utilization of non-conventional energy sources for biomass pretreatment. *Bioresource Technology, 199*, 398–407. doi: 10.1016/j.biortech.2015.08.117.

Suchowska-Kisielewicz, M., Jedrczak, A., Sadecka, Z., & Myszograj, S. (2013). Effect of aerobic pretreatment of waste on the rate of anaerobic treatment processes. *Journal of Material Cycles and Waste Management, 15*(2), 138–145.

Sun, L., Pope, P. B., Eijsink, V. G. H., & Schnürer, A. (2015). Characterization of microbial community structure during continuous anaerobic digestion of straw and cow manure. *Microbial Biotechnology, 8*(5), 815–827.

Sztancs, G., Juhasz, L., Nagy, B. J., Nemeth, A., Selim, A., Andre, A., Toth, A. J., Mizsey, P., & Fozer, D. (2020). Co-Hydrothermal gasification of Chlorella vulgaris and hydrochar: The effects of waste-to-solid biofuel production and blending concentration on biogas generation. *Bioresource Technology, 302*, 122793.

Tang, B., Yu, L., Huang, S., Luo, J., & Zhuo, Y. (2010). Energy efficiency of pre-treating excess sewage sludge with microwave irradiation. *Bioresource Technology, 101*(14), 5092–5097.

Terashima, M., Goel, R., Komatsu, K., Yasui, H., Takahashi, H., Li, Y. Y., & Noike, T. (2009). CFD simulation of mixing in anaerobic digesters. *Bioresource Technology, 100*(7), 2228–2233.

Tian, X., Wang, C., Trzcinski, A. P., Lin, L., & Ng, W. J. (2015). Insights on the solubilization products after combined alkaline and ultrasonic pre-treatment of sewage sludge. *Journal of Environmental Sciences, 29*, 97–105.

Tiehm, A., Nickel, K., Zellhorn, M., Neis, U., & Tiehm, A. (2001). Ultrasonic waste activated sludge disintegration for improving anaerobic stabilization. *Water Research, 35*(8), 2003–2009. doi: 10.1016/S0043-1354(00)00468-1.

Tyagi, V. K., & Lo, S.-L. (2011). Application of physico-chemical pretreatment methods to enhance the sludge disintegration and subsequent anaerobic digestion: An up to date review. *Reviews in Environmental Science and Bio/Technology, 10*(3), 215–242.

Uğuz, G., Atabani, A. E., Mohammed, M. N., Shobana, S., Uğuz, S., Kumar, G., & Ala'a, H. (2019). Fuel stability of biodiesel from waste cooking oil: A comparative evaluation with various antioxidants using FT-IR and DSC techniques. *Biocatalysis and Agricultural Biotechnology, 21*, 101283.

van Lier, J. B., Tilche, A., Ahring, B. K., Macarie, H., Moletta, R., Dohanyos, M., Hulshoff Pol, L. W., Lens, P., & Verstraete, W. (2001). New perspectives in anaerobic digestion. *Water Science and Technology, 43*(1), 1–18. doi: 10.2166/wst.2001.0001.

Wang, H., Cai, W.-W. W., Liu, W.-Z. Z., Li, J.-Q. Q., Wang, B., Yang, S.-C. C., & Wang, A.-J. J. (2018). Application of sulfate radicals from ultrasonic activation: Disintegration of extracellular polymeric substances for enhanced anaerobic fermentation of sulfate-containing waste-activated sludge. *Chemical Engineering Journal, 352*, 380–388. doi: 10.1016/j.cej.2018.07.029.

Wang, Q., Kuninobu, M., Kakimoto, K., Ogawa, H., & Kato, Y. (1999). Upgrading of anaerobic digestion of waste activated sludge by ultrasonic pretreatment. *Bioresource Technology, 68*(3), 309–313. doi: 10.1016/S0960-8524(98)00155-2.

Wawrzynczyk, J., Recktenwald, M., Norrlöw, O., & Dey, E. S. (2008). The function of cation-binding agents in the enzymatic treatment of municipal sludge. *Water Research, 42*(6–7), 1555–1562.

Weiland, P. (2010). Biogas production: current state and perspectives. *Applied Microbiology and Biotechnology, 85*(4), 849–860.

Yan, Y., Feng, L., Zhang, C., Zhu, H., & Zhou, Q. (2010). Effect of ultrasonic specific energy on waste activated sludge solubilization and enzyme activity. *African Journal of Biotechnology, 9*(12), 1776–1782. doi: 10.5897/ajb10.1279.

Yu, B., Xu, J., Yuan, H., Lou, Z., Lin, J., & Zhu, N. (2014). Enhancement of anaerobic digestion of waste activated sludge by electrochemical pretreatment. *Fuel, 130*, 279–285.

Zhang, S., Zhang, P., Zhang, G., Fan, J., & Zhang, Y. (2012). Enhancement of anaerobic sludge digestion by high-pressure homogenization. *Bioresource Technology, 118,* 496–501.

Zhen, G., Lu, X., Kato, H., Zhao, Y., & Li, Y. Y. (2017). Overview of pretreatment strategies for enhancing sewage sludge disintegration and subsequent anaerobic digestion: Current advances, full-scale application and future perspectives. *Renewable and Sustainable Energy Reviews, 69,* 559–577. doi: 10.1016/j.rser.2016.11.187.

Zhou, J., Xu, W., Wong, J. W. C., Yong, X., Yan, B., Zhang, X., & Jia, H. (2015). Ultrasonic and thermal pretreatments on anaerobic digestion of petrochemical sludge: dewaterability and degradation of PAHs. *PLoS One, 10*(9), e0136162.

Zhou, S., Zhang, Y., & Dong, Y. (2012). Pretreatment for biogas production by anaerobic fermentation of mixed corn stover and cow dung. *Energy, 46*(1), 644–648.

Zielewicz, E., & Sorys, P. (2008). The comparison of ultrasonic disintegration in laboratory and technical scale disintegrators. *The European Physical Journal Special Topics, 154*(1), 289–294.

Zupančič, G. D., & Grilc, V. (2012). Anaerobic treatment and biogas production from organic waste. In: S. Kumar (Ed.), *Management of Organic Waste* (pp. 1–28). InTech: London.

14 Overview of Thermal Based Pre-Treatment Methods for Enhancing Methane Production of Sewage Sludge

Gowtham Balasundaram,
Pallavi Gahlot, and Absar Ahmad Kazmi
Indian Institute of Technology, Roorkee

Vinay Kumar Tyagi
National Instiute of Hydrology, Roorkee

CONTENTS

14.1 Introduction 257
14.2 Principles of Anaerobic Digestion...258
14.3 Conventional Thermal Pretreatment 259
14.4 Temperature-Phased Anaerobic Digestion (TPAD) 260
14.5 Microwave Irradiation 261
14.6 Thermal Hydrolysis 262
    14.6.1 The Cambi Thermal Hydrolysis Process (THP)............................263
    14.6.2 The Exelys Thermal Hydrolysis 263
14.7 Thermochemical Methods...264
    14.7.1 Conventional Thermochemical Treatment Methods 265
        14.7.1.1 Alkali Thermal Treatment..265
        14.7.1.2 Acid Thermal Treatment..265
    14.7.2 Microwave-Based Alkali Pretreatment 266
14.8 Conclusions...267
References..267

14.1 INTRODUCTION

A large amount of residual sludge is generated in wastewater treatment plants (WWTPs). Approximately 25%–50% of the influent COD is converted into primary or secondary sludge and hence adequate management of sludge is needed due to

DOI: 10.1201/9781003202431-14

the cost involved which accounts for 40% of all operational input. Anaerobic digestion is considered one of the most important and widely applied sludge management techniques (Turonskiy and Mathai, 2006; Gahlot et al., 2020; Gahlot et al.,
2021). However, the existing anaerobic digestion-based technologies at WWTPs
have shown low productivity in terms of energy. Therefore increasing methane production through the use of municipal sewage sludge could improve the energy self-
sufficiency in WWTPs (Zhang et al., 2014). To improve anaerobic digestion, several
physical, chemical, and biological-based pretreatment methods have been employed.
Since most of the degradable substances are enclosed inside a microbial cell wall or
entangled in an extracellular polymeric matrix, only 35%–45% of the reduction in
volatile solids occurs resulting in limited biodegradability of the sludge (Bolzonella
et al., 2005; Bhattacharya et al., 1996). Pretreatment before anaerobic digestion has
many advantages such as (1) the released soluble substances enhance VFA generation
that results in increased biogas production. (2) Because of the pretreatment, viscosity is reduced that allowing greater solids to be fed into the digester. Because of
the higher solids concentration, increased gas production can be attained using a
lower digester volume (Elliott and Mahmood, 2007). During the past decade, several
studies have been done on the use of thermal-based pretreatment methods (conventional heating, microwave, thermal hydrolysis, and TPAD) that are applied to sewage
sludge that can accelerate the rate-limiting hydrolysis step and increase the biogas
production with an improved biosolid quality (Hasegawa et al., 2000; Pilli et al.,
2015; Ruffino et al., 2015). Thermal pretreatment results in a high level of solubilization and also a reduction in the levels of pathogenic microorganisms (Valo et al.,
2004; Ødegaard et al., 2002). Thermal pretreatments ranging from 50°C to 180°C
have been studied extensively with the period ranging from 15 to 60 minutes; however, temperatures exceeding 160°C result in the production of recalcitrant such as
phenols and furans that may have counter effects on anaerobic digestion (Appels
et al., 2013; Balasundaram et al., 2021; Climent et al., 2007).

This chapter aims to discuss the effects of different thermal-based pretreatment
methods on sludge solubilization and to compare the enhanced methane production.
The chapter starts with the principles behind anaerobic digestion followed by a discussion of several thermal-based pretreatment methods along with underlying mechanisms.

## 14.2	PRINCIPLES OF ANAEROBIC DIGESTION

Anaerobic digestion is a complex process that occurs in a series of steps, namely,
hydrolysis, acidogenesis, acetogenesis, and methanogenesis. During the hydrolysis step, the degradation of insoluble organic material and high-molecular-weight
compounds such as lipids, polysaccharides, proteins, and nucleic acids into soluble
organic substances occurs. During acidogenesis, VFA production takes place along
with ammonia, H_2S, CO_2, and other byproducts. During the next stage, acetogenesis
occurs in which the alcohols and higher organic acids that are produced during acidogenesis give rise to the formation of acetic acid, CO_2, and H_2S. During the last stage
called methanogenesis, two groups of methanogenic bacteria act on it; acetate gets
split into methane and carbon dioxide, and then hydrogen is used as an electron donor
and carbon dioxide is used as an acceptor to produce methane (Appels et al., 2008).

Parameters such as pH, alkalinity, and temperature may significantly affect the above-mentioned digestion steps. A pH between 6.5 and 7.2 is considered optimal for methanogenic bacteria, and they are extremely sensitive to pH changes. Even though VFA production may interfere with the pH of the digester, the effect can be compensated by the increase in alkalinity through the production of carbon dioxide, bicarbonate, and ammonia by methanogens (Boe, 2006).

14.3 CONVENTIONAL THERMAL PRETREATMENT

Thermal pretreatment also known as heat pretreatment is a well-established sludge pretreatment technology. Thermal pretreatment mainly enhances methane production, reduces the volume of sludge, aids in the destruction of pathogens, helps in the removal of odor, and improves dewaterability. Much impact on carbohydrates and proteins is caused by thermal pretreatment (Shrestha et al., 2020). During the process, because of the effects of temperature, pressure, and water, the substrate swells. In addition, the hydrogen bonds between are broken which increases the surface area between microorganisms and substrates which allows the digester to work in shorter Hydraulic Retention Time (HRTs) under reduced volume. Bougrier et al. (2006) attempted to investigate the effect of different temperature-based pretreatment on the semicontinuous anaerobic digestion of waste-activated sludge samples. It was concluded that the thermal pretreatments (130°C, 150°C, and 170°C for 30 minutes) led to an increase in methane production. The highest enhancement in methane yield was reported by the sludge pretreated at 170°C. Chen et al. (2020) investigated the effects of thermal and thermal alkaline pretreatment using continuous mesophilic anaerobic digestion (CMAD). The pretreatment conditions of TP and TAP were 134°C and 30 minutes, pH 12, 134°C ± 1°C and 30 minutes respectively. Compared to the control, the thermally pretreated sludge showed a 1.88 times increase in methane production, and the thermal alkaline pretreated sludge showed 2.2 times increase in methane production. It was concluded that both the treatment processes showed enhanced methane production however TAP showed higher efficiency. Biswal et al. (2020) studied the use of low thermal pretreatment to enhance the anaerobic digestion of waste-activated sludge. It was found that the soluble COD increased by 4.2–11.9 times. When the sludge was pretreated at 60°C and 80°C the methane yield increased by 13.7% and 27.0%. The authors concluded that low-temperature thermal pretreatment can help in enhancing anaerobic digestion. Appels et al. (2013) investigated the influence of low-temperature pretreatment on sludge solubilization and enhanced methane production. It was concluded that the efficiency of anaerobic digestion decreased for the sludge pretreated at 70°C. For the pretreatment conditions (90°C and 60 minutes) the biogas production increased by a factor of 11. When the sludge was pretreated within 200°C a linear trend was observed for COD solubilization versus temperature. However, when the temperature exceeded 200°C, the results seemed to be dispersed. Also, temperatures up to 190°C of resulted in enhanced biogas production. Bougrier et al. (2008) investigated the use of thermal sludge pretreatment at different temperatures and stated that the pretreatment at 190°C was more efficient than the treatment carried out at 135°C in terms of carbohydrate, protein, COD removal, and methane production. When thickened activated

TABLE 14.1

Effects of Conventional Thermal Pretreatment on Anaerobic Digestion

Pretreatment conditions	Effects of pretreatment	Observation on Anaerobic Digestion	References
Thermal 70°C, 9 hours	VS removal efficiency of 45%	$0.523\,m^3\,CH_4$/kg VS	Yu et al. (2014)
Thermal 175°C, 30 minutes	Methane production increased by 14%	234 mL/g COD compared to the 205 mL/g COD (control)	Haug et al. (1978)
Thermal 120°C–210°C, 60 minutes	Methane production increased by 19.5%–63.9%	$0.322\,m^3\,CH_4$/kg VS	Lee et al. (2017)
Thermal 121°C, 60 minutes	Biogas production increased by 20%	420 mL/g VS compared to 350 mL/g VS(control)	Barjenbruch and Kopplow (2003)
Thermal 70°C–90°C, 0.5–4 hours	Methane production increased by 59.82%	$0.286\,m^3$ biogas/kg VS	Gandhi et al. (2018)
Thermal 170°C, 60 minutes	Methane production increased by 61%	142 mL/g COD compared to 88 mL/g COD (control)	Graja et al. (2005)

sludge was pretreated for 60 min at different temperatures (70°C–90°C), a maximum increase in methane production (123%) was observed for the sludge pretreated at 90°C (Appels et al., 2010). Contrary to the above-mentioned results, Climent et al. (2007) stated that no significant increase in methane production was obtained when waste-activated sludge was pretreated at the temperatures of 110°C–134°C. A few studies about the effects of conventional thermal pretreatment on anaerobic digestion are enlisted in Table 14.1.

14.4 TEMPERATURE-PHASED ANAEROBIC DIGESTION (TPAD)

Globally, ~31% of the waste is dumped in open landfills out of which only 19% is recovered (via recycling and composting), and the rest is subjected to incineration. Open dumping is the most common waste management method practiced majorly in South Asia, the Middle East and North Africa, and Sub-Saharan Africa. With the increase in waste sludge generation from WWTPs, its disposal has become an issue. Sludge disposal and transportation bestow almost 50% of the total plant operational cost. Anaerobic digestion (AD) has proven to be a vital process in the treatment and valorization of sewage sludge, thus curtailing solid waste and allowing the recovery of renewable energy in the form of bioenergy/biofuels. However, hydrolysis of waste-activated sludge (WAS) is the rate-limiting step of the AD process (Gavala et al., 2003; Westerholm et al., 2016) and thus requires huge efforts to overcome this limitation as well as enhance waste biomass degradability. The temperature-phased AD (TPAD) method has shown more promising results than the conventional AD process with respect to biogas production, volatile solid reduction, pathogen annihilation, dewaterability, etc. TPAD involves a two-stage reactor procedure where the first stage is driven at a thermophilic temperature (~55°C) and the second stage at a mesophilic temperature (~35°C). Qin et al. (2017) used TPAD using different temperatures, i.e., thermophilic-mesophilic (TM) and hyperthermophilic–mesophilic (HM) conditions. TPAD was initiated from the TM (55°C and 35°C) condition and shifted to the HM (70°C and 35°C) condition on the 134th day within 24 hours. The volatile solid reduction in HM-TPAD was 14.5% higher than the control (mesophilic AD). Eminent segregation of microbial communities was observed in acidogenic and methanogenic phases in comparison to TM-TPAD, where the differences in microbial diversity were not much significant (Qin et al., 2017). Ge et al. (2011) reported enhanced VS reduction and methane yield of WAS by TPAD (70°C TAD and 35°C MAD) process. Also, mathematical modeling using AD Model No. 1 (ADM1) indicated better performance of the system due to an increased apparent hydrolysis rate, rather than an increase in degradability.

TPAD shows better VS reduction, biodegradability, or dewaterability than conventional AD; however, no significant differences were observed between TPAD and AD in terms of the overall formation of volatile sulfur compounds normalized per VS added (Akgul et al., 2016). Pathogen removal is an important factor for obtaining pathogen-free digestate that can be utilized for soil reclamation. Han et al. (1997) reported the complete removal of total and fecal coliforms in TPAD at 11–28 days of sludge retention times (SRT) which was always $\leq$1,000 MPN/g TS in the effluent. Also, the foaming of WAS during the AD process was eliminated. TPAD has been used to investigate different feedstocks like sewage sludge (primary and secondary), food waste, and organic fraction of municipal solid waste (Ge et al., 2010; Borowski, 2015; Akgul et al., 2017). Xiao et al. (2018) compared this two-phased AD of food waste with single-phased mesophilic and thermophilic AD and reported that biogas and CH_4 yields in TPAD were lower (0.759 $\pm$ 0.115 and 0.454 $\pm$ 0.201 L/g added VS, respectively) than that in MAD and TAD but had higher energy conversion efficiency than TAD. TPAD is one of the prime processes that has been explored for enhancing the biodegradability of waste and biogas production with a relatively low energy input and capital cost.

14.5 MICROWAVE IRRADIATION

Microwave operates at the wavelength of 1 mm–1 m and oscillation frequencies ranging from 0.3 to 300 GHz (Toreci et al., 2009). Sludge cell damage may occur in two ways, the rotation of dipoles under oscillating electromagnetic fields heats the liquid that is present inside the cells to its boiling point and causes cell rupture (Tang et al., 2010). The changing dipole orientation of the polar molecules breaks the hydrogen bonds that kill the microorganisms at low temperatures. Microwave has also been used to increase the solubilization of organic matter. Because of the lysing of cell membranes, the intracellular matter shall be exposed and the application of high temperature can lead to the formation of bubbles that results in the rupture of cell membranes (Neumann et al., 2016). When a MW irradiation of 175°C was applied, the solubilization and biogas production increased by 35% and 31%, respectively (Eskicioglu et al., 2009). When MW energy of 336 kJ/kg was applied, the soluble

COD increased by 214% compared to the untreated sludge Appels et al. (2013). Ahn et al. (2009) reported that the sCOD increased from 2% to 22% when MW pretreatment was applied for 15 minutes. MW pretreatment has been effectively shown to remove pathogens from sludge. When a 2,450 MHz MW was used for sludge pretreatment, >2.66 log removal of fecal coliforms was observed (Hong et al., 2006). Experiments conducted by Pino-Jelcic et al. (2006) at 60°C–65°C stated that neither fecal coliforms nor *Salmonella* sp. were detected in microwave-irradiated sludge. Eskicioglu et al. (2007) reported that the MW (1,250 W, 2,450-MHz frequency, and 12.24-cm wavelength) pretreatment causes significant damage to floc structure and releases 3.6, 4.2, and 4.5 times COD, sugars, and proteins. In batch digesters, a 20% increase in biogas production was obtained when compared to the control. In semi-continuous digesters, the SRT has reduced from 20 to 5 days with the implementation of microwave-based pretreatment. When microwave pretreatment was applied, a 3.6-fold increase in soluble to total COD was observed. Doğan and Sanin (2009) applied the use of alkaline solubilization and MW irradiation as a combined method of pretreatment for WAS. An increase of 16.3% in total gas production was obtained in the pretreated sludge (MW + pH-12) when compared to the control. Also, TS, VS, and TCOD reductions improved by 24.9%, 35.4%, and 30.3%. An improved dewaterability of 22% was obtained in the pretreated sludge compared to the control.

14.6 THERMAL HYDROLYSIS

Numerous sewage pretreatment methodologies have been investigated to date to enhance conventional anaerobic processes since the latter is quite a slow and sensitive method to carry out the efficacious digestion of organic waste. Different pretreatments for sludge help in its enhanced hydrolysis and faster degradation before moving to AD. The disposal of sewage sludge is particularly challenging and poses severe environmental hazards due to the high content of organic, toxic, and heavy metal pollutants among its constituents. Thermal hydrolysis has been studied extensively and many full-scale plants around the world are in operation (Kepp et al., 2000). Since hydrolysis is a rate-limiting step in AD thermal hydrolysis helps in overcoming this step by hydrolyzing the substrate prior to AD. This sludge conditioning/thermal treatment between 120°C and 200°C is employed with dewatered sludge cake prior to the AD process. The advantages of thermal hydrolysis are the production of class A biosolids, better digestate dewaterability, easy biomass degradability, reduced viscosity, allowing higher organic loading rates and lower SRTs of digesters, preventing digester souring, higher biogas yield, etc. Commercially, thermal hydrolysis systems utilizing sub-critical wet air oxidation methodologies were applied since the mid-20th century. The Zimpro process followed by the Porteus process was developed in the 1930s and 1960s, respectively. The Zimpro used steam as a source to heat sludge to 150°C and 300°C. While the Porteus used compressed air injected by an air compressor for heating up to similar temperatures, with an energy release of approximately 0.85 kWh/kg air added (Barber, 2020). However, these systems had certain drawbacks like bad odor in the vicinity, high operating costs, and production of wastewater with refractory compounds during the treatment process. The first commercially operating plant was in Hias, Norway in 1995, supplied by the Cambi, Norway. In the mid-2000s, a second supplier, Veolia, UK came with another thermal hydrolysis technology called Thelys, subsequently Biothelys. Other commercial-scale thermal hydrolysis processes in the municipal market included Haaeslev, Turbotech, Lysotherm, Beijing Hinergy, Shenzhen EST, etc.

14.6.1 THE CAMBI THERMAL HYDROLYSIS PROCESS (THP)

The Cambi THP is a more established technology with more than 25 installations globally while the Exelys process is a comparatively naïve technology in thermal hydrolysis. Cambi uses pressure and temperatures of ~120–130 psi and 330°F, respectively, for a duration of around half an hour to hydrolyze biosolids prior to the AD process. It possesses a series of batch reactors. Thermal hydrolysis using Cambi is a three-step process, where the first unit is a preheating tank that eliminates the problem of pumping under pressure and corrosion. The second unit is a steam tank in which the steam is introduced under pressure. Thereafter the sludge goes to the flash tank where the reactor pressure is released rapidly. Initially, the DS content of the sludge is adjusted to 14%–18% before starting the pretreatment. It is then continuously sent to the preheated tank where it reaches a temperature of 100°C and then it is sent to the steam reactor where the reaction times of 20–30 minutes with a temperature of 150°C–160°C and a pressure of 8–9 bars is maintained. Before releasing the sludge into the flash tank, the pressure is reduced to 2 bars which enables the rapid rupture of cells. In the flash tank, the temperature is reduced to 100°C. The main advantage of the Cambi is that the SRT can be reduced to 10–12 days in comparison with the conventional digester (15–30 days), and also the dewaterability can be improved (30%–40% DS) (Pilli et al., 2015). A schematic representation of Cambi THP is shown in Figure 14.1.

14.6.2 THE EXELYS THERMAL HYDROLYSIS

The Exelys™ process is a newer generation of thermal hydrolysis developed by Veolia. A full-scale pilot demonstration plant is located in Hillerod, Denmark. It utilizes a series of continuous plug-flow batch reactors. Usually, thermal hydrolysis is done prior to digestion, i.e., lysis–digestion (LD) mode. However, the Exelys process can also be operated between two digestion steps, i.e., digestion–lysis–digestion (DLD) mode. Dewatered sludge is continuously fed to a static mixer along with steam. Under the high pressure of 9 bars, the sludge is hydrolyzed at ~165°C in a plug-flow

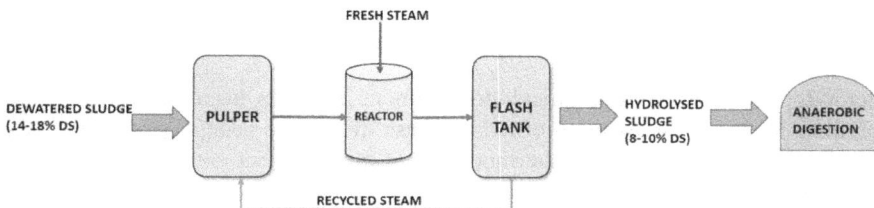

FIGURE 14.1 Schematic representation of the Cambi THP.

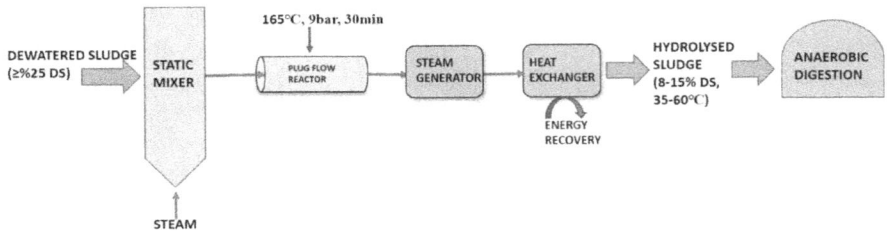

FIGURE 14.2 Schematic representation of the Biothelys process.

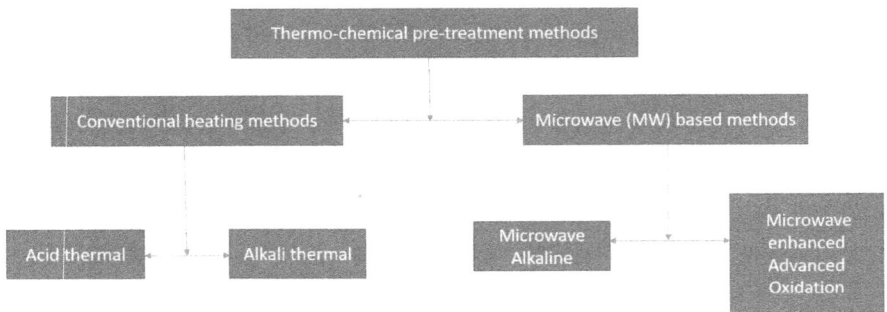

FIGURE 14.3 Classification of thermochemical pretreatment methods.

reactor with sufficient retention time (≥30 minutes). The hydrolyzed biomass is then cooled down in the heat exchangers and then fed to the digesters (Gurieff et al., 2011). Gurieff et al. (2011) compared both batch and continuous thermal hydrolysis pretreatments with both Biothelys and the Exelys processes and reported that the COD solubilization and increase in VSS were almost the same for both, with preferable outcomes for the continuous process. However, unlike Cambi, the steam is not recycled in the Exelys process and thus requires higher steam input/ton DS. However, on the other hand, Exelys is as efficient as Cambi in operating at a high solid content of sewage sludge.

Thus, in thermal hydrolysis, the flash reaction at high pressure causes cell disruptions, leading to higher solubilization of organic matter and the release of exopolysaccharides for microbial activity. A Schematic representation of the Biothelys process is shown in Figure 14.2.

14.7 THERMOCHEMICAL METHODS

Methods such as thermochemical pretreatment methods seem to be promising solutions in which the addition of acid or base avoids the necessity of higher temperatures. A classification of thermochemical pretreatment methods is shown in Figure 14.3. Besides, thermochemical pretreatment methods can be operated at low to moderate temperatures (Uma Rani et al., 2012). Several studies suggest the use of thermochemical pretreatment techniques resulting in increased sludge solubilization, thereby increasing biogas production. Thermochemical pretreatment involves the use of mineral acids, alkalis, and/or oxidation methods in varying temperature ranges for the breakdown of complex organic substances. Since hydrolysis is the rate-limiting step in AD, chemical-based reagents can be used to hydrolyze the cell wall and membrane, thus increasing the solubilization of organics present within the microbial cells (Tyagi and Lo, 2011).

14.7.1 CONVENTIONAL THERMOCHEMICAL TREATMENT METHODS

14.7.1.1 Alkali Thermal Treatment

Various chemical reagents such as NaOH, KOH, $Mg(OH)_2$, and $Ca(OH)_2$ have been used by researchers for the pretreatment of sludge (Kim et al., 2013). At extremely higher pH values of the medium, the cell loses its viability and it cannot maintain its turgor pressure and disrupts. It leads to the leakage of intracellular material out of the cell. Lipids, hydrocarbons, and proteins are decomposed and hydrolyzed into amino acids, polysaccharides, and aliphatic acids (Tyagi and Lo, 2011). The increase in COD solubilization through the use of alkali-based reagent could be explained by different reasons such as (1) degradation of particular materials causes the formation of various acids that results in neutralization, (2) reactions that occur with the free carboxylic groups, and (3) uronic acetyl esters and uronic acid saponification (Jeongsik Kim et al., 2003).

14.7.1.2 Acid Thermal Treatment

Acid-based pretreatment methods make use of HCl, H_2SO_4, H_3PO_4, and HNO_2. Furfural and hydroxyl methyl furfural are some of the inhibitory compounds formed by the use of strong acids and hence diluted acids along with thermal pretreatments are used. Pilli et al. (2020) conducted a thermochemical pretreatment study using waste sulfuric sludge obtained from a chemical Industry. The optimal conditions for pretreatment were deduced to be 30 minutes, 90°C, and 0.175 g waste acid/g dried sludge. Since particle size is one of the major parameters influencing sludge structure, the effect of acid thermal pretreatment on particles was studied and it was found that the average median particle size (D_{50}) of the pretreated sludge was 1,200 μm which was almost ten times larger than that of the untreated sludge. The possible reason for this could be that the addition of sulfuric acid could dissolve the metal ions in the extracellular polymeric substances, thereby releasing more soluble organic biopolymers. These organic biopolymers result in the aggregation of sludge particles because of the bridging flocculation effect thereby increasing its size. Moreover, increasing H^+ ions because of the addition of sulfuric acid could neutralize the negative surface charge of the particles thereby decreasing charge repulsion and causing the particles to agglomerate. At higher temperatures, the random motion of the particles suspended in a liquid medium (Brownian movement) increases the probability of collision causing aggregation. Thus, the increase in particle size leads to the formation of larger aggregates with the loose structure which promotes the passage of water thereby increasing its dewatering ability. Some studies carried out showing the effects of thermochemical pretreatment on AD are enlisted in Table 14.2.

TABLE 14.2

Effects of Thermochemical Pretreatment on Anaerobic Digestion

Pretreatment Conditions	Effects of Pretreatment	Effects on Anaerobic Digestion	References
7 g/NaOHL 121°C, 30 minutes	Biogas production increased by 38%	Biogas production increased from 3,657(control) to 5,037 L/m³	Kim et al. (2003)
45 meq/NaOHL 55°C, 240 minutes	Methane production increased by 88%	CH_4 production from 165 (control) to 310 L/kg VS	Heo et al. (2003)
1.65 g/KOHL pH: 10 130°C, 60 min	Methane production increased by 75%	CH_4 production from 88(control) to 154 mL/g COD_{in}	Valo et al. (2004)
7 g/NaOHL 121°C, 30 minutes	Methane production increased by 79%	CH_4 production from 290(control) to 520 L/kg VS	Park et al. (2005)

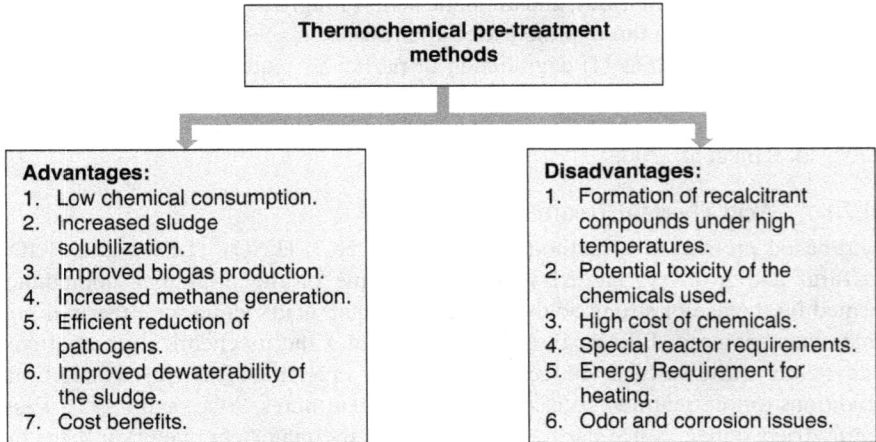

FIGURE 14.4   Advantages and disadvantages of thermochemical pretreatment.

14.7.2 MICROWAVE-BASED ALKALI PRETREATMENT

Microwave irradiation is based on electromagnetic radiation which is in the frequency range of 300 MHz to 300 GHz that corresponds to the wavelength range of 1 m down to 1 mm. The principle of microwave mainly lies in the polar molecules such as water, proteins, and lipids in the medium. The electric field of the microwave causes the polar molecules to rotate leading to friction. Because of this, energy in the form of heat is released into the system. However, if techniques such as alkaline pretreatment and microwave irradiation are combined, the results would be more efficient. Chang et al. (2011) carried out a study using microwave alkali pretreatment and reported an increased COD solubilization of 46%. A comparative study between conventional heating and MW heating was done by Pino-Jelcic et al. (2006), and it was found that the latter showed better results in terms of sludge solubilization, gas production, and sludge dewaterability. Moreover, a few advantages and disadvantages of thermochemical pretreatment are shown in Figure 14.4.

14.8 CONCLUSIONS

Low- and high-temperature thermal pretreatment significantly increases solubility and thereby increases the degradation of complex organic matter that results in increased biogas production. Most studies have suggested the use of temperatures below 160°C for thermal pretreatment, beyond which the formation of toxic recalcitrant has been reported that may have deleterious effects on AD and subsequent methane generation. Commercially existing thermal-based pretreatment processes such as Cambi thermal hydrolysis and Biothelys have been proven to enhance dewaterability. Even though these pretreatment methods for improving AD may incur investment and operational costs they can be compensated by the benefits from the increased gas production due to AD. Therefore, the thermal-based pretreatment methods are a very attractive solution for improving the process viability by reducing the rate-limiting hydrolysis, supporting microbial activity, and increasing the methane yield.

REFERENCES

Ahn, J.H., Shin, S.G. and Hwang, S., 2009. Effect of microwave irradiation on the disintegration and acidogenesis of municipal secondary sludge. *Chemical Engineering Journal*, *153*(1–3), pp. 145–150.

Akgul, D., Cella, M.A. and Eskicioglu, C., 2017. Influences of low-energy input microwave and ultrasonic pretreatments on single-stage and temperature-phased anaerobic digestion (TPAD) of municipal wastewater sludge. *Energy*, *123*, pp. 271–282.

Akgul, D.E.N.İ.Z., Cella, M.A. and Eskicioglu, C., 2016. Temperature phased anaerobic digestion of municipal sewage sludge: A Bardenpho treatment plant study. *Water Practice and Technology*, *11*(3), pp. 569–573.

Appels, L., Baeyens, J., Degrève, J. and Dewil, R., 2008. Principles and potential of the anaerobic digestion of waste-activated sludge. *Progress in Energy and Combustion Science*, *34*(6), pp. 755–781.

Appels, L., Degrève, J., Van der Bruggen, B., Van Impe, J. and Dewil, R., 2010. Influence of low temperature thermal pre-treatment on sludge solubilisation, heavy metal release and anaerobic digestion. *Bioresource Technology*, *101*(15), pp. 5743–5748.

Appels, L., Houtmeyers, S., Degrève, J., Van Impe, J. and Dewil, R., 2013. Influence of microwave pre-treatment on sludge solubilization and pilot scale semi-continuous anaerobic digestion. *Bioresource Technology*, *128*, pp. 598–603.

Balasundaram, G., Banu, R., Varjani, S., Kazmi, A.A. and Tyagi, V.K., 2021. Recalcitrant compounds formation, their toxicity, and mitigation: Key issues in biomass pretreatment and anaerobic digestion. *Chemosphere*, *291*, p. 132930.

Barber, B., 2020. *Sludge Thermal Hydrolysis: Application and Potential*. IWA Publishing: Perth.

Barjenbruch, M. and Kopplow, O., 2003. Enzymatic, mechanical and thermal pre-treatment of surplus sludge. *Advances in Environmental Research*, *7*(3), pp. 715–720.

Bhattacharya, S.K., Madura, R.L., Walling, D.A. and Farrell, J.B., 1996. Volatile solids reduction in two-phase and conventional anaerobic sludge digestion. *Water Research*, *30*(5), pp. 1041–1048.

Biswal, B.K., Huang, H., Dai, J., Chen, G.H. and Wu, D., 2020. Impact of low-thermal pretreatment on physicochemical properties of saline waste activated sludge, hydrolysis of organics and methane yield in anaerobic digestion. *Bioresource Technology*, *297*, p. 122423.

Boe, K., 2006. Online monitoring and control of the biogas process. *Archives of Disease in Childhood*, *59*(6), pp. 36–44.

Bolzonella, D., Pavan, P., Battistoni, P. and Cecchi, F., 2005. Mesophilic anaerobic digestion of waste activated sludge: Influence of the solid retention time in the wastewater treatment process. *Process Biochemistry*, *40*(3–4), pp. 1453–1460.

Borowski, S., 2015. Temperature-phased anaerobic digestion of the hydromechanically separated organic fraction of municipal solid waste with sewage sludge. *International Biodeterioration & Biodegradation*, *105*, pp. 106–113.

Bougrier, C., Delgenes, J.P. and Carrère, H., 2006. Combination of thermal treatments and anaerobic digestion to reduce sewage sludge quantity and improve biogas yield. *Process Safety and Environmental Protection*, *84*(4), pp. 280–284.

Bougrier, C., Delgenès, J.P. and Carrère, H., 2008. Effects of thermal treatments on five different waste activated sludge samples solubilisation, physical properties and anaerobic digestion. *Chemical Engineering Journal*, *139*(2), pp. 236–244.

Chang, C.J., Tyagi, V.K. and Lo, S.L., 2011. Effects of microwave and alkali induced pretreatment on sludge solubilization and subsequent aerobic digestion. *Bioresource Technology*, *102*(17), pp. 7633–7640.

Chen, H., Yi, H., Li, H., Guo, X. and Xiao, B., 2020. Effects of thermal and thermal-alkaline pretreatments on continuous anaerobic sludge digestion: Performance, energy balance and, enhancement mechanism. *Renewable Energy*, *147*, pp. 2409–2416.

Climent, M., Ferrer, I., del Mar Baeza, M., Artola, A., Vázquez, F. and Font, X., 2007. Effects of thermal and mechanical pretreatments of secondary sludge on biogas production under thermophilic conditions. *Chemical Engineering Journal*, *133*(1–3), pp. 335–342.

Doğan, I. and Sanin, F.D., 2009. Alkaline solubilization and microwave irradiation as a combined sludge disintegration and minimization method. *Water Research*, *43*(8), pp. 2139–2148.

Elliott, A. and Mahmood, T., 2007. Pretreatment technologies for advancing anaerobic digestion of pulp and paper biotreatment residues. *Water Research*, *41*(19), pp. 4273–4286.

Eskicioglu, C., Droste, R.L. and Kennedy, K.J., 2007. Performance of anaerobic waste activated sludge digesters after microwave pretreatment. *Water Environment Research*, *79*(11), pp. 2265–2273.

Eskicioglu, C., Kennedy, K.J. and Droste, R.L., 2009. Enhanced disinfection and methane production from sewage sludge by microwave irradiation. *Desalination*, *248*(1–3), pp. 279–285.

Gahlot, P., Aboudi, K., Ahmed, B., Tawfik, A., Khan, A.A., Khursheed, A. and Tyagi, V.K., 2021. Direct interspecies electron transfer (DIET) via conductive materials in anaerobic digestion of organic wastes. In: V.K. Tyagi, M. Kumar, A.K.J. An and Z. Cetecioglu (Eds.), *Clean Energy and Resources Recovery* (pp. 227–252). Elsevier: Amsterdam, Netherlands.

Gahlot, P., Ahmed, B., Tiwari, S.B., Aryal, N., Khursheed, A., Kazmi, A.A. and Tyagi, V.K., 2020. Conductive material engineered direct interspecies electron transfer (DIET) in anaerobic digestion: Mechanism and application. *Environmental Technology & Innovation*, 20, p. 101056.

Gandhi, P., Paritosh, K., Pareek, N., Mathur, S., Lizasoain, J., Gronauer, A., Bauer, A. and Vivekanand, V., 2018. Multicriteria decision model and thermal pretreatment of hotel food waste for robust output to biogas: Case study from city of Jaipur, India. *BioMed Research International*, 2018, p. 9416249.

Gavala, H.N., Yenal, U., Skiadas, I.V., Westermann, P. and Ahring, B.K., 2003. Mesophilic and thermophilic anaerobic digestion of primary and secondary sludge. Effect of pretreatment at elevated temperature. *Water Research*, *37*(19), pp. 4561–4572.

Ge, H., Jensen, P.D. and Batstone, D.J., 2010. Pre-treatment mechanisms during thermophilic–mesophilic temperature phased anaerobic digestion of primary sludge. *Water Research*, *44*(1), pp. 123–130.

Ge, H., Jensen, P.D. and Batstone, D.J., 2011. Temperature phased anaerobic digestion increases apparent hydrolysis rate for waste activated sludge. *Water Research*, *45*(4), pp. 1597–1606.

Graja, S., Chauzy, J., Fernandes, P., Patria, L. and Cretenot, D., 2005. Reduction of sludge production from WWTP using thermal pretreatment and enhanced anaerobic methanisation. *Water Science and Technology*, *52*(1–2), pp. 267–273.

Gurieff, N., Bruus, J., Hoejsgaard, S., Boyd, J. and Kline, M., 2011. Maximizing energy efficiency and biogas production: Exelys™–continuous thermal hydrolysis. *Proceedings of the Water Environment Federation*, *2011*(4), pp. 1103–1116.

Han, Y., Sung, S. and Dague, R.R., 1997. Temperature-phased anaerobic digestion of wastewater sludges. *Water Science and Technology*, *36*(6–7), pp. 367–374.

Hasegawa, S., Shiota, N., Katsura, K. and Akashi, A., 2000. Solubilization of organic sludge by thermophilic aerobic bacteria as a pretreatment for anaerobic digestion. *Water Science and Technology*, *41*(3), pp. 163–169.

Haug, R.T., Stuckey, D.C., Gossett, J.M. and McCarty, P.L., 1978. Effect of thermal pretreatment on digestibility and dewaterability of organic sludges. *Journal (Water Pollution Control Federation)*, *50*, pp. 73–85.

Heo, N.H., Park, S.C., Lee, J.S. and Kang, H., 2003. Solubilization of waste activated sludge by alkaline pretreatment and biochemical methane potential (BMP) tests for anaerobic co-digestion of municipal organic waste. *Water Science and Technology*, *48*(8), pp. 211–219.

Hong, S.M., Park, J.K., Teeradej, N., Lee, Y.O., Cho, Y.K. and Park, C.H., 2006. Pretreatment of sludge with microwaves for pathogen destruction and improved anaerobic digestion performance. *Water Environment Research*, *78*(1), pp. 76–83.

Kepp, U., Machenbach, I., Weisz, N. and Solheim, O.E., 2000. Enhanced stabilisation of sewage sludge through thermal hydrolysis-three years of experience with full scale plant. *Water Science and Technology*, *42*(9), pp. 89–96.

Kim, J., Park, C., Kim, T.H., Lee, M., Kim, S., Kim, S.W. and Lee, J., 2003. Effects of various pretreatments for enhanced anaerobic digestion with waste activated sludge. *Journal of Bioscience and Bioengineering*, *95*(3), pp. 271–275.

Kim, J., Yu, Y. and Lee, C., 2013. Thermo-alkaline pretreatment of waste activated sludge at low-temperatures: effects on sludge disintegration, methane production, and methanogen community structure. *Bioresource Technology*, *144*, pp. 194–201.

Lee, I.J., Lee, Y.S. and Han, G.B., 2017. Effect of two-step thermo-disintegration on the enhancing solubilization and anaerobic biodegradability of excess waste sludge. *Chemical Engineering Journal*, *317*, pp. 742–750.

Neumann, P., Pesante, S., Venegas, M. and Vidal, G., 2016. Developments in pre-treatment methods to improve anaerobic digestion of sewage sludge. *Reviews in Environmental Science and Bio/Technology*, *15*(2), pp. 173–211.

Ødegaard, H., Paulsrud, B. and Karlsson, I., 2002. Wastewater sludge as a resource: sludge disposal strategies and corresponding treatment technologies aimed at sustainable handling of wastewater sludge. *Water Science and Technology*, *46*(10), pp. 295–303.

Park, C., Lee, C., Kim, S., Chen, Y. and Chase, H.A., 2005. Upgrading of anaerobic digestion by incorporating two different hydrolysis processes. *Journal of Bioscience and Bioengineering*, *100*(2), pp. 164–167.

Pilli, S., Yan, S., Tyagi, R. D., & Surampalli, R. Y. 2015. Thermal pretreatment of sewage sludge to enhance anaerobic digestion: A review. *Critical Reviews in Environmental Science and Technology*, *45*(6), 669–702.

Pino-Jelcic, S.A., Hong, S.M. and Park, J.K., 2006. Enhanced anaerobic biodegradability and inactivation of fecal coliforms and Salmonella spp. in wastewater sludge by using microwaves. *Water Environment Research*, *78*(2), pp. 209–216.

Qin, Y., Higashimori, A., Wu, L.J., Hojo, T., Kubota, K. and Li, Y.Y., 2017. Phase separation and microbial distribution in the hyperthermophilic-mesophilic-type temperature-phased anaerobic digestion (TPAD) of waste activated sludge (WAS). *Bioresource Technology*, *245*, pp. 401–410.

Ruffino, B., Campo, G., Genon, G., Lorenzi, E., Novarino, D., Scibilia, G. and Zanetti, M., 2015. Improvement of anaerobic digestion of sewage sludge in a wastewater treatment plant by means of mechanical and thermal pre-treatments: Performance, energy and economical assessment. *Bioresource Technology*, *175*, pp. 298–308.

Shrestha, B., Hernandez, R., Fortela, D.L.B., Sharp, W., Chistoserdov, A., Gang, D., Revellame, E., Holmes, W. and Zappi, M.E., 2020. A review of pretreatment methods to enhance solids reduction during anaerobic digestion of municipal wastewater sludges and the resulting digester performance: Implications to future urban biorefineries. *Applied Sciences*, *10*(24), p. 9141.

Pilli, S., Pandey, A.K., Katiyar, A., Pandey, K. and Tyagi, R.D., 2020. Pre-treatment technologies to enhance anaerobic digestion. In *Sustainable Sewage Sludge Management and Resource Efficiency*. London: IntechOpen.

Tang, B., Yu, L., Huang, S., Luo, J. and Zhuo, Y., 2010. Energy efficiency of pre-treating excess sewage sludge with microwave irradiation. *Bioresource Technology*, *101*(14), pp. 5092–5097.

Toreci, I., Kennedy, K. J. and Droste, R. L. (2009). Evaluation of continuous mesophilic anaerobic sludge digestion after high temperature microwave pretreatment. *Water Research*, *43*(5), pp. 1273–1284. doi: 10.1016/j.watres.2008.12.022.

Turonskiy, I.S. and Mathai, P.K., 2006. *Wastewater Sludge Processing*, Wiley Interscience: Hoboken, NJ.

Tyagi, V.K. and Lo, S.L., 2011. Application of physico-chemical pretreatment methods to enhance the sludge disintegration and subsequent anaerobic digestion: An up to date review. *Reviews in Environmental Science and Bio/Technology*, *10*(3), pp. 215–242.

Uma Rani, R., Adish Kumar, S., Kaliappan, S., Yeom, I. T., & Rajesh Banu, J. (2012). Low temperature thermo-chemical pretreatment of dairy waste activated sludge for anaerobic digestion process. *Bioresource Technology*, *103*(1), pp. 415–424.

Valo, A., Carrère, H. and Delgenès, J.P., 2004. Thermal, chemical and thermo-chemical pre-treatment of waste activated sludge for anaerobic digestion. *Journal of Chemical Technology & Biotechnology: International Research in Process, Environmental & Clean Technology*, *79*(11), pp. 1197–1203.

Westerholm, M., Crauwels, S., Houtmeyers, S., Meerbergen, K., Van Geel, M., Lievens, B. and Appels, L., 2016. Microbial community dynamics linked to enhanced substrate availability and biogas production of electrokinetically pre-treated waste activated sludge. *Bioresource Technology*, *218*, pp. 761–770.

Xiao, B., Qin, Y., Zhang, W., Wu, J., Qiang, H., Liu, J. and Li, Y.Y., 2018. Temperature-phased anaerobic digestion of food waste: A comparison with single-stage digestions based on performance and energy balance. *Bioresource Technology*, *249*, pp. 826–834.

Yu, B., Xu, J., Yuan, H., Lou, Z., Lin, J. and Zhu, N., 2014. Enhancement of anaerobic digestion of waste activated sludge by electrochemical pretreatment. *Fuel*, *130*, pp. 279–285.

Zhang, W., Wei, Q., Wu, S., Qi, D., Li, W., Zuo, Z. and Dong, R., 2014. Batch anaerobic co-digestion of pig manure with dewatered sewage sludge under mesophilic conditions. *Applied Energy*, *128*, pp. 175–183.

15 Management and Disposal of Solid Waste

Practices and Legislations in Different Countries

Arshad Husain
Aligarh Muslim University

CONTENTS

15.1 Introduction 272
15.2 Types of Solid Waste 273
    15.2.1 Municipal Solid Waste 273
    15.2.2 Hazardous Waste 273
    15.2.3 Hospital Waste 274
15.3 Health Impacts of Solid Waste 274
15.4 Life Cycle of Municipal Solid Waste 275
    15.4.1 Determination of Individual Components of Solid Waste
        from MSW 275
15.5 Guidance for Carrying Out Waste Sampling and Analysis 276
    15.5.1 Procedure 277
    15.5.2 Analysis 277
    15.5.3 Energy Equivalent of Solid Waste 278
15.6 Legal Framework of Solid Waste Management in India 278
    15.6.1 The Municipal Solid Waste (Management and Handling)
        Rules 2000 279
        15.6.1.1 Collection of Solid Waste 280
    15.6.2 Environmental Protection – From the Indian Constitution
        Perspective 280
    15.6.3 Hazardous Wastes (Management and Handling) Amendment
        Rules, 2003 281
15.7 Technologies Used for Disposal of Municipal Solid Waste 281
    15.7.1 Composting 281
    15.7.2 Anaerobic Digestion 283
    15.7.3 Incineration 283
    15.7.4 Sanitary Landfills and Landfill Gas Recovery 284
15.8 Municipal Solid Waste Management (MSWM) in the Kingdom of
    Saudi Arabia 284

DOI: 10.1201/9781003202431-15

15.9   Municipal Solid Waste Management (MSWM) in Nigeria 285
15.10   Integrated Waste Management 286
         15.10.1 Rules and Legislation 286
15.11   Municipal Solid Waste Management (MSWM) in Australia................... 287
15.12   Municipal Solid Waste Management (MSWM) in Spain 289
15.13   Municipal Solid Waste Management (MSWM) in Ghana....................... 291
15.14   Municipal Solid Waste Management (MSWM) in Hong Kong............... 292
References.. 293

15.1 INTRODUCTION

The total Indian urban population amounts to approximately 1.38 billion. There are 4,378 cities and towns in India. Of those cities, according to the 2001 census, 423 are considered class I, meaning that the population exceeds 100,000. The class I cities alone contribute to more than 72% of the total municipal solid waste (MSW) generated in urban areas. Class I cities include 7 mega cities (which have a population of more than 4 million), 28 metro cities (which have a population of more than 1 million), and 388 other towns (which have a population of more than 100,000). The population growth rate in urban India is high. The percentage of the total population living in urban areas shows a continuous increase. For 2015, a value of 32.2% is predicted. Although there are no comprehensive data on waste generation rates, collection coverage, storage, transport, and disposal volumes and practices, the Central Public Health and Environmental Engineering Organization (CPHEEO) estimated a per capita waste generation in Indian cities and towns in the range of 0.2–0.6 kg/day. A World Bank publication estimated that in 2000 urban India produces approximately 100,000 metric tonnes of MSW daily or approximately 35 million metric tonnes of MSW annually. Comparing 1996 with 2005 shows how the physical composition of MSW can change over time along with the changing lifestyle and economic growth of the country. Although the typical urban growth rate has been determined at around 2.5% annually, the growth of waste generation is outpacing the urban population growth in Indian cities (Pratap et al. 2021). Therefore, urban population growth as well as increasing per capita waste generation will continue to amplify the waste problem. To prevent future problems, India must take immediate steps to control waste generation, enhance recycling recovery and reuse and ensure better collection and sustainable disposal.

According to the Central Pollution Control Board (CPCB, 1998), the average collection coverage ranges from 50% to 90%. Furthermore, of all collected waste, 94% is disposed of in an unacceptable manner without any consideration of state-of-the-art engineering principles. Hence, there is severe degradation of groundwater and surface water through leachate, as well as degradation of air through uncontrolled burning of waste.

Solid wastes are all the wastes arising from human and animal activities that are normally solid and that are discarded as useless or unwanted. Since the beginning, humankind has been generating waste, be it the bones and other parts of animals they slaughtered for their food or the wood they cut to make their carts. With the progress of civilisation, the waste generated became of a more complex nature. At the end of the 19th century, the industrial revolution saw the rise of the world of consumers. Not only did the air get more and more polluted, but the earth itself became more polluted with the generation of non-biodegradable solid waste. The increase in population and urbanisation was also largely responsible for the increase in solid waste.

15.2   TYPES OF SOLID WASTE

Solid waste can be classified into different types depending on their source:

1. Household waste is generally classified as municipal waste
2. Industrial waste as hazardous waste
3. Biomedical waste or hospital waste as infectious waste

15.2.1   Municipal Solid Waste

It consists of household waste, construction and demolition debris, sanitation residue and waste from streets. This garbage is generated mainly from residential and commercial complexes. With rising urbanisation and changes in lifestyle and food habits, the amount of MSW has been increasing rapidly and its composition changing. In 1947, cities and towns in India generated an estimated 6 million tonnes of solid waste; in 1997, it was about 48 million tonnes; in 2011, it was about 68.8 million tonnes. More than 25% of MSW is not collected at all; 70% of Indian cities lack adequate capacity to transport it and there are no sanitary landfills to dispose of the waste. The existing landfills are neither well equipped nor well managed and are not lined properly to protect against contamination of soil and groundwater. Over the past few years, the consumer market has grown rapidly, leading to products being packed in cans, aluminium foils, plastics and other such non-biodegradable items that cause incalculable harm to the environment. In India, some municipal areas have banned the use of plastics and they seem to have achieved success. For example, today one will not see a single piece of plastic in the entire district of Ladakh where the local authorities imposed a ban on plastics in 1998. Other states should follow the example of this region and ban the use of items that cause harm to the environment. One positive note is that in many large cities, shops have begun packing items in reusable or biodegradable bags.

15.2.2   Hazardous Waste

Industrial and hospital waste are considered hazardous as they may contain toxic substances. Certain types of household waste are also hazardous. Hazardous wastes could be highly toxic to humans, animals and plants; are corrosive, highly inflammable or explosive; and react when exposed to certain things, e.g. gases. India generates around 6.2 million tonnes of hazardous waste every year from 36,165 hazardous waste-generating industries. Household wastes that can be categorised as hazardous waste include old batteries, shoe polish, paint tins, old medicines and medicine bottles. In the industrial sector, the major generators of hazardous waste are metals, direct exposure to chemicals in hazardous waste such as mercury and cyanide can be fatal.

15.2.3 Hospital Waste

Hospital waste is generated during the diagnosis, treatment, or immunisation of human beings or animals or in research activities in these fields or in the production or testing of biological. It may include waste like sharps, soiled waste, disposables, anatomical waste, cultures, discarded medicines, chemical wastes etc. These are in the form of disposable syringes, swabs, bandages, body fluids, human excreta etc. This waste is highly infectious and can be a serious threat to human health if not managed in a scientific and discriminatory manner. It has been roughly estimated that of the 4 kg of waste generated in a hospital at least 1 kg would be infected. Hospital waste contaminated by chemicals used in hospitals is considered hazardous. These chemicals include formaldehyde and phenols, which are used as disinfectants, and mercury, which is used in thermometers or equipment that measure blood pressure. Most hospitals in India do not have proper disposal facilities for these hazardous wastes.

15.3   HEALTH IMPACTS OF SOLID WASTE

Modernisation and progress have had their share of disadvantages, and one of the main aspects of concern is the pollution they cause to the earth – be it land, air or water. With the increase in the global population and the rising demand for food and other essentials, there has been a rise in the amount of waste being generated daily by each household. This waste is ultimately thrown into municipal waste collection centres from where it is collected by the area municipalities to be further thrown into landfills and dumps. However, either due to resource crunch or inefficient infrastructure, not all of this waste gets collected and transported to the final dumpsites. If at this stage the management and disposal are improperly done, it can cause serious impacts on health and problems to the surrounding environment. Waste that is not properly managed, especially excreta and other liquid and solid waste from households and the community, is a serious health hazard and leads to the spread of infectious diseases. Unattended waste lying around attracts flies, rats and other creatures that in turn spread disease. Normally, it is the wet waste that decomposes and releases a bad odour. This leads to unhygienic conditions and thereby to a rise in health problems. The plague outbreak in Surat is a good example. Plastic waste is another cause of ill health. Thus, excessive solid waste that is generated should be controlled by taking certain preventive measures.

The group at risk from the unscientific disposal of solid waste include – the population in areas where there is no proper waste disposal method, especially preschool children, waste workers and workers in facilities producing toxic and infectious material. Another high-risk group includes the population living close to a waste dump and those, whose water supply has become contaminated either due to waste dumping or leakage from landfill sites. Uncollected solid waste also increases the risk of injury and infection.

In particular, organic domestic waste poses a serious threat, since they ferment, creating conditions favourable to the survival and growth of microbial pathogens. Direct handling of solid waste can result in various types of infectious and chronic diseases with the waste workers and the rag pickers being the most vulnerable.

Exposure to hazardous waste can affect human health; children are more vulnerable to these pollutants. In fact, direct exposure can lead to diseases through chemical exposure as the release of chemical waste into the environment leads to chemical poisoning. Many studies have been carried out in various parts of the world to establish a connection between health and hazardous waste.

Waste from agriculture and industries can also cause serious health risks. Other than this, co-disposal of industrial hazardous waste with municipal waste can expose people to chemical and radioactive hazards. Uncollected solid waste can also obstruct stormwater runoff, resulting in the formation of stagnant water bodies that become the breeding ground of disease. Waste dumped near a water source also causes contamination of the water body or the groundwater source. The direct dumping of untreated waste in rivers, seas and lakes results in the accumulation of toxic substances in the food chain through the plants and animals that feed on it directly or indirectly.

The disposal of hospital and other medical waste requires special attention since this can create major health hazards. This waste generated from hospitals, healthcare centres, medical laboratories and research centres such as discarded syringe needles, bandages, swabs, plasters and other types of infectious waste is often disposed of with regular non-infectious waste.

Waste treatment and disposal sites can also create health hazards for the neighbourhood. Improperly operated incineration plants cause air pollution and improperly managed and designed landfills attract all types of insects and rodents that spread disease. Ideally, these sites should be located at a safe distance from all human settlements. Landfill sites should be well-lined and walled to ensure that there is no leakage into the nearby groundwater sources. Recycling too carries health risks if proper precautions are not taken. Workers working with waste containing chemicals and metals may experience toxic exposure.

Disposal of healthcare waste requires special attention since it can create major health hazards, such as hepatitis B and C, through wounds caused by discarded syringes. Rag pickers and others, who are involved in scavenging in the waste dumps for items that can be recycled, may sustain injuries and come into direct contact with these infectious items.

15.4 LIFE CYCLE OF MUNICIPAL SOLID WASTE

Solid waste can be further explored and studied for different economies in terms of its composition as shown in Table 15.1.

15.4.1 DETERMINATION OF INDIVIDUAL COMPONENTS OF SOLID WASTE FROM MSW

As solid wastes are heterogeneous in nature, determination of the composition is not an easy task. For this reason, a more generalised field procedure based on commonsense and random sampling techniques has been developed as determining the composition of solid waste. The procedure involves unloading a quantity of waste

TABLE 15.1

Relative Composition of Household Waste in Low, Medium and High-Income Countries

Parameter		Low-Income Countries	Medium-Income	High-Income Countries
Contents	Organic (putrescible) (%)	40–85	20–65	20–30
	Paper (%)	1–10	15–30	15–40
	Plastics (%)	1–5	2–6	2–10
	Metal (%)	1–5	1–5	3–13
	Glass (%)	1–10	1–10	4–10
	Rubber, leather etc. (%)	1–5	1–5	2–10
	Other (%)	15–60	15–50	2–10
Physical and	Moisture content (%)	40–80	40–60	5–20
chemical	Specific weight (kg/m³)	250–500	170–330	100–170
properties	Calorific value (kcal/kg)	800–1,100	1,000–1,300	1,500–2,700

in a controlled area of a disposal site that is isolated from winds and separate from other operations. A representative residential sample might be a truckload resulting from a typical daily collection in a residential area. To ensure that the result obtained is sound statically, a large-enough sample must be obtained. It has been found that measurements made on a sample size of about 200 lb are very insignificant from the measurement made on samples up to 1,700 lb taken from the same waste load. The following technique is followed for the assessment of solid waste components:

1. Unload a truckload of waste in a controlled area away from other operations.
2. Quarter the waste load.
3. Select one of the quarters, and quarter that quarter.
4. Select one of the quartered quarters and separate the entire individual component of the waste into preselected components such as food waste, paper, cardboard, plastics, rubber, textiles, leather, wood, glass, metals, dirt etc.
5. Place the separated components in a container of known volume and tare mass and measure the volume and mass of each component.
6. Determine the percentage distribution of each component by mass. Typically 100 to 200 kg of waste should be sorted to obtain a representative sample. To obtain a more representative distribution of components, samples should be collected during each season of the year.

15.5 GUIDANCE FOR CARRYING OUT WASTE SAMPLING AND ANALYSIS

Depending on the needs of the client analysis of waste may have different objectives. For instance, waste producers want to know what kind of recovery/disposal is possible, managers of waste treatment plants need to know if they can accept and will be able to treat the waste and authorities are interested in the environmental effects related to a particular waste. The different needs of the concerned actors lead to different testing programmes in which sample taking is required. The strategy of sampling and analysis has to be planned in advance very carefully in order to avoid useless efforts and unnecessary costs. The correct procedure of sampling is very important to get a representative sample of the specific waste subject to testing. A representative sample of a specific waste is important to ensure the reliability of the analysis results obtained, which is the decision basis for the subsequent choice of waste management operations and handling. The objective is to become familiar with the characteristics of municipal solid wastes while investigating techniques used to determine the different fractions of solid waste. Various components of MSW would be separated into ferrous metal, nonferrous metal, glass, paper, plastic, wood, food products and yard trimmings. The volume and mass fractions of each category will be determined. If the source of the solid waste is known, the generation per person per capita day will also be calculated.

15.5.1 Procedure

Weigh the solid waste that is to be used for the laboratory. This will give the total mass (M_t). Estimate the volume of the solid waste by placing (not packing) it in a container and measuring the dimensions. This will give the total volume (Vt). Separate the solid waste into fractions containing ferrous and nonferrous metal, glass, paper, plastic, wood, food products, yard trimmings and unspecified objects. Weigh each fraction. Record each of these individual fractions as M_1, M_2 etc. Using a ruler and tape measure, estimate the volume of each fraction of the solid waste. Volumetric containers such as beakers or buckets can also be used. Keep in mind that solid wastes do not usually conform to geometric shapes. Use your imagination and engineering reasoning. Record each of these individual fractions as V_1, V_2 etc. The sum of these weights should be equal to V_t.

15.5.2 Analysis

Calculate the mass fraction, X_{mi}, of each category of solid waste using the following formula: where M_t is the total mass and M_{iis} is the mass of each individual category (M_1, M_2 etc.).

Once you have calculated X_{mi} for each category of waste, the sum of these fractions should equal 1.00.

$$X_{mi} = \frac{M_1}{M_t}$$

Similarly, calculate the volume fraction, X_{vi}, of each category using the following formula:

$$X_{vi} = \frac{V_i}{V_t}$$

where V_t is the total volume and V_i is the volume of each individual category. As with the mass, the sum of these fractions should equal 1.00.

15.5.3 ENERGY EQUIVALENT OF SOLID WASTE

The energy requirements of a community can be satiated to some extent by energy recovery from wastes as a better alternative to landfilling. Energy recovery is a method of recovering the chemical energy in MSW. Chemical energy stored in wastes is a fraction of the input energy expended in making those materials. Due to the difference in resources (materials/energy) that can be recovered, energy recovery falls below material recovery on the hierarchy of waste management.

15.6 LEGAL FRAMEWORK OF SOLID WASTE MANAGEMENT IN INDIA

In India, SWM is the primary responsibility and duty of the municipal authorities. State legislation and the local acts that govern municipal authorities include special provisions for the collection, transport and disposal of waste. They assign the responsibility for the provision of services to the chief executive of the municipal authority.

Most state legislation does not cover the necessary technical or organisational details of SWM. Laws talk about sweeping streets, providing receptacles in various parts of the city for storage of waste and transporting waste to disposal sites in general terms, but they do not clarify how this cleaning shall or can be done. The municipal acts do not specify in clear terms which responsibilities belong to the citizens (for example, the responsibility not to litter or the accountability for storing waste at its source). Moreover, they do not mention specific collection systems (such as the door-to-door collection of waste), do not mandate appropriate types of waste storage depots, do not require covered waste transport issues and do not mention aspects of waste treatment or sanitary landfills. Thus, most state legislation, with the exception of that of Kerala, does not fulfil the requirements for an efficient SWM service. Given the absence of appropriate legislation or of any monitoring mechanism on the performance of municipal authorities, the system of waste management has remained severely deficient and outdated. Inappropriate and unhygienic systems are used. At disposal sites, municipal authorities dump municipal waste, human excreta from slum settlements, industrial waste from small industrial establishments within the city and biomedical waste without imposing any restrictions, thus provoking serious problems of health and environmental degradation. A public interest litigation was filed in the Supreme Court in 1996 (Special Civil Application No. 888 of 1996) against the government of India, state governments and municipal authorities for their failure to perform their duty of managing MSW adequately. The Supreme Court then appointed an expert committee to look into all aspects of SWM and to make recommendations to improve the situation. After consulting around 300 municipal authorities, as well as other stakeholders, the committee submitted a final report to the Supreme Court in March 1999. The report included detailed recommendations regarding the actions to be taken by class 1 cities, by the state governments and by the central government to address all the issues of Municipal Solid Waste Management (MSWM) effectively.

On the basis of the report, the Supreme Court directed the government of India, state governments and municipal authorities to take the necessary actions. The Ministry of Environment and Forests was directed to expeditiously issue rules regarding MSW management and handling. Such rules were already under development and had been under consideration for quite some time. Thus, in September 2000, the ministry issued the MSW (Management and Handling) Rules 2000 under the Environment Protection Act 1986.

15.6.1 THE MUNICIPAL SOLID WASTE (MANAGEMENT AND HANDLING) RULES 2000

The steps to be taken by all municipal authorities to ensure the management of solid waste according to best practices, the municipal authorities must meet the deadlines laid down in schedule I of the rules and must follow the compliance criteria and procedure laid down in Schedule II. Hence, municipal authorities are responsible for implementing provisions of the 2,000 rules. They must provide the infrastructure and services with regard to the collection, storage, segregation, transport, treatment and disposal of MSW. Municipal authorities are requested to obtain authorisation (that is, permission or technical clearance) from the state pollution control board or committee to set up waste processing and disposal facilities, and they must deliver annual reports of compliance. The state pollution control boards are directed to process the application of municipal authorities and to issue an authorisation to the municipalities within 45 days of the application's submission. The CPCB is responsible for coordinating the implementation of the rules among the state boards. The municipalities were mandated to implement the rules by December 2003, with punishment for municipal authorities that failed to meet the standards prescribed; nevertheless, most municipalities did not meet the deadline.

The urban development departments of the respective state governments are responsible for enforcing the provisions of the rules in metropolitan cities. The district magistrates or deputy commissioners of the concerned districts are responsible for enforcing the provisions within the territorial limits of their jurisdictions. The state pollution control boards are responsible for monitoring compliance with the standards on groundwater, ambient air and leachate pollution. They must also monitor compliance with compost quality standards and incineration standards as specified in the rules.

The deadline for implementing schedule I of the 2,000 rules has already passed, and compliance is far from effective. Some cities and towns have not even started implementing measures that could lead to compliance with the rules. Enforcement and sanctioning mechanisms remain weak. Other cities and towns have moved somewhat forward, either of their own accords or because of pressure from the Supreme Court, their state government, or their state pollution control board. Under Schedule II of the rules, municipal authorities have been further directed to set up and implement improved waste management practices and services for waste processing and disposal facilities. They can do so on their own or through an operator of a facility (as described in Schedules III and IV of the rules).

15.6.1.1 Collection of Solid Waste

To prohibit littering and to facilitate compliance, municipal authorities must take the following steps:

- Organise collection of MSW at household level by using methods such as door-to-door, house-to-house, or community bin service. Collection must be on a regular pre-informed schedule or by the acoustic announcement (without exceeding permissible noise levels).
- Give special consideration to devising waste collection in slums and squatter areas, as well as to commercial areas such as areas with hotels, restaurants and office complexes.
- Segregate at the source all recyclable waste, as well as biomedical waste and industrial waste, to prevent special waste from being mixed with ordinary MSW.
- Collect separately all horticultural waste and construction or demolition waste or debris, and dispose of it following proper norms. Similarly, waste generated at dairies will be regulated in accordance with state laws.
- Prohibit the burning of waste.
- Do not permit stray animals at waste storage facilities.

15.6.2 ENVIRONMENTAL PROTECTION – FROM THE INDIAN CONSTITUTION PERSPECTIVE

a. The state's responsibility with regard to environmental protection has been laid down under Article 48-A of the Constitution, which reads as follows:
    The State shall endeavor to protect and improve the environment and to safeguard the forests and wildlife of the country.
b. Environmental protection is a fundamental duty of every citizen of this country under Article 51-A(g) of our Constitution which reads as follows:
    It shall be the duty of every citizen of India to protect and improve the natural environment including forests, lakes, rivers and wildlife and to have compassion for living creatures.
c. Article 21 of the Constitution is a fundamental right which reads as follows:
    No person shall be deprived of his life or personal liberty except according to procedure established by law.
d. Article 48-A of the Constitution comes under Directive Principles of State Policy and Article 51 A(g) of the Constitution comes under Fundamental Duties.
e. The State's responsibility with regard to raising the level of nutrition and the standard of living and to improve public health has been laid down under Article 47 of the Constitution which reads as follows:
    The State shall regard the raising of the level of nutrition and the standard of living of its people and the improvement of public health as among its primary duties and, in particular, the State shall endeavor to bring about prohibition of the consumption except for medicinal purposes of intoxicating drinks and of drugs which are injurious to health.

h. The 42nd amendment to the Constitution was brought about in the year 1974 makes it the responsibility of the State Government to protect and improve the environment and to safeguard the forests and wildlife of the country. The latter, under Fundamental Duties, makes it the fundamental duty of every citizen to protect and improve the natural environment including forests, lakes, rivers and wildlife and to have compassion for living creatures.

As conferred by Article 246(1), while the Union is supreme to make any law over the subjects enumerated in List I, the states, under Article 246 (3), enjoy competence to legislate on the entries contained in List II, and both the Union and the states under Article 246(2) have concurrent jurisdiction on entries contained in List III. In the event of a clash, the Union enjoys primacy over states in that its legislation in the Union and the Concurrent List prevails over state legislation. Also, the Parliament has residuary powers to legislate on any matter not covered in the three lists (Art. 248).

15.6.3 HAZARDOUS WASTES (MANAGEMENT AND HANDLING) AMENDMENT RULES, 2003

These rules classify used mineral oil as hazardous waste under the hazardous waste (Management and Handling) Rules, 2003 which requires proper handling and disposal. The organisation will seek authorisation for the disposal of hazardous waste from concerned state pollution control boards (SPCB) as and when required.

15.7 TECHNOLOGIES USED FOR DISPOSAL OF MUNICIPAL SOLID WASTE

The main methods available for processing, treatment and disposal of MSW are composting, vermicomposting, anaerobic digestion, incineration, gasification, pyrolysis, plasma pyrolysis and production of refuse derived fuel.

15.7.1 COMPOSTING

The MSW generated in India constitutes a large amount of organic matter in it. Composting is a natural process of decomposition of organic waste that yields manure or compost, having high nutrient value. Composting is a biological process involving microorganisms (mainly fungi and bacteria); they convert degradable organic waste into humus-like substances. This finished product is rich in carbon and nitrogen and is an excellent medium for growing plants. Composting also helps in increasing the moisture-holding capacity of the soil and maintains soil health. It improves the soil texture and counters micronutrient deficiencies. No large capital investment is required in comparison to other methods available for waste treatment. The nutrient value is higher in the compost made of heterogeneous MSW of urban areas as compared to agrowaste.

Segregation is very important before using composting pits for minimising the risks of contamination. Contaminated compost should be dumped as it is also a pollutant.

The foul odour, insects, stray animals and chances of fire can be minimised by covering the waste. Also, the waste should not be stored for long periods and inflammable wastes such as plastic and PVC plastic should be kept away. Otherwise, in case of breaking of fire, they will release dioxins, one of the most toxic chemicals.

Earthworms added in the compost which are fed upon scientifically semi-decomposed organic wastes can produce natural organic manure; this special method of composting is called vermicomposting. It requires less mechanical effort, is easy to operate and is preferred over simple composting in small towns. Toxic matters which can kill the earthworms should be necessarily be ceased to enter the chain.

Organic matter constitutes 35%–40% of the MSW generated in India. This waste can be recycled by the method of composting, one of the oldest forms of disposal. It is the natural process of decomposition of organic waste that yields manure or compost, which is very rich in nutrients. Composting is a biological process in which microorganisms, mainly fungi and bacteria, convert degradable organic waste into humus-like substances. This finished product, which looks like soil, is high in carbon and nitrogen and is an excellent medium for growing plants.

The process of composting ensures the waste that is produced in the kitchens is not carelessly thrown and left to rot. It recycles the nutrients and returns them as nutrients. Apart from being clean, cheap and safe, composting can significantly reduce the amount of disposable garbage. Organic fertilisers can be used instead of chemical fertilisers and are better, especially when used for vegetables. They increase the soil's ability to hold water and make the soil easier to cultivate. They help the soil retain more of the plant nutrients.

Vermicomposting has become very popular in the past few years. In this method, worms are added to the compost. This helps to break the waste, and the added excreta of the worms make the compost very rich in nutrients.

To make a compost pit, you have to select a cool, shaded corner of the garden or the school compound and dig a pit, which ideally should be 3 feet deep. This depth is convenient for aerobic composting as the compost has to be turned at regular intervals in this process. Preferably, the pit should be lined with granite or brick to prevent nitrite pollution of the subsoil water, which is known to be highly toxic. Each time the organic matter is added to the pit, it should be covered with a layer of dried leaves or a thin layer of soil, which allows air to enter the pit, thereby preventing bad odour. At the end of 45 days, the rich pure organic matter is ready to be used.

Similar to the recycling of inorganic materials, source-separated organic wastes can be composted and the compost obtained can be used as an organic fertiliser on agricultural fields. Organic compost is rich in plant macronutrients like nitrogen, phosphorous, potassium and other essential micronutrients. The advantages of using organic manure in agriculture are well established and are a part of public knowledge.

The United Nations Environment Program (UNEP) defines composting as the biological decomposition of biodegradable solid waste under predominantly aerobic conditions to a state that is sufficiently stable for nuisance-free storage and handling and is satisfactorily matured for safe use in agriculture. Composting can also be defined as human intervention in the natural process of decomposition as noted by the Cornell Waste Management Institute. The biological decomposition accomplished by microbes during the process involves the oxidation of carbon present in the organic waste. The energy released during oxidation is the cause of the rise in temperatures in windrows during composting. Due to this energy loss, aerobic composting falls below anaerobic composting on the hierarchy of waste management.

Life cycle impacts of extracting virgin raw materials and manufacturing make material recovery options like recycling and composting the most environment-friendly methods to handle the waste. They are positioned higher on the hierarchy compared to other beneficial waste-handling options like energy recovery. However, the quality of the compost product depends upon the quality of input waste. Composting mixed wastes results in low-quality compost, which is less beneficial and has the potential to introduce heavy metals into the human food chain.

Aerobic composting of mixed waste results in a compost contaminated by organic and inorganic materials, mainly heavy metals. Contamination of MSW compost by heavy metals can cause harm to public health and the environment and is a major concern, leading to its restricted agricultural use. Mixed waste composting is therefore not an option for sustainable waste management, but this issue is not a part of public knowledge. Mixed waste composting is widely practised and is considered better (if not best) (8) in countries like India where more than 91% of MSW is landfilled, and there are no other alternatives. It is considered better probably because public health and environmental impacts of unsanitary landfilling are more firmly established by research than those impacts due to heavy metal contamination of MSW compost.

15.7.2 ANAEROBIC DIGESTION

The United States Environmental Protection Agency (USEPA) defines anaerobic digestion (AD) as a process where microorganisms break down organic materials, such as food scraps, manure and sewage sludge, in the absence of oxygen. In the context of SWM, AD (also called anaerobic composting or biomethanation) is a method to treat source-separated organic waste to recover energy in the form of biogas and compost in the form of a liquid residual. Biogas consists of methane and carbon dioxide and can be used as fuel or by using a generator it can be converted to electricity on-site. The liquid slurry can be used as organic fertiliser. The ability to recover energy and compost from organics puts AD above aerobic composting on the hierarchy of waste management.

Similar to aerobic composting, AD needs a feed stream of source-separated organic wastes. AD of mixed wastes is not recommended because contaminants in the feed can upset the process. Lack of source-separated collection systems and public awareness and involvement strike off large-scale AD from feasible SWM options in India. However, AD on a small scale (called small-scale biogas) has emerged as an efficient and decentralised method of renewable energy generation and waste diversion from landfills. It also reduces greenhouse gas emissions by using methane as an energy source which would otherwise be emitted from landfilling waste.

15.7.3 INCINERATION

The waste with high calorific value mainly composed of large amounts of rags, papers, plastics and pathological wastes is advantageous for incineration. The incineration method is usually adopted for the disposal of waste in developed countries. It easily reduces the volume of waste by up to 90% with a handsome amount of recovered energy. The amount spent on waste transportation can be reduced by maintaining the plant within the city, which causes no problem for civilians as this method is noiseless, with no foul odour and also the land required is also minimum. Initial and maintenance cost is, however, high in comparison to conventional olden methods because of the installation of equipment and the requirement of skilled labour. With low net energy recovery for the disposal of low calorific waste i.e. chlorinated waste and high moisture waste, the plant is not economical. Toxic metals in ash and the presence of SO_x and NO_x chlorinated compounds in the smoke emitted is a serious issue to be taken care.

15.7.4   SANITARY LANDFILLS AND LANDFILL GAS RECOVERY

All types of MSW, non-recyclable waste and other types of inorganic waste generated can be finally disposed of by sanitary landfill. All types of commercial and institutional waste can also be disposed of by this method in an economical way with very low cost with no requirement of skilled labour. Net environmental gains can be achieved if organic wastes are landfilled, having the potential to recover landfill gas. The gas can be used directly for heating purposes in domestic uses and can be used for power generation also. The main disadvantage involved is pollution that can be caused by polluted leachate flowing down the gradient areas in case of a poor drainage system; the other is the emission of greenhouse gases, i.e., carbon dioxide and methane in case of a less efficient gas recovery process. Also, in case of poor gas ventilation, there will always be a risk of breaking of fire and explosion due to the buildup of methane concentration.

15.8   MUNICIPAL SOLID WASTE MANAGEMENT (MSWM) IN THE KINGDOM OF SAUDI ARABIA

The increasing rate of industrialisation, construction activities and expansion of fast growth in the Kingdom of Saudi Arabia not only pose problems related to the allocation of resources and powers but also severely challenges the natural environment severely which has led to environmental degradation such as contaminated water, sinking groundwater levels, unhealthy soils and polluted air seen in various parts of the world also. Solid waste (SW) is generated from households, offices, shops, markets, restaurants, public institutions, industrial installations, waterworks and sewage facilities, construction or demolition sites and agricultural activities. SW generation rates and composition varies from country to country depending on the economic situation, industrial structure, waste management regulations and lifestyle. Although statistics on waste generation and treatment have been improved substantially in many countries during the past decade, at present, only a small number of countries have comprehensive waste data management systems covering all waste types and treatment techniques.

Sustainable waste management aims at the global environmental quality which is a pre-requisite for a rise in per capita welfare over a period of time. "Efficient management of waste is a global concern requiring extensive research and development works towards exploring newer applications for a sustainable and environmentally sound management". In our daily life, hundreds of objects pass through our hands, but we rarely think of the processes preceding their production and following their use. Being aware and knowing what to buy and how to recycle and/or dispose of them has a tremendous impact on our environment.

Landfilling is the simplest and normally cheapest method for disposing of waste. In most low-to medium-income developing nations and many developed countries too, almost all generated SW goes to landfill. European Union also follows policies of reduction, reuse and diversion from landfill which is strongly encouraged; more than half of the member states still send an excess of 75% of their waste to landfill. However, such disposal of waste to landfill will decrease in future as large volumes of MSW are being produced and are still increasing significantly for many developed countries. Therefore, the landfill will then be a relevant source of groundwater pollutants for the foreseeable future. A report by UNEP states that all over the world nearly 3,000 million people live in urban areas and every day approximately 160,000 people join the Global Environment Outlook (2000). Current global MSW generation levels are approximately 1.3 billion tonnes per year and are expected to increase to approximately 2.2 billion tonnes per year by 2025. This represents a significant increase in per capita waste generation rates from 1.2 to 1.42 kg per person per day in the next 15 years. However, global averages are broad estimates only as rates vary considerably by region, country, city and even within cities. The annual waste generation in East Asia and the Pacific region is approximately 270 million tonnes per year. This quantity is mainly influenced by waste generation in China, which makes up 70% of the regional total. Per capita waste generation ranges from 0.44 to 4.3 kg per person per day for the region, with an average of 0.95 kg/capita/day.

15.9 MUNICIPAL SOLID WASTE MANAGEMENT (MSWM) IN NIGERIA

The MSW in Nigeria contain all sources of unsorted wastes, such as commercial refuse, construction and demolition debris, garbage, electronic wastes etc., which are dumped indiscriminately on roadsides and any available open pits irrespective of the health implication on people. The aim is to emphasise various waste management options, which integrated waste management disclosed the hierarchy of waste management options, environmental impacts of those options were studied under health and social effects, and the legislation of Extended Producer Responsibility was suggested where by-products are taken back by manufacturers, especially when remanufacturing and reuse is available to ensure sound management practice in developing country Nigeria. In Nigeria, the common method of disposal is an open dump. The amount of trash that accumulates in a matter of hours would be more than waste collectors could haul in a day; these garbage "dumps" are located on the side of the highway at the fringe of cities and slums (personal experience). In most developing nations such as Nigeria with about 140 million population, waste is dumped indiscriminately on roadsides and any available open pits irrespective of the health implication for people. All classes of SW are collected and dumped together without much effect at segregating and differentiating the different components of solid waste. There are cases where this waste is dumped in streams or river channels. Due to deposits in the rivers, Abulude et al. (2006) reported that developing countries are witnessing changes in groundwater, which constitute another source of potable water. Comparing waste produced in developing countries and developed countries such as America, where an average American produces about 4.4 pounds of MSW each day, resulting in roughly 210 million tonnes per year for the nation and that of developing countries, waste is disposed of in open space or roadside; scavenging at dump sites is still common. Metals in raw surface water reflect erosion from natural sources, the fallout from the atmosphere and additions from industrial activities. These metals in soil and water may enter the food chain; further potential sources of human exposure include consumer products and industrial waste, as well as the working environment.

15.10   INTEGRATED WASTE MANAGEMENT

Integrated waste management, IWM, is defined as "the application of suitable techniques and technologies and management programmes to achieve specific wastes management objectives and goals".

15.10.1   RULES AND LEGISLATION

The industry should realise that the impact their product has on the environment does not start and end with the manufacture of the product. The impact a product has on the world starts with the design and ends at the ultimate disposal of the product after its useful life. The Society for Environmental Toxicity and Chemistry (SETAC) defined the period of the analysis that is Life Cycle Assessment (LCA) as an objective process to evaluate the environmental burden associated with a product, process or activity by identifying and quantifying energy and materials usage and environmental releases, to assess the impact of those energy and material uses and releases on the environment and to evaluate and implement opportunities to effect environmental improvements. The assessment includes the entire life cycle of the product from "cradle to grave" encompassing:

1. Raw material extraction and processing
2. Manufacturing
3. Transportation, distribution and trade
4. Use, reuse and maintenance
5. Recycling
6. Final disposal

LCA address the environmental impacts of the system under study in the areas of ecological health, human health and resource depletion. The law and policies concerning the proper management of MSW are continuing to evolve. However, in some cases, efforts to divert waste from landfills and incinerators have resulted in hazardous dismantling, shredding, burning, exporting and other unsafe or irresponsible disposal methods. Most of the electrical and electronics equipment (EEE) imported into developing countries have to undergo their end of life (EOL), before being imported. Backlogged demand for EEE in developing countries as well as the lack of national regulation and/or lax enforcement of existing laws can also promote the growth of semiformal or informal waste EEE recycling economies that are poorly controlled and involved extremely risky techniques. Often the participants in these sectors are not aware of the risks, do not know of better practices, or simply have no access to investment capital to finance profitable improvements. According to United Nations Basel Convention 1989, the convention puts the onus on exporting countries to ensure that hazardous waste is managed in an environmentally sound manner in the country of import. If the waste items are proclaimed to be recycled by the Environment Protection Agency (EPA), the corresponding responsible enterprises are enforced to pay recycling-clearance-disposal fees to the Recycling Management Fund (RMF) of the EPA for recycling the waste items. From this fact, the following rules are needed to implement in the developing country Nigeria to ensure better management practices or a sound recycling system:

1. Pollution Pay Principle (PPP),
2. Extended Producer Responsibility (EPR), and
3. Integrated Product Policy (IPP).

In PPP, producers and importers have only the obligation for paying the recycling-clearance-disposal fees to the EPA but do not assume any responsibility for recycling work. Also in EPR, the system requires producers to have a full obligation for recycling the products they produce both within and outside the manufactured country. While in the IPP, seek to improve environmental performance by looking at all phases of the product's life cycle and taking action where it is most effective. All these strategies depend on the nature of the product and waste. The introduction of EPR with well-defined roles for all participants, producers, users, authorities and waste managers is essential for designing an effective waste management system, especially in developing countries like Nigeria, where there are no recycling facilities by Environmental Protection Agencies in all states. This EPA ensure product takeback by manufacturers, especially when remanufacturing and reuse is available, it is then the responsibility of the manufacturers to pay in deposit/refund systems and in kerbside collection of reverse logistic.

15.11 MUNICIPAL SOLID WASTE MANAGEMENT (MSWM) IN AUSTRALIA

Australia's urban waste streams are an untapped renewable energy resource. The space available for landfill is decreasing in our major cities, and the methane produced by landfilled MSW, green waste and biosolids is now recognised as a significant, long-term source of greenhouse gas emissions. Local authorities, state and federal governments and the waste management industry now recognise opportunities in converting the energy in urban waste streams to renewable power and other energy products. There is a clear international precedent that modern waste-to-energy (WtE) plants are clean and efficient. With appropriate technology choice, it is technically feasible for a similar industry to be developed in Australia. This is evident by plans for two major projects in Western Australia to convert more than 200,000 tonnes of MSW into electricity annually. Despite this encouraging activity, knowledge of the thermochemical properties of Australian waste streams is considerably lacking. Such knowledge is critical for the effective planning and development of WtE projects. The SW generated in Australia is usually classified into three main categories: municipal, commercial and industrial and construction and demolition waste. An estimated 53 million tonnes of SW was generated from all sources in Australia during 2010–2011. Of this, 27% was a municipal (household) waste, equivalent to 14.3 million tonnes (with a per capita MSW generation rate of 660 kg/year). This is a significant increase compared with the per capita rate of 447 kg/year from just eight years earlier, in 2002–2003. Globally, MSW is usually managed in four major ways: recycling, composting, landfilling and WtE. Despite the significant amount of energy that could be recovered from urban waste streams as renewable energy, Australia uses only the first three methods to manage MSW. The country has no large-scale thermal treatment facilities for the disposal of non-hazardous MSW; the last MSW incineration plant shut down in 1997, and an attempt to develop a SW energy recycling facility in Wollongong, New South Wales, failed, with the plant shut down in 2004. Information about feedstock chemical properties and energy content is essential to the design and operation of any type of thermochemical conversion system, whether combustion or gasification-based. The energy content of MSW can be estimated based on average physical compositions using empirical models (Chang et al., 2007; Lee et al., 2017; Kathiravale et al., 2003; Wang et al., 2020). While this approach is quick and inexpensive, the downside is that the energy content of the type of organic waste in the country where the empirical model was developed is likely to differ significantly from that in the country where the model is applied. This variation is directly related to sociocultural properties; for example, differences in the amount and type of food waste. Geographical and seasonal considerations also influence the quantity and type of waste generated in different countries. To avoid this uncertainty, waste samples should be systematically collected and prepared, and the energy content should be directly measured using standard laboratory apparatus such as a bomb calorimeter. Energy content can also be calculated from a sample's ultimate analysis, which usually lists the carbon, hydrogen, oxygen, nitrogen, sulphur and ash content of the dry fuel on a weight percentage basis.

Regular surveys have been conducted in Australia's major cities to understand the physical composition of MSW, which is routinely managed by local councils. For example, Swales (2013) reported that an average MSW stream (samples collected from Brisbane City Council's transfer stations in 2013) contained 53.3% of organic matter, 14.7% of plastic, 13% of paper, 4.2% of glass, 2.7% of metal, 11.6% of others and 0.5% of household hazardous. As MSW waste streams are landfilled according to the current waste management system, the surveys focus only on the quantity and distribution of wastes; the determination of thermochemical characteristics is out of their scope. If WtE is to feature in strategic thinking and future planning, then the chemical characteristics of waste – in particular, the calorific value – become important. Researchers from developing and developed countries have reported their findings of chemical characteristics, including caloric values of MSW samples, via direct measurement (e.g. Algeria Guermoud et al., 2009; Zhou et al., 2014; Komilis et al., 2012; Eisted and Christensen, 2011; Kumar and Goel, 2009; Abu-Qudais and Abu-Qdais, 2000; Lee et al., 2017.

However, Australian data for this research area are scarce. One of the challenges in analysing the chemical characteristics of MSW is the lack of a standard method for sample collection and preparation. While most researchers categorised and analysed the different physical components of MSW (e.g. food, paper, plastics, textiles, wood, glass, metals etc.), Agrawal (1988) analysed only two fractions of MSW: combustible and non-combustible. Most researchers collected MSW samples directly from transfer stations, community bins and final disposal sites, sorting them into different categories later on (Brunner and Ernst, 1986; Chang et al., 2007; Gidarakos et al., 2006; Kumar and Goel, 2009), while some collected each category separately from different locations (Hanc et al., 2011; Katiyar et al., 2013; Komilis et al., 2012). The work presented here begins to address the lack of data for Australian waste streams by developing a method to characterise waste in terms of chemical composition and energy content, and applying the method to MSW and green waste from Brisbane, Australia.

15.12 MUNICIPAL SOLID WASTE MANAGEMENT (MSWM) IN SPAIN

A typical municipal solid waste treatment plant consists of various sites, including a landfill site, leachate pond, tipping floor (area to discharge and classify fresh MSW) and sites for AD of MSW and the composting process. These can be sources of odours. Odours from MSW treatment plants consist of a complex mixture of organic compounds, hydrogen sulphide and ammonia. These odours can be an important problem for MSW plant operators and also a source of annoyance to nearby urban populations. The chemical characterisation of the emissions from an MSW treatment plant has proven to be helpful in order to determine which compound or group of compounds are characteristic in each of the point sources of the treatment plant. This characterisation is fundamental to improving the waste treatment system in order to reduce odorous emissions (Gallego et al., 2012). Landfill sites are one of the most important sources of gaseous emissions. Landfill emissions contain mainly methane carbon dioxide (from 45% to 55% and 30% to 40%, respectively Jönsson et al., 2003), both of which are considered to be greenhouse gases contributing to global climate change. Landfill gas may also contain nitrogen, oxygen, hydrogen sulphide and ammonia and various organic compounds being the main contributors to the odour. Among the organic compounds, it is possible to find saturated and unsaturated hydrocarbons, acidic hydrocarbons and organic alcohols, aromatic hydrocarbons, halogenated compounds, sulphur compounds such as carbon disulphide and mercaptans, and other compounds such as siloxanes (Keller, 1988). Although VOC emissions are <1% of total gaseous emissions, they are considered to be among the most hazardous air pollutants (Allen et al., 1997). Some of them are known to be toxic or carcinogenic, such as benzene, toluene, ethylbenzene, and m-, p- and o-xylene (xylenes) (commonly called BTEX). Therefore, the control of BTEX concentrations in the vicinity of landfills is really important. In fact, several studies found BTEX in landfills and in the surrounding areas, although these emissions do not pose a health risk, owing to their low concentrations as a result of air dilution (Dincer et al., 2006; Durmusoglu et al., 2010). In terms of odours, some VOCs have been identified as responsible for the odour, such as sulphur compounds, alkyl benzenes, limonene and certain esters. In addition, VOCs are known precursors of photochemical smog formation.

A previous study carried out by Davoli et al. (2003) reports that p-cymene is a characteristic compound of emissions from landfill. In more recent work, this result was also confirmed in the digested waste from AD (Orzi et al., 2010). There are progressively more MSW treatment plants that collect and even use landfill gas for electricity production in gas engines, but not all the gas can be collected by these extraction systems, and leakage can still pose a problem. Even with well-designed covers, few landfills are thought to recover more than 60% of the available gas, and the losses to the atmosphere through the surface are one of the most important causes (Johannessen, 1999). Therefore, strict control and maintenance of landfills are essential if gas leaks are to be prevented. For this purpose, recent studies suggest landfill biocovers as an effective solution for odour reduction (Capanema et al., 2014; Lakhouit et al., 2014).

With regard to composting, it is known that ammonia is one of the main components of the emissions that result from this practice, especially when the nitrogen content of the composting waste is high. Ammonia emissions from composting depend on the following parameters: the C/N ratio of the waste, composting temperature and pH. Hydrogen sulphide, methyl mercaptan, dimethyl sulphide, carbon disulphide and dimethyl disulphide are volatile sulphur compounds produced by composting. These compounds are generated as a result of incomplete or insufficient aeration during this process. After testing the emissions at various stages of the composting process, many studies have concluded that most VOCs are emitted during the early stages of the process (Eitzer, 1995; Gallego et al., 2012; Scaglia et al., 2011). Eitzer (1995) performed a survey of emissions of volatile organic compounds from MSW composting facilities. The classification includes different locations on the facility into tipping floor, shredder, fresh from newly formed active compost, mid-aged from one-fifth to four-fifths of the way through the active composting region, old if it was at the end of the active composting region, curing if it was still on-site but out of the active composting area. The highest concentrations were found in samples taken from the tipping floors, near the shredders and in the freshest active composting regions, while more mature compost samples exhibited lower concentrations. Another composting study was carried out in the laboratory by Komilis et al. (2012), but in this case on three organic components of MSW (food wastes, yard wastes and mixed paper wastes) and mixtures of these. This study concluded that all VOCs were emitted early during the composting process and that their production rates decreased with time at thermophilic temperatures. The mixed paper primarily produced alkylated benzenes, alcohols and alkanes. Yard wastes primarily produced terpenes, alkylated benzenes, ketones and alkanes, while food wastes primarily produced sulphides, acids and alcohols.

15.13   MUNICIPAL SOLID WASTE MANAGEMENT (MSWM) IN GHANA

Ghana is located in West Africa and has a total area of 238,533 km² with a coastal line of 550 km. It has a tropical climate with two major seasons; the rainy season (May–October) and the dry season (November–April). The average temperature is 30°C and annual rainfall is between 1,100 mm in the north and 2,100 mm in the south. Data on MSW generation and composition are available in only a few selected cities, most of which are over a decade old. Nationwide waste statistics, in general, are lacking; field study on household waste composition and generation has not been conducted holistically in the ten regions of the country, hence lack of reliable data which could provide information to the local and national waste management authorities for decision-making. Human and resource capacity to carry out these studies which involve the collection of informative data on waste composition and quantity that is hauled to treatment sites or recycling centres or disposal sites is lacking (Kanat, 2010; Pichtel, 2005). Municipal or household wastes are often generated from several sources where variable human activities are encountered. Several studies indicate that much of the MSW from developing countries is generated from households (55%–80%), followed by commercial or market areas (10%–30%) with varying quantities from streets, industries and institutions among others (Okot-Okumu, 2012). Waste from these sources is highly heterogeneous in nature and has variable physical characteristics depending on their sources; notably in their composition are food waste, yard waste, wood, plastics, papers, metals, leather, rubbers, inert materials, batteries, paint containers, textiles, construction and demolishing materials and many others which would be difficult to classify. The heterogeneity of the generated waste is a major setback in its utilisation as a raw material. There is therefore the need for fractionation of the waste before it can be subjected to any meaningful treatment process. Source sorting and separation of waste is one of the traditional fractionation methods and fundamental steps in an integrated waste management system with the potential to provide data on waste generation and the quality of the fractions. However, the success of any designed waste segregation system will depend largely on the active participation of the waste generators in the various communities and how they comply with the principles of sorting and separation of the waste. The generation of waste from commercial outfits in Ghana is difficult to quantify on a per capita basis since all the generators are not known. Assessment is mostly done on the bulk of the waste collected. The composition may depend on the business activities; hence the household is the right source to obtain correct data for managing waste. The organic fraction in the waste was the highest in the waste stream and ranged from 48% to 69%. Paper increased the percentage of biodegradables to 58%–76% which could be used as raw material for biological conversion processes like composting, biogas and bioethanol refinery process. The organic composition varied among the various socioeconomic areas but these differences were not significant ($p < 0.05$). The geographical locations recorded decreases in organic waste from the coastal regions through the forest to the Northern savanna. Plastic waste was the second highest fraction and decreased from the Northern savanna through the forest regions to the coastal regions. Paper waste as a percentage of the total waste stream was less in the

Northern savanna but almost the same for the coastal and forest regions. Nationally, the waste generation rate was 0.47 kg/person/day. A total of 12,710 tonnes of household waste was generated from households in Ghana. The 8,389 tonnes of the waste are biodegradables and available for bioconversion processes and 2,754 tonnes are for recycling. National sorting and separation efficiency was 84% for biodegradables and 76% for other waste. The one-way separation system was effective.

15.14 MUNICIPAL SOLID WASTE MANAGEMENT (MSWM) IN HONG KONG

With the growth of population, urbanisation and affluence, disposal of MSW has become a major environmental challenge affecting people throughout the world (Bogner et al., 2007). Hong Kong, as a world-class metropolis, is also inevitably faced with this pervasive issue. The Hong Kong Environmental Protection Department (HKEPD) has listed waste reduction and management policies as an intractable environmental issue to be resolved. At present, Hong Kong relies solely on landfills for MSW disposal. Approximately 9,000 tonnes of unrecoverable MSW are still discarded in the landfills every day, albeit Hong Kong has achieved a recycling rate of 52% in 2010. Hong Kong is experiencing a serious shortage of MSW disposal sites with an anticipation that the current three strategic landfills, namely South East New Territories (SENT), North East New Territories (NENT) and West New Territories (WENT) will be exhausted in 2014, 2016 and 2018, respectively. In response to this acute problem, there is a pressing need for the HKEPD to identify a comprehensive solution. A policy framework for the management of MSW was introduced by the HKEPD in late 2005 to address this problem. One of the approaches applied in this policy framework is bulk reduction and disposal, in which the HKEPD has proposed to implement landfill extension (LFE) and Integrated Waste Management Facility (IWMF), with the advanced incineration facility (AIF) as the core technology in this IWMF. The implementation of LFE and AIF, however, has triggered a strong dispute among stakeholders such as Hong Kong's citizens and green groups (Ng, 2011, 2012). It spurs an intense debate as to whether these waste disposal facilities are truly suitable and sustainable for Hong Kong MSW management practices. Perhaps, it could not be told merely based on the general perceptions and good experiences of the public and executive authorities. Apart from this issue, the inconvenient truth about the unprecedented challenge of climate change has created observable changes in various weather patterns and drawn extensive concerns from the public, climate panels and policymakers. The waste management sector accounted for approximately 3%–5% of total anthropogenic greenhouse gas (GHG) emissions at a global scale in 2005. The maximum, minimum and annual average shares of GHG emissions from the waste sector in Hong Kong from 1990 to 2006 were 5.9%, 3.2% and 4.5%, respectively, which was the third largest sector after electricity generation and transportation. Hong Kong, as a responsible international community, always takes initiative to reduce GHG emissions and combat climate change. These initiatives include using cleaner fuel and renewable energy for power generation, promoting energy efficiency and carbon audits in buildings, and using energy-efficient transport and cleaner vehicles in the city. In 2003, the Kyoto Protocol was extended to Hong Kong, where Hong Kong is attached to mainland China (defined as a non-Annex 1 Party), and assists the Central People's Government in fulfilling the obligations under the Kyoto Protocol. While mainland China announced a target of cutting carbon intensity, which is defined as the total mass of carbon dioxide equivalent emissions per gross domestic product (GDP), by 40%–45% by 2020 from the 2005 level, the Hong Kong Special Administrative Region (HKSAR) Government has set a more aggressive target to reduce carbon intensity by 50%–60% by 2020 from the 2005 level for its own region. Besides enhancing energy efficiency and revamping the fuel mix for electricity generation, one should take action with the waste sector as there is plenty of room for GHG emission reductions by employing cleaner waste management practices or converting waste to wealth through displaced energy from fossil fuels (Bogner et al., 2007). The accounting of GHG emissions on various MSW disposal methods provides a conceptual framework with which to describe a carbon footprint concept to the public and policymakers for understanding the issues surrounding the need to reduce GHG emissions (Hammond, 2007). The association of MSW disposal issue and GHG abatement arouses a challenge that how to manage the MSW effectively without adversely impacting the environment. Several studies have recently been conducted to examine the GHG emissions from landfills and incineration facilities, particularly focused on European and North American regions (Christensen et al., 2009; Kaplan et al., 2009; Morris, 2010; Vergara et al., 2011; Assamoi and Lawryshyn, 2012). Morris (2010) reported that the multi-criteria complexity of the landfills and incineration facilities (e.g. performance factors, waste characteristics and methodology issues) affected local preferred waste technology decisions. The results are difficult to generalise and represent the local environmental conditions. Also, most studies did not investigate the GHG emissions from the individual subprocesses of landfill and incineration.

REFERENCES

Abu-Qudais, M., and Abu-Qdais, H.A. Energy content of municipal solid waste in Jordan and its potential utilization. *Energy conversion and Management*, 2000, 41, 983–991.

Abulude, F.O., Obidiran, G.O., and Orungbemi, S. Determination of physico-chemical parameter and trace metal contents of drink water sample in Akure Nigeria. *Electronic Journal of Environmental, Agricultural and Food Chemistry*, 2006, 6(8), 2298–2299.

Agrawal, K. R. A rapid technique for characterization and proximate analysis of refuse-derived fuels and its implications for thermal conversion. *Waste Management & Research*, 1988, 6(3), 271–280.

Allen, M.R., Braithwaite, A., and Hills, C.C. Trace organic compounds in landfill gas at seven U.K. waste disposal sites. *Environmental Science & Technology*, 1997, 31, 1054–1061.

Assamoi, B., and Lawryshyn, Y. The environmental comparison of landfilling vs. incineration of MSW accounting for waste diversion. *Waste Management*, 2012, 32, 1019–1030.

Bogner, J., Ahmed, M.A., Diaz, C., Faaij, A., Gao, Q., Hashimoto, S., et al. Waste management. In: Metz, B., Davidson, O.R., Boesch, P.R., Dave, R., and Meyer, L.A. (Eds.), *Climate Change 2007: Mitigation Contribution of Working Group III to the Fourth Assessment Report of the International Panel on Climate Change*. Cambridge, UK: Cambridge University Press, 2007, pp. 585–618.

Brunner, P.H., and Ernst, W. R. Alternative methods for the analysis of municipal solid waste. *Waste Management & Research*, 1986, 4(2), 147–160.

Capanema, M.A., Cabana, H., and Cabral, A.R. Reduction of odours in pilot-scale landfill biocovers. *Waste Management*, 2014, 34, 770–777.

Central Pollution Control Board (CPCB). Collection, transportation and disposal of municipal solid wastes in Delhi, 1998.

Chang, Y. F., Lin, C. J., Chyan, J. M., Chen, I. M., and Chang, J. E. Multiple regression models for the lower heating value of municipal solid waste in Taiwan. *Journal of Environmental Management*, 2007, 85(4), 891–899.

Christensen, T.H., Simion, F., Tonini, D., and Moller, J. Global warming factors modelled for 40 generic waste management scenarios. *Waste Management & Research*, 2009, 27, 871–884.

Davoli, E., Gangai, M.L., Morselli, L., and Tonelli, D. Characterisation of odorants emissions from landfills by SPME and GC/MS. *Chemosphere*, 2003, 51, 357–368.

Dincer, F., Odabasi, M., and Muezzinoglu, A. Chemical characterization of odorous gases at a landfill site by gas chromatography–mass spectrometry. *Journal of Chromatography A*, 2006, 1122, 222–229.

Durmusoglu, E., Taspinar, F., and Karademir, A. Health risk assessment of BTEX emissions in the landfill environment. *Journal of Hazardous Materials*, 2010, 176, 870–877.

Eitzer, B.D. Emissions of volatile organic chemicals from municipal solid waste composting facilities. *Environmental Science & Technology*, 1995, 29, 896–902.

Gallego, E., Roca, F.J., Perales, J.F., Sánchez, G., and Esplugas, P. Characterization and determination of the odorous charge in the indoor air of a waste treatment facility through the evaluation of volatile organic compounds (VOCs) using TD–GC/MS. *Waste Management*, 2012, 32, 2469–2481.

Gidarakos, E., Havas, G., and Ntzamilis, P. Municipal solid waste composition determination supporting the integrated solid waste management system in the island of Crete. *Waste Management*, 2006, 26, 668–679.

Guermoud, N., Ouadjnia, F., Abdelmalek, F., and Taleb, F. Municipal solid waste in Mostaganem city (Western Algeria). *Waste Management*, 2009, 29(2), 896–902.

Hammond, G., Time to give due weight to the carbon footprint issue. *Nature*, 2007, 445(7125), 256.

Hanc, A., Novak, P., Dvorak, M., Habart, J., and Svehla, P. Composition and parameters of household bio-waste in four seasons. *Waste Management*, 2011, 31, 1450–1460.

Johannessen, L.M. Guidance Note on Recuperation of Landfill Gas from Municipal Solid Waste Landfills, Working Paper series 4. Urban Development Division, Urban Waste Management Thematic Group, The World Bank, Washington, 1999.

Jönsson, O., Polman, E., Jensen, J.K., Eklund, R., Schyl, H., and Ivarsson, S. Sustainable Gas Enters the European Gas Distribution System. Danish Gas Technology Center, 2003. Available at: http://www.dgc.eu/sites/default/files/filarkiv/documents/C0301_sustainable_gas.pdf.

Kanat, G. Municipal solid-waste management in Istanbul. *Waste Management*, 2010, 30(8–9), 1737–1745.

Kaplan, P.O., Decarolis, J., and Thorneloe, S. Is it better to burn or bury waste for clean electricity generation? *Environmental Science & Technology*, 2009, 43(6), 1711–1117.

Kathiravale, S., Yunus, M.N.M., Sopian, K., Samsuddin, A.H., and Rahman, R.A. Modeling the heating value of Municipal solid waste. *Fuel*, 2003, 82(9), 1119–1125.

Katiyar, R.B., Suresh, S., and Sharma, A.K. Characterisation of municipal solid waste generated by the City of Bhopal, India. *International Journal of ChemTech Research*, 2013, 5, 623–628.

Keller, A.P. Trace Constituents in Landfill Gas. Task Report on Inventory and Assessment of Cleaning Technologies. Final report, May 1984–February 1987, 1988. Other Information: See also PB–88-217013, p. Medium: X; Size: p. 140.

Komilis, D., Evangelou, A., Giannakis, G., and Lymperis, C. Revisiting the elemental composition and the calorific value of the organic fraction of municipal solid wastes. *Waste Management*, 2012, 32, 372–381.

Kumar, N., and Goel, S. Characterization of Municipal solid waste (MSW) and a proposed management plan for Kharagpur, West Bengal, India. *Resources Conservation and Recycling*, 2009, 53(3), 166–174.

Lakhouit, A., Schirmer, W.N., Johnson, T.R., Cabana, H., and Cabral, A.R. Evaluation of the efficiency of an experimental biocover to reduce BTEX emissions from landfill biogas. *Chemosphere*, 2014, 97, 98–101.

Lee, J., Choi, D., Tsang, Y. F., Oh, J. I., and Kwon, E. E. Employing CO2 as reaction medium for insitu suppression of the formation of benzene derivatives and polycyclic aromatic hydrocarbons during pyrolysis of simulated municipal solid waste. *Environmental Pollution*, 2017, 224, 476–483.

Morris, J. Bury or burn North America MSW? LCAs provide answers for climate impacts and carbon neutral power. *Environmental Science & Technology*, 2010, 44, 7944–7949.

Okot-Okumu, J. Solid waste management in African cities–East Africa. In: Marmolejo Rebellon, L.F. (Ed.), *Waste Management–An Integrated Vision*. London: Intech, 2012, pp. 3–20.

Orzi, V., Cadena, E., D'Imporzano, G., Artola, A., Davoli, E., Crivelli, M., and Adani, F. Potential odour emission measurement in organic fraction of municipal solid waste during anaerobic digestion: Relationship with process and biological stability parameters. *Bioresource Technology*, 2010, 101, 7330–7337.

Pichtel, J. *Waste Management Practices: Municipal, Hazardous, and Industrial*. Boca Raton, FL: CRC Press, 2005.

Pratap, V., Bombaywala, S., Mandpe, A., and Khan, S. U. Solid waste treatment: Technological advancements and challenges. In: *Soft Computing Techniques in Solid Waste and Wastewater Management*. Amsterdam: Elsevier, 2021, pp. 215–231.

Scaglia, B., Orzi, V., Artola, A., Font, X., Davoli, E., Sanchez, A., and Adani, F. Odours and volatile organic compounds emitted from municipal solid waste at different stage of decomposition and relationship with biological stability. *Bioresource Technology*, 2011, 102, 4638–4645.

Swales, L. 2013 Annual Waste Characterisation Survey. Waste Minimisation Team, Brisbane City Council, 2013.

United Nations Environment Program (UNEP). Urban areas. GEO-2000, Global Environment Outlook, 2000. https://www.unep.org/resources/global-environment-outlook-2000.

Vergara, S.E., Damgaard, A., and Horvath, A. Boundaries matter: Greenhouse gas emission reductions from alternative waste treatment strategies for California's municipal solid waste. *Resources, Conservation & Recycling*, 2011, 57, 87–97.

Wang, H., Liu, X., Wang, N., Zhang, K., Wang, F., Zhang, S., ... and Matsushita, M. Key factors influencing public awareness of household solid waste recycling in urban areas of china: A case study. *Resources, Conservation and Recycling*, 2020, 158, 883–896.

Zhou, H., Meng, A., Long, Y., Li, Q., and Zhang, Y. Classification and comparison of municipal solid waste based on thermochemical characteristics. *Journal of the Air and Waste Management Association*, 2014, 64(5), 597–616.

16 Sources, Characteristics, Treatment Technologies and Disposal Methods for Faecal Sludge

Gulafshan Tasnim, Mohd Dawood Khan, Izharul Haq Farooqi, and Farrukh Basheer
Aligarh Muslim University

CONTENTS

16.1 Introduction 298
    16.1.1 Characteristics 299
    16.1.2 Sampling and Analysis 299
16.2 Guidelines for Handling and Disposal of Faecal Sludge 301
    16.2.1 Regulations and Guidelines for Faecal Sludge
        Capture and Containment 302
    16.2.2 Regulations and Guidelines for Faecal Sludge Emptying and
        Transportation 304
        16.2.2.1 Fee Structures 307
        16.2.2.2 Specific Guidelines for Faecal Sludge Desludging
                Trucks and Accessories.. 308
        16.2.2.3 Regulations on Faecal Sludge Emptying and
                Transportation: Examples from Selected Countries ... 308
        16.2.2.4 Other General Health Requirements for Operators 309
    16.2.3 Regulations and Guidelines for Faecal Sludge
        Treatment and Disposal 310
        16.2.3.1 Need and Objective of FS Treatment......................... 310
    16.2.4 Regulations and Guidelines for Faecal Sludge Use................... 314
    16.2.5 Guidelines for Agricultural and Non-Agricultural Land
        Application 316
    16.2.6 Lime Stabilisations for Agricultural Reuse 317
    16.2.7 Occupational Risks on Farms 318
    16.2.8 Soil Application of (Co-)Composted Faecal Sludge 319
    16.2.9 Compost Application Guidelines (Nikiema et al., 2014) 319
    16.2.10 Faecal Sludge Reuse... 320
    16.2.11 Faecal Sludge Reuse in Aquaculture 320
    16.2.12 Effluent Discharge and Reuse for the STP 321

DOI: 10.1201/9781003202431-16

297

16.3   Treatment Technologies for Faecal Sludge 321
        16.3.1   Faecal Sludge Treatment Practices: Developed vs
                  Developing Countries 321
        16.3.2   Factors to Be Considered While Deciding Overall Treatment
                  Process or an Individual Technology........................... 323
                  16.3.2.1   Faecal Sludge Characteristics 323
                  16.3.2.2   Technical and Economic Feasibility 323
                  16.3.2.3   Local Context, Regulations and Existing Faecal
                              Management Practices in the Area........ 324
                  16.3.2.4   Treatment Objective, End Goal and Reuse Options 324
        16.3.3   Steps for Choosing Appropriate Treatment Processes and
                  Technologies for a Treatment Plant............................. 324
                  16.3.3.1   Technical Features and Specifications for Optimum
                              Performance of Technologies........................ 325
                  16.3.3.2   Co-Treatment of Faecal Sludge at WWTPs................. 333
                  16.3.3.3   Land Application.. 333
16.4   Conclusion 334
References... 335

16.1   INTRODUCTION

Faecal sludge or FS is basically, highly variable with a concentrated mixture of faeces, urine and flush water combined with grey water, diluted cleansing liquids, menstrual hygiene products, hair, grease, sand, grit, debris, heavy metals, pathogenic microorganisms and various harmful chemical compounds. This raw or partially digested slurry containing a series of solid, liquid and semi-solid waste, gets accumulated in pits, tanks and vaults of onsite sanitation systems. Once the containment system or storage tank reaches its functional capacity, it needs to be emptied. These non-sewered systems are the predominant form of excreta disposal installations in urban, peri-urban and rural areas of low- and middle-income countries where sewerage networks are either absent or limited to a small portion of urban areas. Households in these areas rely heavily on improperly designed or substandard Onsite sanitation services (OSSs) like dry or wet pit latrines, non-sewered public ablution blocks, flush or pour-flush toilets with septic tanks, aqua privies, dry toilets, ventilated improved pit latrines (VIPs), bucket latrines etc. The column chart (Figure 16.1) depicts the percentage of urban households using these sanitation systems in various cities of developing economies. As reported in different works of literature and research studies, figures from individual cities vary from 47% for Delhi, India, 51% for Santa Cruz, Bolivia, 65% for Thailand, Bangkok, through 79% for Bangladesh, Dhaka, 85% for Ghana and Tanzania, 88% for Manila, Philippines, 89% for Maputo, Mozambique, 90% for Kampala, Uganda, to 98% for Bamako, Mali (Ingallinella et al., 2002; Strauss et al., 2000; Census of India, 2011). This means that over 50%–90% of dwellers in towns and cities of different regions of Africa, Asia and Latin America are linked to OSSs making it the major source of faecal sludge generation.

16.1.1 CHARACTERISTICS

As stated earlier, compared to wastewater and sewage, FS is several times more concentrated and widely differs in quality and quantity from place to place. It has an offensive odour and appearance and contains three main components, i.e., scum, effluent and sludge. FS parameters are influenced by a variety of factors like differing localities, climate, topographies, soil conditions, depth of groundwater table, intrusion of surface water, admixtures to FS retention/storage durations, storage, collection, emptying and transport practices, design, construction and performance of OSSs and household usage patterns (Heinss et al., 1998). Based on physicochemical characteristics, FS can be classified as low-, medium- and high-strength sludge (Tayler, 2018). Another classification is done on the basis of the retention time of sludge in the containment systems under which FS is categorised as either fresh or digested sludge (Strauss and Montangero, 2002). Fresh sludge is also known as raw sludge or unstabilised sludge, while digested sludge is also called stabilised sludge (also called septage). Broadly, septage is a combination of scum, sludge and liquid that accumulates in the septic tank.

In case the faecal sludge is to be used as a bio-solid, it has to be checked for nematode eggs which are the indicators of choice to determine hygienic quality and safety. It provides us with an understanding that treating FS can be much more challenging than compared to sewage. Along with the high amount of COD, biological oxygen demand (BOD), helminth eggs and solid content, faecal sludge may also contain considerable amounts of heavy metals like cadmium (Cd), lead (Pb), copper (Cu), zinc (Zn) and chromium (Cr). Surprisingly, FSs are usually "cleaner" than sewage treatment plant sludges as they tend to contain fewer heavy metals or refractory organics, but exceptions are always a possibility (Heinss et al., 1998). Table 16.1 gives a brief introduction to the extent to which FS characteristics vary when compared to sewage. Be it pathogen concentration, nutrient content or organic, dissolved and suspended solids, FS values are 10–100 times greater than sewage values in all cases. Another important thing to note is that the calorific value of FS is also quite comparable to sewage, which means that, just like sewage and sewage sludge, FS also has the capability to be utilised as a biofuel.

The bar graph shown in Figure 16.1 depicts the variability of FS from city to city and country to country. That is why, although, most methods are readily adaptable to other locations and contexts (based on past case studies and ongoing research), some methods must be adapted specifically to suit the local context to achieve the best results.

16.1.2 SAMPLING AND ANALYSIS

Now another question pops into our heads as to how to determine FS characteristics. Are there standard methodologies just like wastewater or sewage sludge for sampling and analysis in place for FS too? The answer is not yet! While standard methods for the examination of water and wastewater do exist for environmental disciplines like waste and wastewater treatment, being a relatively new area to the research community, this is not the case with FS treatment. No standard reference exists for methods for the analysis

TABLE 16.1

Comparison of Faecal Sludge (from OSSs) and Sewage Characteristics (Strauss et al., 1997; Koné and Strauss, 2004; Heinss et al., 1998; Semiyaga et al., 2015; Seck et al., 2015; Tsai, 2012; Syed-Hassan et al., 2017; Sedlak, 1991; Kengne et al., 2011; Ingallinella et al., 2002)

Analytical Parameters Used to Describe FS	Type "A"	Type "B"	Type "C"
Example	Frequently emptied sludge from public toilets or bucket Latrines	Septage/sludge from septic tanks and/or pit latrines	Wastewater and excrement conveyed in sewers
Characterisation	Mostly fresh FS; sludge retained only for several days or a week	Digested FS; sludge retained in the collection containers for several months or years	Tropical sewage
Concentrations	High strength, exhibits high concentrations of organics, ammonium and solids	Low strength, undergone biochemical stabilisation to a considerable extent	
pH	6.55–9.34		6–8
COD (mg/L)	20,000–50,000	<15,000	500–2,500
BOD (mg/L)	7,600	840–2,600	150–1,250
COD/BOD	5:1 to 10:1	5:1 to 10:1	2:1
NH_4-N (mg/L)	2,000–5,000	<1,000	30–70
Total solids, TS (%)	≥3.5%	<3%	<1%
TS (mg/L)	30,000–50,000	30,000–50,000	
Suspended solids, SS (mg/L)	≥30,000	≅7,000	200–700
Helminth eggs (no./litre)	20,000–60,000, 2,500 (lowest reported value)	4,000–6,000, 600 and 16,000 (lowest and highest reported value)	300–2,000
Faecal coliforms (cfu/100 mL)	1×10^5	1×10^5	2×10^5 to 7×10^5
Total phosphorus, TP (mg P/L)	400	150	5–20
Total Kjeldahl Nitrogen, TKN (mg N/L)	3,400–3,750	1,000	20–70
Total volatile solids (%TS)	60–70 (VS)	45–73 (VS) 60–75 (VSS)	76–79 (VSS)
Calorific values (MJ/kg DM)	12.2–19.1		11.10–22.10 (dried sewage sludge)

of FS. That is why many of the methods that are used to characterise wastewater and soils are frequently adapted for FSs too. However, since FS is also much more variable in consistency, quantity and concentration as compared to sewage and wastewater, methods have to be adapted on an individual case-to-case basis (Bassan et al., 2016).

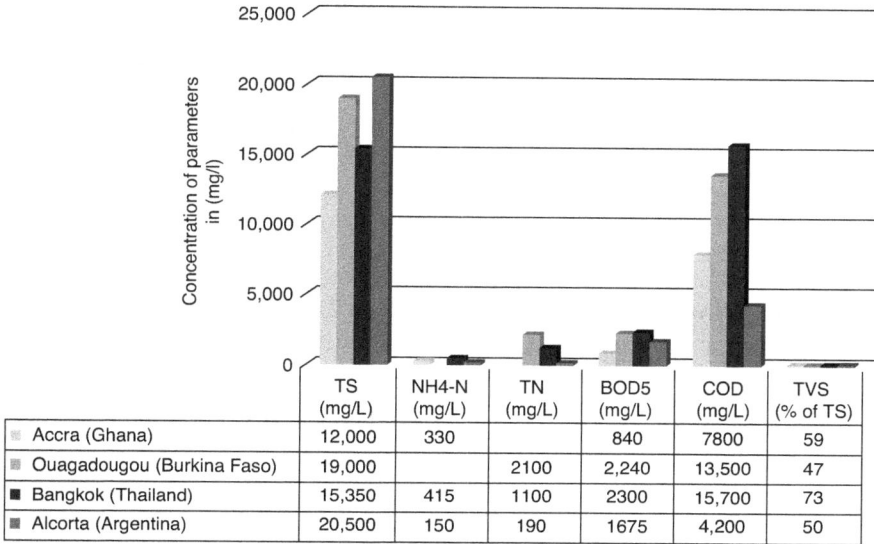

	TS (mg/L)	NH4-N (mg/L)	TN (mg/L)	BOD5 (mg/L)	COD (mg/L)	TVS (% of TS)
Accra (Ghana)	12,000	330		840	7800	59
Ouagadougou (Burkina Faso)	19,000		2100	2,240	13,500	47
Bangkok (Thailand)	15,350	415	1100	2300	15,700	73
Alcorta (Argentina)	20,500	150	190	1675	4,200	50

FIGURE 16.1   The bar graph shown depicts the variability of FS from city to city and country to country.

Analytical parameters (physical, chemical and biological) are determined based on the objectives of the characterisation study. Temperature and pH (potential hydrogen) are two parameters that are detected immediately onsite at the sampling point. Solid and organic content (i.e., total solids (TSs), volatile solids (VSs), total and volatile suspended solids (TSS and VSS), BOD, chemical oxygen demand (COD) etc.), as well as the nutrient content of FSs (i.e., total nitrogen (TN), total phosphorus (TP), ammonium (NH_4^+), ortho-phosphate etc.), are analysed in the laboratory. Apart from that, parameters like volatile fatty acids (VFAs) and heavy metals like iron (Fe), zinc (Zn), nickel (Ni), lead (Pb) etc. which influence the anaerobic digester operation, are also determined. Last but not the least, disease-causing pathogens in FS are determined using two indicator organisms, i.e., *Salmonella spp.* (an indicator for bacterial contamination) and *Ascaris lumbricoides* (an indicator for faecal contamination of water) (Bassan et al., 2016).

16.2   GUIDELINES FOR HANDLING AND DISPOSAL OF FAECAL SLUDGE

Different organisations use different versions of the sanitation service chain. The chain drawn in Figure 16.2 is a combination of what is put forward by the World Bank (WB) and the Bill and Melinda Gates Foundation (BMGF). While onsite sanitation is prevalent across towns and cities in the state, there are major gaps across the sanitation service chain.

FIGURE 16.2   Sanitation service chain as put forward by WB.

16.2.1   Regulations and Guidelines for Faecal Sludge Capture and Containment

The design and construction of containment systems are an essential part of faecal sludge management guidelines. Improper and substandard design of OSSs contaminates groundwater, produces unbearable odour and leads to a variety of diseases.

Many guidelines have been formed by various international organisations like the World Health Organization (WHO), Netherlands Development Organization (SNV) and Water Aid, and they have been compiled for OSSs.

At the household level, factors like expense, ease of construction, emptying frequency, cost of the desludging services as well as area, comfort and privacy requirements, affect the decision-making process. It is essential that the regulations should opt for something which is locally appropriate, common, safe to use, technically feasible and, last but not least, easy to access for routine desludging.

Table 16.2 shows various OSSs in households that are adopted based on traditional, economic, environmental and health-related aspects (Franceys et al., 1992).

These are the various OSSs options along with their main features as discussed (Tilley et al., 2014; Reed et al., 2014; Franceys et al., 1992).

A. **Single-pit latrine**: It is the simplest, cheapest and most widely used technology. It should be at least 3 m in depth and 1 m in diameter with a minimum design capacity of 1,000 L. It has a low space requirement and can be constructed using timber, concrete, stones or mortar plastered into the soil. The dual pit is always prioritised over single pit ones when it comes to health and safety reasons. In case, capital cost and space availability is an issue then, single-pit latrines are preferred. The design life of these latrines varies between 15 and 30 years.

   Limitations: Excreta is visible, and there is a possibility of odour and groundwater contamination. It cannot incorporate an offset pit.

B. **Ventilated improved latrine**: It is a simple and low-cost technology which is a modified version of a single pit latrine. There is continuous airflow through the ventilation pipes vents. The vent pipe should have an internal diameter of at least 11 cm.

   Limitations: Excreta is visible and there is a possibility of odour and groundwater contamination. The construction process is a little complex and an extra cost is required for vent pipes and superstructure. It cannot incorporate an offset pit.

TABLE 16.2
Comparison of OSS Options (Tilley et al., 2014; Kalbermatten et al., 1980)

Type of Latrine	Required Soil Condition	Water Table	Construction Cost	O&M Cost	Water Requirement	Self-Building Potential	Ease of Construction	Reuse Potential (in Agriculture)
Single pit latrine	Stable permeable soil	Bottom of the pit should be at least 2 m above the groundwater level	Low	Low	None	High	Very easy; except on rocky or wet grounds	Low
Ventilated improved latrine	Stable permeable soil	Same as above	Low	Low	None	High	Very easy; except on the rocky or wet ground	Low
Pour flush latrine	Stable permeable soil	Same as above	Low	Low	Water near toilet	High	Easy	Low
Double vault composting toilet	None	None	Medium	Low	None	High	Skilled labour is required	High
Self-topping aqua privy	Permeable soil	Bottom of the pit should be at least 2 m above the groundwater level	Medium	Low	Water near toilet	High	Skilled labour is required	Medium
Septic tank	Permeable soil	None	High	High	Water piped to the toilet	Low	Skilled labour is required	Medium
Container-based sanitation	Permeable soil	None	Medium	High	None	NA	Externally provided	High

C. **Pour flush latrine**: It is affordable, easy to install and clean with reduced levels of odour and excreta visibility. The pit can be placed outside the house while the toilet can be installed inside the house. It can incorporate an offset pit.

   Limitations: Excreta is visible and there is a possibility of odour and groundwater contamination. The construction process is a little complex, and extra cost is required for vent pipes and superstructure

D. **Septic tank**: It is a watertight chamber made from PVC, plastic, concrete, fibreglass etc. It is usually divided into two chambers with the first chamber being half of the total length. It can incorporate an offset pit and has reduced levels of odour and excreta visibility. It has the convenience of a conventional flush toilet and if properly sealed, possesses no fear of water contamination

   Limitations: Large space and water supply requirement with the challenge of effluent disposal.

E. **Aqua privy**: It is a simple storage and settling tank, much like a cheaper septic tank where the excreta falls directly into the tank. Permeable land and an ample amount of water supply are a must in this case. A water seal needs to be maintained for reduced odour.

The FS accumulation rate is a critical parameter while designing and selecting the most appropriate OSSs. These rates vary from place to place and are influenced by factors like toilet use frequency, tank size, local decomposition rate or type of anal cleansing materials as shown in Table 16.3. If local data for accumulation rates are not available, then the suggested general values can be used, which are different for different countries, as shown in Table 16.4.

16.2.2   REGULATIONS AND GUIDELINES FOR FAECAL SLUDGE EMPTYING AND TRANSPORTATION

OSSs need to be emptied periodically through mechanical (using vacuum trucks etc.) or manual desludging. While there can be no technical alternative to simple and inexpensive manual cleaning, the biggest drawback, in this case, is that it puts its staff under hazardous health risk issues. Often, due to several complexities in different onsite technologies on the basis of economic status and access, several operators are found working in any given setting (Strande et al., 2014).

TABLE 16.3
General Data for Faecal Sludge Accumulation Rates (Franceys et al., 1992)

Type of Waste Retained	Maximum Accumulation Rate (l capita⁻¹year⁻¹) When Waste Is Retained in Water	Maximum Accumulation Rate (l capita⁻¹year⁻¹) When Waste Is Retained under Dry Conditions
Where degradable materials are used (e.g. paper and leaves)	40	60
Where non-degradable materials are used (e.g. mud and sticks)	60	90

TABLE 16.4
Examples of OSS Building Codes and Standards from Different Countries

Country	Referenced Guidelines and Legislations
India	• IS 2470:1985 Indian code for the installation of septic tanks – Construction of facilities for containing sludge. The code has two parts: (1) design criteria for construction and (2) secondary treatment and disposal of effluents. (www.indiawaterportal.org/articles/indian-standard-code-practice-installation-septictanks-2470-bureau-indian-standards-1986) • This is also present in the handbook on technical options for onsite sanitation (2012) by the Ministry of Drinking Water and Sanitation. (https://mdm.nic.in/mdm_website/Files/WASH/handbookon-_technical-options-for-on-site-sanitation-modows-2012_0.pdf)
Malaysia	• MS1228:1991 – Malaysian Standard Code of Practice for the Design and Installation of Sewerage Systems 1991. Available at https://kupdf.net/download/ms-1228-1991_58c77 bccdc0d600452339028_pdf • Malaysian Sewerage Industry Guidelines Vol.5: Septic tanks (2008). Available at: www.scribd.com/document/378170193/Malaysia-Sewerage-Industry-GuidelineVolume-5 • Malaysian Standard (MS) 2441-1:2012- Onsite sewage treatment units, Part 1: Prefabricated septic tanks specifications. Listed at www.jeces.or.jp/spread/pdf/02S PAN5ws.pdf
Vietnam	• Manual for septic tank design, installation and O&M – Ministry of Health • Draft Design Code for Septic Tank Design and Construction — Ministry of Construction Both cited in www.susana.org/_resources/documents/default/2-1673-vietnam-fsmstudy.pdf
USA	• Manual of septic tank practice, U.S. Public Health Service, revised edition 1967. Available at https://nepis.epa.gov/Exe/ZyPDF.cgi/9101V1SI.PDF?Dockey=9101V1SI.PDF • For state-specific guidelines see also www.epa.gov/septic/advanced-technology-onsitetreatment-wastewater-products-approved-state
The Philippines	• Revised National Plumbing Code of the Philippines. Available at www.itnphil.org.ph/docs/sanitation%20-%20wastewater%20-magtibay.pdf
South Australia	• Standard for the construction, installation and operation of septic tank systems in South Australia. Available at www.lga.sa.gov.au/page.aspx?u=6640&c=59014
Canada	• The Ministry of Municipal Affairs and Housing is responsible for administering septic system approvals as outlined in the Building Code Act. See www.ontario.ca/laws/regulation/ r12332
Ghana	• Ministry of Water Resources, Works and Housing; Community Water and Sanitation Agency. Small towns sector guidelines (Design Guidelines) Vol. III, 2010. Available at http://lgs.gov.gh/index.php/protocols/ (under CWSA's Operational Documents and Guidelines) • Latrine technology manual 2016 (UNICEF supported). Available at www.unicef.org/ghana/Latrine_technology_option_manual_final__a4_size.pdf

(Continued)

TABLE 16.4 (*Continued*)

Examples of OSS Building Codes and Standards from Different Countries

Country	Referenced Guidelines and Legislations
Sri Lanka	• SLS 745 part 1: 2004: Code of Practice for Design and Construction of Septic Tanks and Associated Effluent Disposal Systems. Part 1 — Small Systems Disposing to Ground • SLS 745 part 2: 2009: Code of Practice for the Design and Construction of Septic Tanks and Associated Effluent Disposal Systems. Part 2: Systems Disposing to Surface, Systems for Onsite Effluent Reuse and Larger Systems Disposing to Ground. Available at www.slsi.lk/index.php?lang=en (Search Standards with keyword Septic Tanks)

TABLE 16.5

Septage Emptying Frequency as Reported by the Water Environment Federation, USA

Sanitation systems	Desludging interval	Accessibility	Inspection of the filled level
Septic tanks	2–6 years	Access can be provided through a manhole that should not be covered by concrete, roads, or flooring and must have a cover that can be removed by an adult only	12–18 months
Cesspool	2–10 years		
Privies/portable toilets	1 week to a few months		
Aerobic tanks	Up to 1 year		
Dry pits (associated with septic fields)	2–6 years		
Pit latrines	4–9 years	It should not be covered by concrete, roads, or flooring and must have a cover that can be removed by an adult only	Same methodology as used in septic tanks.

Source: Adapted from USEPA (1999)

Depending upon emptying frequencies as shown in Table 16.5 and the presence or absence of water for flushing, the concentration of collected FS varies considerably from place to place. It also impacts the type and costs of FS treatment to be provided. The desludging frequency is based on the OSS type, containment capacity and incoming flow. In Table 16.6, several desludging guidelines are given which were issued by the USA, which demonstrates after how much time from installation and by what methods faecal sludge should be desludged.

TABLE 16.6
Tank Desludge Guidelines (USEPA, 1994)

When to Desludge?

An inspection for sludge layer thickness can determine whether desludging is required. It is recommended that a septic tank must be pumped out:

- when sludge and scum occupy half to two-thirds of the tank's working capacity (the tank volume below the outlet pipe invert level);
- every 3–5 years (sludge hardens over time and is difficult to remove by suction);
- if the bottom of the scum mat is <8 cm above the bottom of the baffle/outlet pipe;
- if the minimum working capacity is reached; or
- if the anticipated accumulation rate would result in one of these conditions by the time of the next inspection.

How to Desludge?

Firstly, the scum mat is manually broken to facilitate pumping. If the liquid level in the tank is higher than the outlet pipe, the liquid level has to be lowered below the invert of the outlet, which prevents grease and scum from being washed into the drain. Normally, the vacuum/suction hose draws air at a point where 1–2 inches (2.5–5 cm) of sludge remains over the tank bottom, and this material should be retained in the tank. Washing down the inside of the tank is not required unless leakage is suspected and it needs to be inspected. If internal inspection is warranted, fresh air should be continuously blown into the tank for at least 10 minutes before a worker enters

TABLE 16.7
Monthly Charges for OSSs in Malaysia

Category (For Domestic Customers)	Connected Charge per Month in USD (MYR 1.00 = USD 0.24)
Low-cost houses, houses with an annual value of less than MYR 600 and government 0.48 quarters in categories F, G, H and I (receiving either an individual septic tank or connected sewerage services)	0.48
Premises and government quarters with individual septic tanks	1.44
Houses in traditional and new villages and estates (receiving either individual septic tanks or connected sewerage services)	0.72
Premises and government quarters in categories A, B, C, D and E receiving connected 1.92 sewerage services	1.92
Category (For Industrial Customers)	**Rate Based on Number of Employees in USD**
Premises receiving individual septic tank service	0.48 capita^{-1} month^{-1}
Premises with connected sewerage services	0.60 capita^{-1} month

16.2.2.1 Fee Structures

The Malaysian Water Association (2017) provides a good example of fee structures for domestic, industrial, governmental and commercial stakeholders. The fee system covers connection to sewage networks as well as OSS. Table 16.7, gives an idea about the fee structure in Malaysia for OSSs.

16.2.2.2   Specific Guidelines for Faecal Sludge Desludging Trucks and Accessories

The service providers of transportation and collection of septage either public or private have to follow the rules and regulations to ensure that the business operators are satisfying the public, social and environmental safety regulations. For desludging services, local government can devise the operating structure in two ways:

- First, they can make a contract with the licensed contractors that the local entities will be given their own trucks for cleaning and emptying services at scheduled times.
- Second is that they encourage the contractors to invest in a good cause, meanwhile the local entities will provide them access to their parking and cleaning stations [at wastewater treatment plants (WWTPs) usually].

The main aim of these regulations in the transportation section is the minimisation of public and environmental health risks.

16.2.2.3   Regulations on Faecal Sludge Emptying and Transportation: Examples from Selected Countries

Many low-income countries do not have any regulations on utility operators. In countries where such regulations exist, requirements are often based on existing standards from other countries, adapted to local implementation capacity and other applicable standards from the existing regulatory systems.

Two major aspects of regulations are:

a. Occupational health and safety (OHS) standards and
b. Service standards.

The following examples show what such guidelines can entail, from truck registration to health care:

i. **Malaysia**: Malaysian guidelines require Faecal Sludge Management (FSM) service providers to obtain various licences, permits or levies from different authorities. These licences are used as tools to monitor the industry.

   The relevant agencies responsible for handling different authorisation procedures, while conforming to the requested standards in Malaysia (Ho et al., 2011), are listed below:
   - The National Water Service Commission – desludging services permit.
   - Department of Occupational Safety and Health – operation of pressure vessels permit.
   - Construction Industry Development Board – registered operators. Moreover, there should be a system in which these authorities should regularly inspect the permit holders and services licences holders based on performance and activities. Supervision of fair pricing is part of the regulation.

ii. **Australia**: In accordance with the South Australian guidelines, the contractor must have records in satisfaction of the local council, including the details of when the OSS system has been desludged and the location for the disposal of FS. The contractors responsible for the transport of FS should be licensed by the state Environmental Protection Agency (EPA). Vehicles used in the transport of FS must be cleaned in a location where there are wastewater treatment facilities or at a location approved by the EPA for reprocessing of FS so that the wash-down water down not enter stormwater systems. Any transport leakage or spills must be cleaned immediately and the methods used for that in general dry-clean-up methods. Any disposal of these overflows FS in water streams is prohibited by EPA (Brown et al., 2017).

iii. **India**: In accordance with the policy on FS and septage management in India in 2017, there is development in state-specific policies, strategies and guidelines. According to the Tamil Nadu guidelines, the local bodies should be empowered with the knowledge of procedures and facilities for effective septage management. Only certified and licensed septage operators are allowed to desludge and transport septage to the location of sewage treatment plants (STPs) and these operators should be selected in accordance with the Tamil Nadu Transparency in Tenders Act, 1998. Moreover, the operators of septage vehicles must be well-trained and equipped with protective safety gear, uniforms, tools and proper vacuum trucks to ensure safety while handling septage. The guidelines also order the management facility to maintain a record for the location of septage generation, the type of OSS, the in-charge of operating locations and the name and location of STP. From these guidelines, the most important initiative is the use of geographical information system (GIS) for routing the roadmap for vehicles and keep a track of them accordingly (Operative guidelines for septage management, Government of Tamil Nadu, 2017).

16.2.2.4   Other General Health Requirements for Operators

FS is an infectious organic matter, which can cause several diseases if inhaled, ingested or exposed to pierced or broken skin.

- Hands should be washed as soon as possible after coming in contact with FS even after using proper gear.
- Gear and equipment must be washed and sanitised properly after use.
- The workers should be immunised at least against tetanus, hepatitis A and hepatitis B and dewormed (via pills).
- Caution around septic tanks is essential and workers should never enter a septic tank without due precaution. These tanks are confined spaces that may contain toxic or oxygen-limited atmospheres and deaths from careless entry occur every year.
- Moreover, accidents can occur if excessive weight is placed on the lid or utility hole cover (Robbins, 2007).

16.2.3   REGULATIONS AND GUIDELINES FOR FAECAL SLUDGE TREATMENT AND DISPOSAL

The characteristics of FS show that it is harmful to the environment and the health of living beings if it is disposed off without treatment. Therefore, it is necessary that it is collected from various points in the city and is disposed of at an appropriate treatment facility. Linda Strande, a leading scientist at SANDEC (Swiss Federal Institute of Aquatic Sciences and Technology) FSM, suggests an integrated approach of planning, technology and management for solving global sanitation challenges. The definition as given in her book is as – "Faecal sludge treatment involves dewatering of sludge transported from the onsite sanitation systems, their treatment, reuse or end use of resulting (liquid/solid) end products as well as safe disposal of treated matter" (Strande et al., 2014). To operate sustainably in low- and middle-income countries, the septage treatment plants should have low O&M costs, low energy consumption and if possible the capacity to support operational costs. The adopted systems must be built in such a way that it fits with the place conditions and also goes along with the institutional/entrepreneurial set-up responsible for scheme implementation. Both FS and effluent need to be treated. Small bore systems can be used to convey effluent to small distances and then treat it at a decentralised scale. To choose the best combination of technologies, things like population density, water usage, type of onsite system prevalent in the city, soil strata, groundwater table, land available, the topography of the city, characteristics of the FS, the demand of the end product, capital cost and operation cost should be considered.

16.2.3.1   Need and Objective of FS Treatment

As compared to wastewater, FS is highly variable in characteristics and in most countries, there are no specific discharge or reuse standards for their treatment. Inefficient FS collection and haulage systems, the non-affordability of mechanised pit emptying by the urban populace, the difficult-to-access OSSs installations for emptying vehicles, excessive haulage distances to designated disposal or treatment sites, lack or absence of fully operational treatment sites and non-availability of low-cost treatment options at affordable haulage distances etc. are some of the main reasons due to which a large portion of FS goes untreated (Koné and Strauss, 2004). That is why it is important to have a comprehensive and sustainable FS treatment and disposal system, meeting the needs of the local area to eliminate these harmful impacts. This becomes more essential for countries where safe sanitation and drinking water availability are still a huge challenge in the 21st century.

The treatment process focuses on the following things:

- Decreasing the water content of sludge to a great extent. This will lead to volume reduction which will make it less cumbersome to handle and then transport to the local treatment site. Dewatering saves space as well as simplifies and optimises the performance of successive treatment units.
- Stabilisation of FS for reducing the suspended solid content, excess ammonia content and high oxygen demand, i.e., BOD and COD of the faecal sludge to ensure its safe use and disposal.

- Depending upon the objective of treatment, complete/partial removal of pathogens like faecal coliforms and helminth eggs from the dried sludge and the liquid effluent.
- Detect and remove heavy metals and other organic and inorganic pollutants and toxic substances to prevent soil or water contamination during end use or disposal of faecal sludge.
- Reducing excess nutrient content of the liquid effluent before discharging them into water bodies to avoid Eutrophication and water contamination.

As per the United States Environmental Protection Agency (USEPA) (1994), FS treatment can be achieved in three ways which are as mentioned below.

16.2.3.1.1   Co-Treatment with Wastewater at Sewage Treatment Plant

Discharge of septage into publicly operated WWTP (Wastewater Treatment Plants) is common practice in large cities.

WWTP can constitute a low-cost and environmentally sound option for septage treatment. However, if the WWTPs are not designed to take the load of septage, the overall load will affect its cleaning, working and operational costs negatively.

In the case of a conventional activated sludge treatment plant with a primary clarifier, which is designed for 7,500 m³/day and that is operating on an efficiency of 50% will be able to take 100 m³ sepatage/day. The different options for handling FS at WWTP are stated below:

- Septage addition to upstream sewer utility holes
- Septage addition to plant head works
- Septage addition to the sludge handling process
- Septage addition to both liquid stream and sludge handling process

Key points of regulation on the co-treatment of septage at WWTPs could be:

- If a person is engaged in septage collection, that person should dispose of the septage at the facility within the operating area.
- A facility may charge a little fee for receiving septage.
- Authorities may issue an order prohibiting delivery of FS to a WWTP if there is excessive hydraulic or organic load, odour or any other environmental or public health issue.

16.2.3.1.2   Separate Treatment at FS/Septage Treatment Plant (FSTPs)

In those situations where there is a need for WWTP but these are located at far distant places or have insufficient capacity, this creates a need for independent separate treatment facilities. Such treatment facilities have separate units to handle liquid and solid portions of septage (USEPA, 1994). The determination of the facility or technology that should be used depends on the type of septage volume, odour, land availability, reuse option as well as local governing aspects.

There are three types of mechanisms that are generally used for treating faecal sludge i.e., physical (for solid-liquid separation), biological (for stabilisation and

Activity diagram

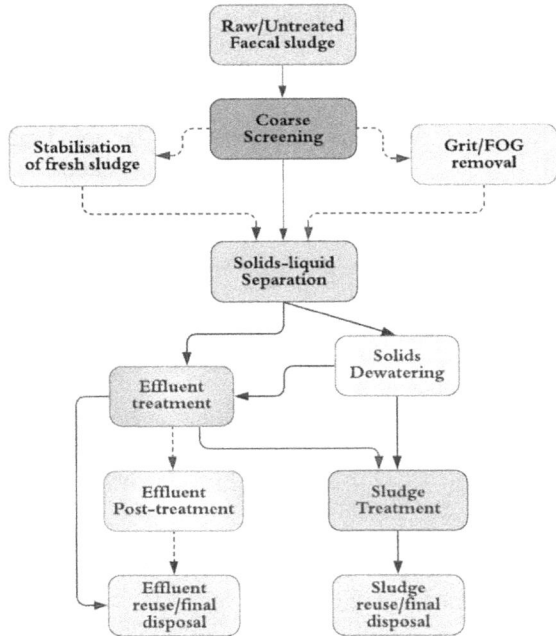

FIGURE 16.3 Mechanisms generally used for treating faecal sludge, i.e., physical, biological and chemical.

liquid treatment.) and Chemical (stabilisation and pathogen inactivation) as shown in Figure 16.3. The steps given below demonstrate how FS is treated in FSTPs.

- **Preliminary treatment**: Here, the main objective is to eliminate bulky objects, large solids and non-organic items like grit, fats, oil and grease (FOG). It also partially stabilises fresh or high-strength sludge. This step of treatment is important to prevent blockages and ensure the smooth and optimum functioning of all units in the whole treatment plant. Desludging vehicles transport the collected FS to the FSTPs where it is transferred to the reception chambers/tanks. In most cases, the tank is equipped with screen chambers, grit chambers and an equalisation facility (if needed). Equalisation/attenuation facility function as buffering tank and prevents flow fluctuations which can be problematic for the whole treatment plant's performance. Sludge after coarse screening can be sent for grit and FOG removal. For these cases, Parabolic grit channels or Vortex separators are generally used.
- **Stabilisation**: Unlike sludge collected from septic tanks which are partly or completely stabilised, raw and fresh sludge collected from pit latrines or public toilets requires some degree of stabilisation because of its unbearable odour, low settleability and difficulty to handle nature.

Lime stabilisation, Anaerobic stabilisation reactors, Anaerobic baffled reactors with integrated settler and filter chambers and small-scale bio-digesters are some of the most commonly used options for this purpose. Aerobic digestion can also be used for this purpose, but it has high energy requirements and operational costs, and thus, it is not very favourable for low-income countries.

- **Solid–liquid separation**: In WWTPs, sedimentation tanks are used for these purposes. Here, the objective is to separate easily settleable solids from the liquid fraction to reduce the organic and suspended loads in the liquid flow. This step reduces the bulk volume of solids to be handled and saves extra space and hassle required in the dewatering and drying processes. It also leads to the reduction of a considerable amount of COD, BOD and TSS and aids in a small amount of pathogen removal. This step helps in saving space and power requirements for the subsequent steps. The separation process is sometimes coupled with dewatering or biological treatment for treating FS. One important thing to note here is that this step should always be considered for septage rather than freshly digested sludge which has considerably high solid content, thus not requiring solid-liquid separation. Gravity-based settling–thickening tanks/ponds, Imhoff tanks, decanting drying beds, gravity thickeners, centrifuges, geobags, waste sta-bilisation ponds etc. are some of the technologies that are currently being widely used for this purpose in many countries. In the case of septage with very low solid content i.e., about 1%–2%, anaerobic ponds can be used whereas stabilised FS with high solid content, i.e., 5%–6%, can be best suited for sludge drying beds.

- **Solid dewatering/drying**: The aim here is to reduce the water content or moisture content present in the sludge so that it becomes easier to handle and transport it. In the long run, it helps in decreasing the plant and treat-ment unit area and, thus, increased efficiency for treatment and resource recovery. This can be achieved through a variety of means, i.e., gravity, evaporation, evapotranspiration, heat application or mechanical means. Unplanted and planted drying beds and mechanical dewatering methods like belt presses, geobags/geotubes, centrifuges etc. are some of the most commonly used treatment methods for dewatering. Chemicals like coagu-lants, complex mixtures of polymers, or locally produced conditioners like chitosan can be used to condition the sludge for mechanical dewatering.

- **Effluent treatment**: After solid–liquid separation and dewatering of FS, supernatant or effluent from those chambers is conveyed to the effluent treatment unit. It is essential for this liquid to meet the designated effluent discharge and reuse standards that exist for sewage or wastewater in that area. Once treated, the liquid can either be safely reused for irrigation or land reclamation purposes or it can be discharged off into a nearby water body. The focus here is to reduce the excessive nutrient content, oxygen and nitrogenous demand, suspended solid content and, last but not least, the pathogen fraction of the FS supernatant to enhance its resource recov-ery options.

For this purpose, inexpensive anaerobic processes are often followed by a last aerobic process and a final polishing step. Anaerobic baffled reactors, waste stabilisation ponds like aerobic facultative and maturation ponds, aerated lagoons, constructed wetlands etc. are some of the low-cost and frequently used options for effluent treatment. The co-treatment of FS effluent at STPs/WWTPs using biological treatment technologies like Upflow Anaerobic Sludge Blanket Reactor (UASB), trickling filters, extended aeration and activated sludge is also a great alternative. It is a great option for FSTPs situated in peri-urban areas which are connected to STP through sewerage network systems. Note that it is essential to assess the compatibility of FS load with the WWTP/STP for the smooth operation of the plant as well as the co-treatment process. As per the end use, treated liquid can further undergo an additional last treatment step, i.e., disinfection. This additional step is called post-treatment of the effluent.

- **Sludge treatment**: Thickened or dried sludge from solid-liquid separation, dewatering and effluent treatment is collected to carry out a final treatment step to make it suitable for safe end use and reuse. The focus here is on stabilising and hygienising the biosolids for optimum resource recovery and negligible risk of contamination. Vermicomposting and co-composting with an organic portion of municipal solid waste are some of the most commonly adopted options for treating dried sludge from FS treatment processes. Other innovative and resourceful treatment options are thermal drying, sludge incineration, co-combustion or pyrolysis/gasification of dried sludge, Ladepa technology for wet sludge, pelletising, solar drying beds, solar sludge ovens, black soldier fly larvae treated FS, trenching, deep row entrenchment, lime/ammonia addition, hydrothermal carbonisation (HTC) etc. (Strande et al., 2014).

16.2.3.1.3 Land Application

In smaller communities or areas that are not close to a treatment plant, transfers are not practical and septage is typically land applied. In many places, septage disposal in landfills is not allowed because it is in a liquid form and waste landfills cannot accept materials containing free liquids. Still, land application is one of the most widely practiced methods for FS disposal in low- and middle-income countries. Although this method was once considered useful because of the nutrient and carbon-rich nature of FS, later on, due to its harmful impacts, it was either completely banned or severely restricted in many countries. Other methods like deep trench burial or subsurface incorporation of FSs (with controlled application rates) or disposal in landfills which are actively being practiced in South Africa and Malaysia are considered comparatively better options for disposal.

16.2.4 REGULATIONS AND GUIDELINES FOR FAECAL SLUDGE USE

End use of FS/septage refers to the safe, beneficial use of human excreta, i.e., faeces and wastewater from onsite sanitation technologies. The type of end use should decide the level of treatment. Considering the nutrients, organic matter and energy contained in FS/septage, it can be used as a soil conditioner or fertiliser in agriculture, aquaculture, or horticultural activities. Other uses include use as a fuel source, building material, or for protein food production. Closing the loop would not only help in reducing freshwater demand and chemical fertiliser demand but also prove to be a source of revenue; in other words, it can help improve the business model. As shown in Figure 16.4, there are many treatment products of FS with the potential for reuse. In many FSTPs around the world, biogas generated from anaerobic digesters or reactors is used as a source of power in the plant area. Treated effluent can be safely reused in the plant vicinity for gardening and irrigation or it can be discharged in the nearest watercourse after meeting the discharge standards. Compost can be safely used as a soil conditioner in agriculture, while dried sludge can be used as a source of biofuel and building material. FS can also be used to generate biofuel (gas and char briquettes) as well as in the production of protein (e.g. animal feed and via the black soldier fly). Figure 16.4 lists all the known resource recovery end products that can be obtained through FS treatment. If these products can be commercially sold, they can be a great source to generate revenue for the FSTPs.

Other than land application for FS disposal rules and guidelines, the regulations for the use of FS in agriculture and in aquaculture are rare. The following sections are only regarding the use of FS in aquaculture, agriculture and the use of faecal sludge as dry fuel in households and in cement factories.

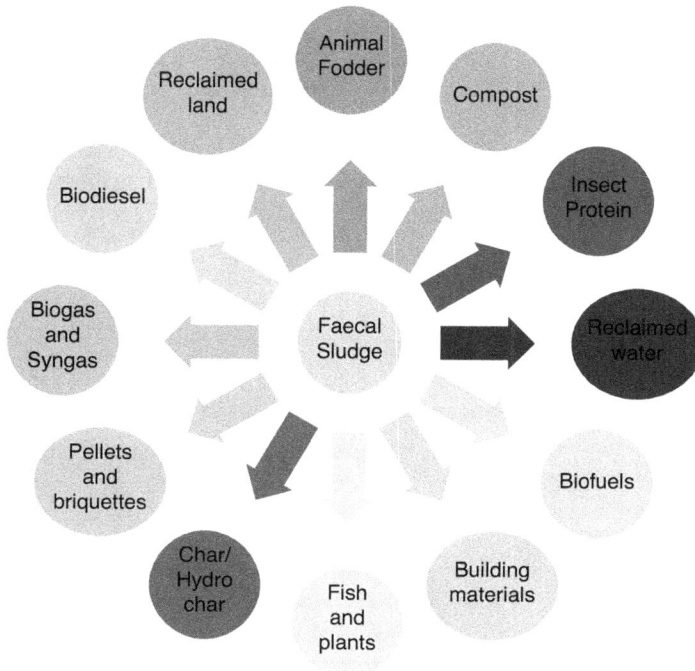

FIGURE 16.4 All the known resource recovery end products that can be obtained through FS treatment.

16.2.5 Guidelines for Agricultural and Non-Agricultural Land Application

There are several international guidelines on public health, for example, in the European Community, the application of highly organic sewage sludge content in agricultural fields (Directive 86/278/EEC), which set limits on the concentration level of FS in soil. The USEPA's standards for the use or disposal of sewage sludge are a thorough regulation in this area (Title 40 of the Code of Federal Regulations [CFR], Part 503). In this regulation other than domestic septage, there is a separate portion on the requirements of its application on nonpublic contact sites which includes agricultural land, forest and reclamation areas (USEPA, 1993a). Part 503 describes the ceiling concentrations (mg kg dry weight^{-1}, mg ha^{-1} and the load in kg ha^{-1} and year^{-1}) for heavy metals in biosolids which applied to land and cannot be exceeded specifically for arsenic, cadmium, chromium, copper, lead, mercury, molybdenum, nickel, selenium and zinc as shown in Tables 16.8–16.13.

According to the USEPA (1993b), the methods should be applied on the septage before using them in the land application (also see USAID, 2008):

- Aerobic digestion for 40 days at 20°C or 60 days at 15°C.
- Anaerobic digestion for 15 days at 35°C–55°C or 60 days at 20°C.
- Air drying for at least three months. Two of the months must have average daily temperatures above freezing.
- Composting or co-composting at temperatures >40°C for 5 days. The temperature of all of the material being composted must be >55°C–65°C for at least four hours during the five days.
- Lime stabilisation to bring the pH higher than 12 for 30 minutes or bring the pH higher than 9 during more than six months if the temperature is above 35°C and/or moisture is below 25%.

TABLE 16.8
Pollutant Ceiling Concentrations in Biosolids (USEPA, 1993a, b)

Pollutant	Ceiling Concentration (mg/kg Dry Weight)
Arsenic	75
Cadmium	85
Copper	4,300
Lead	840
Mercury	57
Molybdenum	75
Nickel	420
Selenium	100
Zinc	7,500

TABLE 16.9
Guidelines and Regulations from Different Countries on FS Land Application (USAID, 2008)

Country	Land Application Methods
USA	Subsurface injection, spraying or spreading on the soil surface or ploughing, disking or injecting into the soil
Canada	It is a common practice to apply sludge and biosolids to agricultural lands as a nutrient source while strictly following governmental regulations and guidelines
Philippines	Surface application

TABLE 16.10
Suitable Soil Conditions for Land Application Sites (Minnesota Pollution Control Agency, 2015)

Characteristic	Minimum Requirement
Soil texture	At the zone of FS application (surface horizon or injection depth), the soil texture must be one of (US system) the following: fine sand, loamy sand, sandy loam, loam, silt, silty loam, sandy clay loam, clay loam, sandy clay, silty clay or clay
Surface horizon	If 0.2 inches per hour or less, this soil is suitable only for surface application with the incorporation
Permeability	within 48 hours or injection
Depth to bedrock	3 feet
Depth to seasonally saturated soil	3 feet

TABLE 16.11
Specifications for Fertilisers and Compost/Soil Conditioner (USAID, 2008)

	Plain Organic Fertiliser	Compost and Soil Conditioner	Fortified Organic Fertiliser
Total NPK	5%–7%	3%–4%	8% minimum
C:N	12:1	12:1	12:1
Moisture content	<35%	<35%	<35%
Organic matter	>20%	>20%	>20%

16.2.6   Lime Stabilisations for Agricultural Reuse

As mentioned previously, the application of lime is one of the most effective methods to stabilise sludge for disposal or reuse. Following the USEPA recommendations, from a case study in the Philippines, a dosage of about 10–20 kg is enough to stabilise 4,001 kg of septage, and then it was ready to use as fertilisers. The material can then

TABLE 16.12

Test for Pathogens for Organic Fertiliser/Soil Conditioner (USAID, 2008)

Pathogen	Safe Limit
Faecal streptococci	$<5 \times 10^3 g^{-1}$ compost
Total coliforms	$<5 \times 10^2 g^{-1}$ compost
Salmonella	0
Infective parasitic	0

TABLE 16.13

Allowable Levels of Heavy Metals in Organic Fertiliser/Compost Soil Conditioner (USAID, 2008)

Heavy Metal	mg/kg Dry Weight
Zn	1,000
Pb	750
Cu	300
Cr	150
Ni	50
Hg	5
Cd	5

be more easily handled for final disposal. There are two common ways to perform lime stabilisation:

I. Lime is simply poured into the (stainless steel) vacuum truck. Lime can be added before or after the septage is pumped, but not in the toilet or septic tank. The lime and septage can then be mixed using the truck's pump.

II. Adding lime to a land-based pit that receives the sludge load on a daily or weekly basis.

16.2.7 OCCUPATIONAL RISKS ON FARMS

All those people who are in contact with FS along the service chain must be checked for proper protective gear and hygiene. In the case of the risk of farmers who work with unsterilised septage in some countries, Seidu (2010) recommends extending the drying periods from 1 to 3 months in the temperature conditions same as northern Ghana to meet the WHO microbial monitoring benchmark for *E. coli*, helminth eggs, Ascaris and rotavirus infections.

Treated sludge is defined as having undergone "biological, chemical or heat treatment, long-term storage or any other appropriate process so as significantly to reduce its fermentability and the health hazards resulting from its use." The WHO (2006) provides an example of storage and treatment options as shown in Tables 16.14 and 16.15.

TABLE 16.14

Recommendations for Storage Treatment of Dry Excreta and FS before Use at Household and Municipal Levels (WHO, 2006)

Treatment	Criteria
Storage; ambient temperature of 2°C–20°C	1.5–2 years
Storage; ambient temperature of 2°C–20°C	>1 year
Alkaline treatment	pH >9 over 6 months

TABLE 16.15

Guideline Value for Verification Monitoring in Excreta and FS Use in Agriculture (WHO, 2006)

	Helminth Eggs (Number)	*E. coli* (Number)
Treatment of faeces and FS	<1 per g total solids	<1,000 per g TS

16.2.8 Soil Application of (Co-)Composted Faecal Sludge

FS is usually co-composted with other organic waste, like food waste, other organic MSW, sawdust and so forth. The mix of materials improves the carbon–nitrogen balance of the material, which again supports microbial activities during the composting process. The better the composting is performing, the higher the temperatures in the compost pile and the elimination of pathogens (Cofie et al., 2016). The resulting FS-based co-compost is a hybrid between:

a. an organic soil ameliorant, which helps to improve soil's physical characteristics
b. an organic fertiliser, which provides plants with crop nutrients

16.2.9 Compost Application Guidelines (Nikiema et al., 2014)

- FS-based composts may be used in different ways, either as a growing media, alone or combined with soil or bio-char or as a soil conditioner-cum-fertiliser (hybrid), to boost and sustain crop yields.
- Only matured (well-composted) FS products should be applied to soils in order to eliminate possible negative side effects ('burning') on crop growth.
- Application rates for FS-based composts vary with soils, crops and compost enrichment and range commonly between 5 and 25 t/ha.
- Compost or compost pellets can be applied in different ways (broadcasting, placement etc.). If applied directly to a plant, the pellets should be placed about a 5 cm radius from the base of the plant, not closer, either on the soil surface or buried at a depth of about 5 cm.
- Non-pelletised compost can be mixed with the soil or planting media. It can be applied on the soil surface or ploughed into the soil, e.g. 1 week before planting.

- The quantity of fertiliser to apply is in general defined by crop requirements and soil fertility. Nutrient requirements for specific crops can usually be obtained from the Ministry of Agriculture. Although crops need a range of different nutrients, it is common practice to calculate the fertiliser application rates based (only) on the crop nitrogen (N) demand.
- On very poor (sandy) soils, a compost can have a high impact by alleviating many soil structure- and soil fertility-related shortcomings, but it can only contribute to some extent to immediate crop nutrient requirements.
- An N-enriched compost will for some crops still require an extra application of other nutrients, like phosphorus (P) or potassium (K). This might, for example, be the case on sandy or highly weathered soils and can be addressed through the use of additional fertiliser.

16.2.10 FAECAL SLUDGE REUSE

The use of FS as dry fuel or biogas in full filling household energy needs is increasingly common. Pyrolysis is a thermochemical process that is used in converting biomass into solid (char), gaseous and liquid compounds in the absence of oxygen. The temperature used for this process is 300°C–700°C, from the past it is seen that slow pyrolysis is used for producing cooking charcoal (Cunningham et al., 2016).

Pyrolysis is the thermochemical process of converting biomass into solid (char), gaseous and liquid compounds in the absence of oxygen (liquids can be recovered from the condensable fraction of gas). Using a heating temperature of 300°C–700° C, slow pyrolysis has historically been employed to produce cooking charcoal (Cunningham et al., 2016). Faecal chars made at low temperatures can be briquetted, for example, with molasses/lime and starch binders, generating thermal efficiency of around 25-mega joule kg^{-1}, which are comparable to those of commercial charcoal briquettes (Ward et al., 2014).

Septage as dry fuel for cement kilns: The use of alternative fuel sources, including sewage sludge, is common in the cement industry (WBCSD, 2014). As septage use is less common, compared to sewage sludge use, in many developed cities a few certain rules and regulations have to be followed before its use, although it has been tested at scale (Wald, 2017). Regulations and standards are needed in five key areas (Hasanbeigi et al., 2012):

- Environmental performance
- product quality,
- Waste quality,
- operational practices and
- Safety and health requirements for employees and local residents

16.2.11 FAECAL SLUDGE REUSE IN AQUACULTURE

Waste-fed aquaculture is centuries old in various countries in East, South and Southeast Asia, especially China. It has been developed mainly by farmers and local communities to use nutrients contained in waste to produce aquatic food. There is a great diversity in current waste-fed aquaculture practices involving septage, fish and aquatic plants, including high-protein plants grown in wastewater as feed for fish grown in freshwater systems (e.g. duckweed). Wastewater may also be used in aquaculture nurseries to produce seeds or fingerlings, which are then grown out into full-size table fish in separate systems without the use of waste. The practice is largely a grey area untouched by regulations and policies, although most waste-fed aquaculture involves the direct addition of waste with little or no prior treatment, resulting in a range of potential hazards: excreta-related pathogens (bacteria, helminths, protozoans and viruses), skin irritants, vectors that transmit pathogens and toxic chemicals. However, only a few risks are considered high. Microbial contaminants, for example, rarely penetrate into edible fish flesh or muscle except for trematodes (parasitic tissue flukes). In fact, the transmission of trematode parasites is of particular concern in aquaculture as trematode-associated diseases are associated with high morbidity. The risk can be reduced through FS storage prior to application. Protection is achieved by a combination of different measures, including cooking fish thoroughly prior to consumption. Microbial quality targets for pond water have been established that can be used to facilitate compliance with the WHO's health-based targets, e.g.,

  i. Viable trematode eggs not detectable (per 100 mL or per gram of TS);
  ii. $\leq 10^4$ *E. coli* (arithmetic mean per 100 mL or per g of TS) and
  iii. ≤ 1 helminth eggs (per litre or per g TS) to protect consumers.

Finally, Annex 1 of WHO (2006) presents design criteria for wastewater treatment ponds that can support microbially safe fish farming and could be referenced in national FS reuse regulations.

16.2.12   Effluent Discharge and Reuse for the STP

In the majority of less-industrialised countries, effluent discharge legislation and standards have been enacted. The standards usually apply for both wastewater and faecal sludge treatment. They are often too strict to be attained under the unfavourable economic and institutional conditions prevailing in many countries or regions. Quite commonly, effluent standards are neither controlled nor enforced. Examples of faecal sludge treatment standards are known from China and Ghana. In the Province of Santa Fé, Argentina, e.g. current WWTP effluent standards also apply to FS treatment. For sludges used in agriculture, a helminth egg standard has been specified. Except for a few countries like China and Ghana, most countries around the world do not have specific discharge standards for FS. In most countries including India, general effluent discharge standards for sewage or wastewater are also applicable to FS. Table 16.16 shows the comparison of effluents from a FSTP with national effluent discharge standards in India.

16.3   TREATMENT TECHNOLOGIES FOR FAECAL SLUDGE

16.3.1   Faecal Sludge Treatment Practices: Developed vs Developing Countries

Globally, over the last decade, a shift can be noticed in the acceptance of the concept of FSM. In developed economies like the USA where barely one-third of households use septic tanks, most of the septage is either co-treated in WWTPs or treated separately in pond systems before discharging in surface water. These pond systems typically consist of an Anaerobic sedimentation pond followed by an Infiltration

TABLE 16.16

Comparison of Effluent of an Indian FSTP with National Effluent Discharge Standards Given by CPCB (New Delhi, India)

Parameters	Raw FS Characteristics	Treated Water Characteristics	General Effluent Standards Set Up by CPCB, India (for Inland Surface Water)	General Effluent Standards Set Up by CPCB, India (for Irrigation)
pH	7.4	8.0	5.5–9.0	5.5–9.0
BOD_5 (mg/L)	5,000	≤9	30	100
COD (mg/L)	15,000	≤67.5	250	-
TSS (mg/L)	10,000	≤96	100	200
Faecal coliform count/100 mL	107	≤1,000	500 (desirable) 2,500 (maximum permissible limit)	1,000
NH_3-N (mg/L)	700	≤18	50	-
PO_4 (mg/L)	400	≤10	5	-

Pond (Strauss and Montangero, 2002). In these countries, FS treatment is based on mechanised sewage and sewage sludge treatment technologies like Extended aeration systems, anaerobic digestion, sludge thickeners, centrifuges, presses and thermal drying technologies. With established mandatory guidelines, stringent regulations and standards, the availability of government support, funds and infrastructure, and a well-connected and functioning sanitation service chain, almost all of the FS or septage is treated and safely discharged. The highly mechanised and expensive technologies used in industrialised countries are not considered adaptable for solving the challenges of FSM in low- and middle-income countries.

Unlike before, when only sewer-based systems were considered appropriate for achieving citywide sanitation, now decentralised options like FSTPs are also being considered as a likely alternative. Newly designed technologies are first assessed in pilot-scale or community-scale plants to check their applicability and efficiency. These plants can then further be upscale based on their performance, govt. support and availability of funds. Noteworthy efforts in this direction have been observed in many developing economies of South Asia, West, Central and South Africa where the sanitation situation is extremely grim. Over the years, many pilot-scale research studies, small-scale treatment facilities and innovative technologies have been developed in countries like Malaysia, China, India, Thailand, Indonesia, Philippines, Bangladesh, Nepal, Vietnam, Argentina, Ghana, Benin, Botswana, South Africa, Senegal, Mali, Ivory Coast, Burkina Faso etc.

Some of the most commonly adopted low-cost practices for FS treatment and disposal in these countries include landfill disposal, land application, co-treatment at WWTPs/STPs/sludge treatment facilities, trenching; settling–thickening tanks or ponds, Imhoff tanks, lime stabilisation, waste stabilisation ponds; stabilisation reactors/tanks, anaerobic digestion with biogas utilisation, anaerobic baffled reactors, gravel filter beds, unplanted drying beds and HF/VF constructed wetlands. In many places, dried sludge from FS treatment units is used for vermicomposting or co-composting with MSW to produce hygienic compost. Most of these methods are employed in small-scale decentralised plants where biological and gravity-based treatment is done. They have low operation and maintenance requirements with no need for chemicals and specially skilled labours. They also have comparatively low capital and O&M and a negligible power requirement. Recent research on FS has also investigated various thermochemical treatment technologies that can be used for changing FS into a solid fuel. These include combustion, pyrolysis, gasification, hydrothermal carbonisation, co-incineration, solar sludge ovens, solar drying beds, Ladepa, pelletisers etc. While these technologies do save space and optimise resource recovery from FS, it should be noted that most of these technologies are not well established and are still undergoing research trials. They are also comparatively expensive and energy-intensive and need expert design and maintenance.

16.3.2 FACTORS TO BE CONSIDERED WHILE DECIDING OVERALL TREATMENT PROCESS OR AN INDIVIDUAL TECHNOLOGY

16.3.2.1 Faecal Sludge Characteristics

The variability in FS characteristics is mainly dependent on the location, the type of OSSs in use, and the emptying practices prevalent in the area. Characteristics like dewaterability, solids concentration, stabilisation, spreading ability etc. are important. For example, Sludge collected from the septic tank which has longer storage periods (two to three years) is usually more diluted and biochemically stable as compared to sludge from pit latrines or unsewered public toilets which is unstable and has high ammonia concentration. Since FS is highly variable, the design of a treatment system or choice of the best-suited alternative must be determined on a case-to-case basis results rather than standard characteristics.

16.3.2.2 Technical and Economic Feasibility

We know that each technology is unique in its function and has a distinct application. While some technologies are suitable for treating undigested FS, some others may work best for digested or pre-treated FS. For example: Unlike fresh FS, undigested sludge is difficult to dewater and needs to be digested first, so we need to select a treatment technology best suiting our goals. For every technology, firstly, we need to find the following details:

- Land requirement
- Power requirement
- Capital cost (for land acquisition, infrastructure, human resources, capacity, building, training etc.)

- Operation and maintenance cost (for the daily running of FSTP, accidents, repair, desludging services etc.)
- Local environmental impact
- Technical knowledge and skills required for construction, management, operation, maintenance, repair and monitoring
- Adequacy of the supply chain for the materials and spare parts
- Treatment efficiency and end products
- Existing and possible future institutional structures, systems and capacities
- Based on our assessment we then select the most suitable set of technologies in the long run, prioritising constraints (varying on a case-to-case basis) like space, fund, viability, end use, efficiency, effluent standards etc.

16.3.2.3 Local Context, Regulations and Existing Faecal Management Practices in the Area

Several local conditions like regional sanitation priorities, priorities of the region with regard to sanitation, traditional FSM approach prevalent in the area, government + public support and awareness, interest in end use, availability of land, skilled labour, power input, capital and funding, existing institutional and regulatory framework, population density, temperature, topography and climate, water usage and availability, soil type and water table, type of OSSs and emptying practices, presence/absence of STPs/FSTPs etc. play small or a big role in influencing the decision-making of a treatment methodology. Since these local conditions change on a case-to-case basis, having sufficient knowledge about them will help in identifying the gaps in the service chain, eliminate harmful practices, ensure the use of local resources in the best way, and aid in selecting the best alternative.

16.3.2.4 Treatment Objective, End Goal and Reuse Options

Having a clear idea as to what we want, as a product, is key to selecting any particular set of technologies. For example, if our goal is dewatering, we can use Settling-thickening tanks while for digestion and biogas generation, we have anaerobic digesters. If the final goal is to discharge the treated effluent in water bodies, then FS effluent can be co-treated in the STP/ETP after receiving pre-treatment. Now, for reuse or end-use options, we also have a number of technologies that can be used. If we want to use FS as a soil conditioner in agriculture or for land application, then we need to ensure proper dewatering, pathogen removal and composting. Similarly, if we want to produce biofuels for energy production, we can employ technologies like Ladepa, incineration, pyrolysis, pelletising etc. In this case, pathogen removal is not important, but dryness is important to achieve the end goal.

16.3.3 Steps for Choosing Appropriate Treatment Processes and Technologies for a Treatment Plant

Conventional treatment plants employ standard treatment procedures as shown in Figure 16.5.

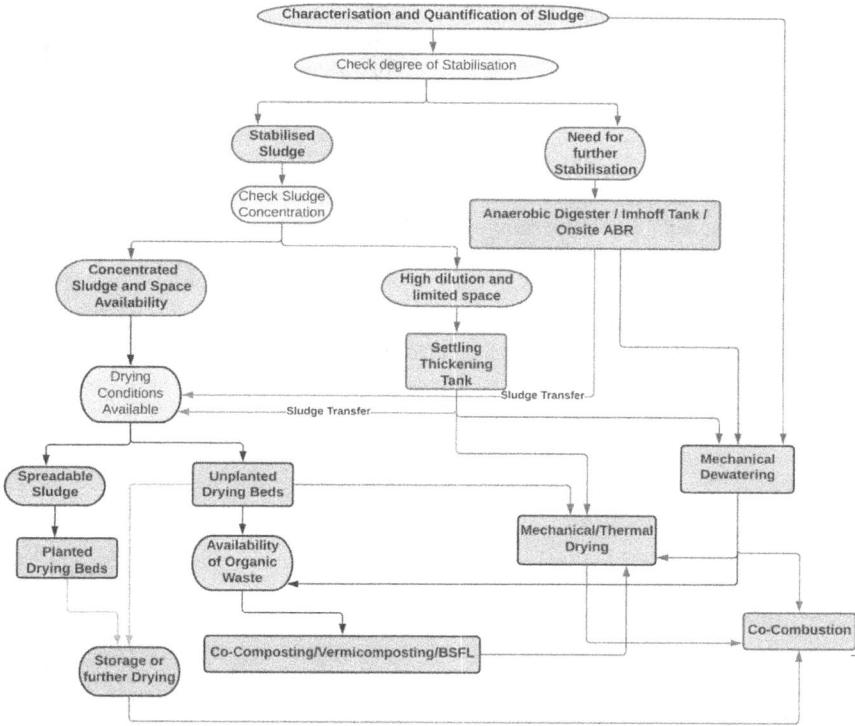

FIGURE 16.5   Conventional treatment procedures employed according to the characterisation of FS.

16.3.3.1   Technical Features and Specifications for Optimum Performance of Technologies

16.3.3.1.1   Stabilisation of Fresh FS

16.3.3.1.1.1   Anaerobic baffled reactor (ABR)

Role in the treatment process: Settling, anaerobic digestion and stabilisation of FS; high reduction of BOD; low reduction of pathogens and nutrients.

Required input characteristics of FS: Suitable for both raw and digested FS, particularly suited for influents with a high percentage of non-settleable suspended solids and a small (COD/BOD) ratio.

When to use or suitable areas of application: Peri-urban/rural area with limited space; easily adaptable and can be applied at the household or community level; should not be installed in areas with a high groundwater table or prone to flooding as infiltration; easily adaptable technology.

End products and post-treatment: Produces biogas; low sludge production; effluent and sludge require further treatment and/or appropriate discharge; pathogen and nutrient removal are generally not satisfactory for reuse in agriculture (Table 16.17).

16.3.3.1.1.2 Anaerobic Digester (AD)

Role in the treatment process: Digests and stabilises FS; good pathogen reduction; designed to capture biogas for resource recovery; nutrients are also conserved; anaerobic mono digestion or co-digestion with MSW, both can be done here.

Required input characteristics of FS: Good for stabilisation of fresh FS; Only fresh FS (like from public toilets) is appropriate for biogas production.

When to use or suitable areas of application: Peri-urban/rural areas with limited space; small-scale bio-digesters can be easily installed at the household or community level.

End products and post-treatment: Produces biogas and slurry; sludge production is low; sludge and effluent from here require further settling and treatment (Table 16.17).

TABLE 16.17
Critical Design Parameters and Treatment Performance of ABR and AD

Treatment Technology	Design Criteria	Treatment Efficiency	Pathogen Removal	Case studies and References of Mentioned Data
Anaerobic baffled reactors	Upflow velocity, Peak: (0.10–1) m/hour (over 8 hours inflow period); Average: 0.03 m/hour (over a 24-hour period); HRT: 48–72 hours; Organic loading rate: 2–6 kg COD/m^3 day; Depth of compartment: 1.8–2.5 m No. of compartments: 4–8 Length of compartment: Minimum of 0.75 m between wall and baffle and up to half of compartment depth	When influent COD ranges between (1,000–7,000) mg/L COD: 40%–80%; BOD: 70%–95%; TSS: 40%–90%; TN: 20%; Oil and grease: 70%; TS in separated sludge: 2%–25%; NH_3 and P removal –not satisfactory	Low pathogen reduction	Example: Devnahalli, India (Morel, 2006; Gutterer et al., 2009)
Anaerobic digesters (small-scale fixed dome or geobag bio-digesters)	Floating gas dome digester for FS digestion: TS in influent: 3%–10%; C:N – 16–25: 1; HRT: 30–50 days; Organic loading rate: 1.6–2.2 kg VS/m^3 day	When Influent COD ranges between (80,000 and 20,000 mg/L) - BOD: 20%–50%; COD: 20%–40%; VSS: 50%–60%; TSS: 60%–70%; Biogas production-10 l – 63 litre biogas/kg dry solids	<50%–60%	Example: Dar e salaam, Tanzania India, Nepal (Tayler, 2018)

16.3.3.1.2   Solid-Liquid Separation

16.3.3.1.2.1   Settling-thickening tank

Role in treatment process: Easily settleable solids are settled, slightly digested and thickened; helps in flow equalisation, limited stabilisation, pathogen reduction and volume reduction of FS.

Required Input characteristics of FS: Since it is a primary treatment, this method is best for stabilised or partly stabilised FS with 1%–2% solid content.

When to use or suitable areas of application: Peri-urban and rural areas with any climate, temperate or rainy climates (desirable); best for areas where limited space is available.

End products and post-treatment: Needs further dewatering, organic, suspended solids and pathogen removal for thickened sludge and scum; further effluent treatment for supernatant for safe reuse or discharge.

16.3.3.1.2.2   Imhoff Tank

Role in the treatment process: Settling, thickening and then digestion of settled sludge, all in a single unit; a small amount of biogas is also produced; high organic load removal and low pathogen reduction.

Required input characteristics of FS: Since it is a primary treatment or pre-treatment, this method is best for partly stabilised and screened sludge.

When to use or suitable areas of application: Best for small settlements in both rural and urban areas with either warm or cold climates; should be used when space is limited and conditions are unfavourable for biogas digesters or stabilisation ponds.

End products and post-treatment: Effluent is almost odourless and requires further treatment for safe reuse and discharge; needs further dewatering, organic, suspended solids and pathogen removal for thickened sludge; biogas can be captured for use.

16.3.3.1.3   Dewatering

16.3.3.1.3.1   Geobags

Role in the treatment process: Time-efficient dewatering as compared to drying beds.

Required input characteristics of FS: Suitable for partly stabilised FS.

When to use or suitable areas of application: Suitable technology for temporary dewatering when space is limited; suitable for Peri-urban and rural areas.

End products and post-treatment: Dewatered sludge can be trenched or further treated; leachate needs further treatment too.

16.3.3.1.3.2   Mechanical Presses

Role in the treatment process: Fast dewatering and sludge separation; solid recovery efficiency is same as centrifuges; achieves high cake dryness.

Required input characteristics of FS: Can be used for both fresh and digested FS with 1%–2% and more solids content.

When to use or suitable areas of application: Suitable for urban areas; the
need for electricity limits their widespread application and makes their use
quite expensive; has much lesser power and capital requirement as well as
easy maintenance than centrifuges.

End products and post-treatment: Further treatment for organic, suspended
solids and pathogen removal is a must, both for thickened sludge and
percolate.

16.3.3.1.3.3 Unplanted drying beds

Role in the treatment process: Partial biodegradation; limited pathogen
reduction; can provide combined solid-liquid separation and dewatering
through percolation and evaporation.

Required input characteristics of FS: Preferred when the flow rate is low, i.e.,
FS is partly digested or pre-settled; best for screened septage with 3%–5%
or more solids content (high).

When to use or suitable areas of application: Peri-urban and rural areas
with hot, dry and arid tropical climates; land available is sufficient, inexpensive and limited operational skills are available; can be used both as
a primary treatment of screened FS or secondary treatment for settled
sludge.

End products and post-treatment: Dried sludge needs to be at least vermi-
composted or co-composted for satisfactory pathogen removal; leachate
requires further effluent treatment for safe reuse or discharge.

16.3.3.1.3.4 Centrifuges

Role in the treatment process: Fast dewatering and sludge separation; better solid recovery efficiency than mechanical presses; achieves high cake
dryness.

Required input characteristics of FS: Can be used for both fresh and digested
FS with 1%–2% or more solids content.

When to use or suitable areas of application: Suitable for urban areas; need
for electricity limits their widespread application; high energy and O&M
cost; best for processing large sludge volumes.

End products and post-treatment: Further treatment for organic, suspended
solids and pathogen removal is a must, both for thickened sludge and
percolate.

16.3.3.1.4 Effluent Treatment

16.3.3.1.4.1 Planted Drying Bed/Constructed wetlands

Role in the treatment process: Enhanced sludge treatment due to the plants;
dewatering, stabilisation and hygienisation through evapotranspiration and
percolation; also used for effluent treatment.

Required input characteristics of FS: Can be used as a primary, secondary and tertiary treatment for FS.

When to use or suitable areas of application: Peri-urban and rural areas with warm tropical as well as colder climates; best for Dry tropical climates; land available is sufficient and inexpensive; local marshland non-invasive species of plants are available.

End products and post-treatment: Dried sludge can be directly used as a soil conditioner; leachate may require further effluent treatment; percolate quality compares favourably to other primary treatments.

16.3.3.1.4.2   Aerated Lagoons

Role in the treatment process: Good stabilisation capacity; high reduction of suspended solids, organic solids and pathogens.

Required input characteristics of FS: Suitable for treated FS from anaerobic processes for reduced cost of treatment; can also be used for co-treatment of the liquid fraction of FS and Wastewater.

When to use or suitable areas of application: Suitable for warm climates; because of high power and land requirements and increased costs, it should only be used when there is no space for facultative ponds and a reliable power supply is available; requires two-thirds of land required by facultative ponds.

End products and post-treatment: Produces relatively good effluent quality as compared to Anaerobic counterparts; effluent needs further pathogen and nutrient removal before discharge and can be utilised for irrigation purposes.

16.3.3.1.4.3   Waste Stabilisation ponds

Role in the treatment process: Better sedimentation properties than settling tanks; good stabilisation capacity; high reduction of solids, nutrients, pathogens and BOD.

Required input characteristics of FS: Can receive fresh or partly digested FS with <1% solid content; anaerobic pond can be used as a first-stage liquid treatment process while facultative and maturation ponds can be used as a secondary and tertiary treatment for effluent.

When to use or suitable areas of application: Used if large inexpensive land area is easily available; best suited for rural or peri-urban areas in tropical and subtropical countries; anaerobic pond has reduced land area and power requirement as compared to facultative ponds and other aerobic counterparts; can also be used for co-treatment of septage and wastewater.

End products and post-treatment: Both liquid and sediments require further treatment; treated water cannot be discharged in surface water but can be used for irrigation or agriculture; biogas produced can be captured for use.

16.3.3.1.5   Sludge Treatment and Reuse

16.3.3.1.5.1   Extended Storage of dried sludge

Role in the treatment process: Reduces pathogen concentration over time; retention time varies between 12 and 16 months.

Required input characteristics of FS: FS has to be pre-treated and dewatered and solid content has to be more than 20%.

When to use or suitable areas of application: Can be used in dry climates; sludge should be stored on permeable material.

End products and post-treatment: Treated sludge can be opted for restricted use; leachate produced will undergo further effluent treatment.

16.3.3.1.5.2   Composting (Co-composting or Vermicomposting)

Role in the treatment process: Produces a good and pathogen-free soil conditioner in a relatively short time, waste materials can be converted into useful products like hygienically safe soil conditioner-cum-fertiliser.

Required input characteristics of FS: FS has to be pre-treated and dewatered solid content has to be in the range of 40%–45% and moisture content between 50% and 60% needs to be mixed with a bulking agent. The design parameters and treatment performance for co-composting of FS with MSW are shown in Table 16.18.

TABLE 16.18

Critical Design Parameters and Treatment Performance for Co-Composting of FS with MSW

Treatment Technology	Design Criteria	Treatment Efficiency	Pathogen Removal	Case Studies and References of Mentioned Data
Co-composting with solid wastes	Mixing Ratio (FS/SW) For dewatered FS, (TS: 20%–30%) - 1:2 to 1:4; For fresh, non-dewatered FS - 1:5 – 1:10; Temp: >50°C; Moisture content = 50%–60%; Pile size (width:height:length) = (2:1.6:Length) m C:N ratio = 20–35; Co-compost cycle: 6–12 weeks; Turning frequency: 3–6 turnings/3 months	C/N ratio of Final product: 13; NO_3/ NH_4:7.8; TVS: 21% TS; NH_4–N: 0.01%	HE:1 viable egg/g TS	Examples: Kumasi (Ghana) and Kushtia (Bangladesh) (Cofie et al., 2016)

When to use or suitable areas of application: Used as an intermediate solution for management of FSM in developing countries if land, skilled labour and market demand for compost are available; Suitable for small and medium-sized towns and rural areas.

End products and post-treatment: Treated compost can be directly used as a soil amendment.

16.3.3.1.5.3   Solar drying

Role in the treatment process: Much faster Drying or Dewatering of Sludge as compared to drying beds; Takes less time, area and pasteurisation.

Required input characteristics of FS: Required input of dried FS (20% ds) (Andriessen et al., 2019).

When to use or suitable areas of application: Used when faster drying is required and lesser space is required compared to drying beds; suitable for hot and dry tropical climates; not suitable in places with variable weather conditions.

End products and post-treatment: 90% output dryness for dried sludge (Andriessen et al., 2019), pathogens are not completely removed. End use depends on quality; percolate needs further treatment.

16.3.3.1.5.4   Solar Sludge Oven

Role in the treatment process: Much faster drying or dewatering of sludge as compared to drying beds; produces hygienic dried sludge.

Required input characteristics of FS: Required input of dried FS (20% ds) (Andriessen et al., 2019).

When to use or suitable areas of application: Used when faster drying is required and lesser space is required compared to drying beds; suitable for hot and dry tropical climates; the system levels out any climatic variations and the sludge can be removed with targeted dry matter content.

End products and post-treatment: 90% output dryness for dried sludge (Andriessen et al., 2019), pathogens are not completely removed. End use depends on quality; percolate needs further treatment.

16.3.3.1.5.5   Pelletisers

Role in the treatment process: Produces dried pellets which can be used as a solid fuel, compost, animal feed, soil amendment or for combustion as a biofuel.

Required input characteristics of FS: Required input of dried FS (70% ds, for conventional) 30%–60% ds, for bio-burn pelletisers) (Andriessen et al., 2019).

When to use or suitable areas of application: It is compatible with a range of moisture contents and sludge properties; pellets can be used as industrial fuel in nearby areas.

End products and post-treatment: Same output dryness as input; pathogens may not be completely removed depending on the level of treatment.

16.3.3.1.5.6 Ladepa (Latrine dehydration and pasteurisation)

Role in the treatment process: Removes detritus from FS, followed by drying and infrared radiation; produces sanitised pellets.

Required input characteristics of FS: Required input of dried FS (20%–30% ds) (Andriessen et al., 2019); best suited for thicker or dewatered sludge or where a sanitised final product is required.

When to use or suitable areas of application: When only limited space is available for drying and skilled personnel is available for safe operation; then, LaDePa could be a solution.

End products and post-treatment: No need, 80% output dryness (Andriessen et al., 2019); Pellets can be used as a soil amendment, and percolate needs further treatment.

16.3.3.1.5.7 Black Soldier Fly Larvae (BSFL)

Role in the treatment process: Decreases the dry mass of waste; insect protein could be used in aquaculture; they can reduce the volume of organic waste by about 55%.

Required input characteristics of FS: Black Soldier fly (BSF) larvae grown only on FS with a dry matter content of 40%; the larvae grow while feeding on organic matter in FS and organic wastes.

When to use or suitable areas of application: The BSF larvae can be used as a conventional protein and fat source for poultry and fish feed.

End products and post-treatment: The residue remaining after digestion can be composted or anaerobically digested to produce a soil conditioner; Ammonia sanitisation for treated sludge can also be done.

16.3.3.1.5.8 Co-incineration or Co-Combustion

Role in the treatment process: Reduces the volume of waste to a high degree making it easy for disposal.

Required input characteristics of FS: FS has to be dewatered and dried and solid content has to be around 80% or more.

When to use or suitable areas of application: When an incineration facility for MSW is available in the area.

End products and post-treatment: Heat energy is recovered and ash and exhaust gases are produced as by-products.

16.3.3.1.5.9 Pyrolysis

Role in the treatment process: Thermochemical treatment of biomass by heating to temperatures between 300°C and 700°C in the absence (or near absence) of oxygen which leads to a reduction in sludge volume.

Required input characteristics of FS: Desirable input of dried FS (70%–90% ds); High moisture sludge (20% ds) can also be used (Andriessen et al., 2019).

When to use or suitable areas of application: When no space is available for drying and skilled personnel is available for safe operation, then, pyrolysis could be a solution; fuel quality is not very satisfactory compared to when other biomass fuels.

End products and post-treatment: Heat energy is recovered and ash and exhaust gases are produced as by-products; produces char or carbonised FS in the form of powder or chunks or briquettes; 100% output dryness (Andriessen et al., 2019); liquid by-products need further treatment.

16.3.3.1.5.10   Hydrothermal Carbonisation

Role in the treatment process: Thermochemical conversion of wet biomass at temperatures ranging from 180°C to 250°C for 1–12 hours of reaction time under pressure (>30 bar).

Required input characteristics of FS: Required input of dried FS (20% ds)

When to use or suitable areas of application: When no space is available for drying and skilled personnel is available for safe operation, then, HTC could be a solution.

End products and post-treatment: Produces hydro-char or carbonised FS in the form of powder or chunks or briquettes; 100% output dryness for sludge (Andriessen et al., 2019).

16.3.3.2   Co-Treatment of Faecal Sludge at WWTPs

Role in the treatment process: Usually biological treatment based (i.e., activated sludge process, trickling filters, upflow anaerobic sludge blanket reactors etc.), aerated lagoons or stabilisation ponds based STPs can be utilised for co-treatment; high reduction of suspended solids, organic solids and pathogens can be achieved.

Required input characteristics of FS: Suitable for treated FS from anaerobic processes for reduced cost of treatment; pre-treatment of FS is a must for proper application at an appropriate point at STP.

When to use or suitable areas of application: Used in areas where an underloaded sewage treatment plant exists. The STP/WWTP needs to have adequate capacity to receive the extra load from FS.

End products and post-treatment: Sludge is treated before reuse or final disposal; Produces relatively good effluent quality as compared to anaerobic counterparts; effluent needs further pathogen and nutrient removal before discharge and can be utilised for irrigation purposes.

16.3.3.3   Land Application

16.3.3.3.1   Trenching (Deep row/Shallow)

Role in the treatment process: Helps in land remediation and safe disposal of FS without creating odours and much lesser risk of pathogen exposure

Required Input characteristics of FS: Pre-treated Sludge or even raw FS can be used.

When to use or Suitable areas of application: Suitable for warm climates; Deep row entrenchments can be used in forestry applications or in areas where inexpensive and unlimited spare land area is available.

End products and post-treatment: It is a cheap method of disposal of FS; helps in Land reclamation.

16.3.3.3.2   Landfill Disposal

Role in the treatment process: Helps in unsafe disposal of FS in open spaces which can lead to soil and water contamination.

Required input characteristics of FS: Suitable for treated FS from anaerobic processes for reduced cost of treatment; pre-treated sludge or even raw FS can be used, although can create odour problems and health hazards; generally, not recommended unless necessary.

When to use or suitable areas of application: Can be used when designated treatment and discharge facilities are not available.

End products and post-treatment: It is a cheap method of disposal of FS with no end products.

16.4   CONCLUSION

With no specific discharge standards for the treatment of faecal sludge, methods need to be adapted on an individual case-to-case basis. Moreover, since faecal sludge remains much more variable in consistency, quantity and concentration as compared to sewage and wastewater; therefore, specific attention is needed to avoid harmful health consequences. Unlike before, when only sewer-based systems were considered appropriate for achieving citywide sanitation, now decentralised options like FSTPs are also being considered as a likely alternative. For lower-income and lower- to middle-income countries with highly variable climates, topography and high population density, adopting decentralised technologies for FS treatment (as a part of an integrated citywide approach to FSM) seem to be the only viable way ahead. Some of the most commonly adopted low-cost practices for FS treatment and disposal in these countries include landfill disposal, land application, co-treatment at WWTPs/STPs/sludge treatment facilities, trenching, settling–thickening tanks or ponds, Imhoff tanks, lime stabilisation, waste stabilisation ponds, stabilisation reactors/tanks, anaerobic digestion with biogas utilisation, anaerobic baffled reactors, gravel filter beds, unplanted drying beds and HF/VF constructed wetlands. Moreover, utilising existing STPs to their full capacity, promoting improved household and community-level sanitation systems, maintaining public sanitation facilities, government and public awareness, private sector involvement, a well-connected FSM chain, innovation and application of the latest technologies, availability of continuous financial support and strictly enforced monitored legislations and discharge standards are the strategies to work upon. The adopted treatment methods for this purpose are mostly biological and gravity-based with low capital, power and O&M requirements.

REFERENCES

Andriessen, N., Ward, B.J. and Strande, L., 2019. To char or not to char? Review of technologies to produce solid fuels for resource recovery from faecal sludge. *Journal of Water, Sanitation and Hygiene for Development*, 9(2), pp. 210–224.

Bassan, M., Ferré, A., Hoai, A.V., Nguyen, V.A. and Strande, L., 2016. Methods for the characterization of faecal sludge in Vietnam. Eawag: Swiss Federal Institute of Aquatic Science and Technology: Dübendorf, Switzerland.

Brown, C., Ellson, A., Ledger, R., Sorensen, G., Cunliffe, D., Simon, D., McManus, M., Schrale, G., McLaughlin, M., Warne, M. and Liston, C., 2017. Draft-South Australian biosolids guidelines for the safe handling and reuse of biosolids.

Cofie, O., Nikiema, J., Impraim, R., Adamtey, N., Paul, J. and Koné, D., 2016. *Co-Composting of Solid Waste and Fecal Sludge for Nutrient and Organic Matter Recovery* (Vol. 3). IWMI: Colombo.

Cunningham, M., Gold, M. and Strande, L., 2016. Literature review: Slow-pyrolysis of faecal sludge.

Franceys, R., Pickford, J., Reed, R. and World Health Organization, 1992. *A Guide to the Development of On-Site Sanitation*. World Health Organization: Geneva, Switzerland.

Government of Tamil Nadu. 2017. Operative guidelines for septage management for local bodies in Tamil Nadu. Fort St George, Chennai, India: Municipal Administration and Water Supply Department, Government of Tamil Nadu. Available at http://muzhusugad-haram.co.in/wp-content/uploads/2017/07/english-septage-operative-guidelines-tn.pdf (accessed on January 9, 2019).

Gutterer, B., Sasse, L., Panzerbieter, T. and Reckerzügel, T., 2009. *Decentralised Wastewater Treatment Systems (DEWATS) and Sanitation in Developing Countries*. BORDA: Bremen.

Hasanbeigi, A., Lu, H., Williams, C. and Price, L., 2012. *International Best Practices for Pre-Processing and Co-Processing Municipal Solid Waste and Sewage Sludge in the Cement Industry* (No. LBNL-5581E). Lawrence Berkeley National Lab (LBNL): Berkeley, CA.

Heinss, U., Larmie, S.A. and Strauss, M., 1998. Solids separation and pond systems for the treatment of faecal sludges in the tropics: Lessons learnt and recommendations for preliminary design, EAWAG/SANDEC, Report No. 05/98.

Ho, P.Y.C., The, T.H., Yassin, Z.N., Lean, C.L., Tan, S.H., and Sasidharan, V., 2011. Landscape analysis and business model assessment in fecal sludge management: Extraction and transportation model in Malaysia. Consultancy report by ERE Consulting Group in collaboration with Indah Water Konsortium (IWK) commissioned by the Bill & Melinda Gates Foundation: Seattle, USA, 76p. Available at https://www.susana.org/_resources/documents/default/2-1670-malaysia-final-report-revised-31-march2012v4.pdf (accessed on February 6, 2019).

Ingallinella, A.M., Sanguinetti, G., Koottatep, T., Montangero, A. and Strauss, M., 2002. The challenge of faecal sludge management in urban areas-strategies, regulations and treatment options. *Water Science and Technology*, 46(10), pp. 285–294.

Kalbermatten, J.M., Julius, D.S. and Gunnerson, C.G., 1980. *Appropriate Technology for Water Supply and Sanitation*. World Bank: Washington, DC.

Kengne, I.M., Kengne, E.S., Akoa, A., Bemmo, N., Dodane, P.H. and Koné, D., 2011. Vertical-flow constructed wetlands as an emerging solution for faecal sludge dewatering in developing countries. *Journal of Water, Sanitation and Hygiene for Development*, 1(1), pp. 13–19.

Koné, D. and Strauss, M., 2004. Low-cost options for treating faecal sludges (FS) in developing countries: Challenges and performance. *In 9th International IWA Specialist Group Conference on Wetlands Systems for Water Pollution Control and to the 6th International IWA Specialist Group Conference on Waste Stabilisation Ponds*, Avignon, France, Vol. 27.

Malaysian Water Association, 2017. Malaysia water industry guide 2017: Extract on sewerage charges. Malaysian Water Association: Kuala Lumpur, Malaysia. Available at https://www.span.gov.my/document/upload/rLpAXOJorApj6VirbDkN9eG1PkUe1Ell.pdf (accessed on January 10, 2019).

Minnesota Pollution Control Agency, 2015. Septage and restaurant grease trap waste management guidelines, 22 p. Available at https://www.pca.state.mn.us/sites/default/files/wqwwists4-20.pdf (accessed on January 10, 2019).

Morel, A., 2006. *Greywater Management in Low and Middle-Income Countries* (No. 628.2 G842g). Swiss Federal Institute of Aquatic Science and Technology: Dübendorf, Switzerland.

Nikiema, J., Cofie, O., and Impraim, R., 2014. Technological options for safe resource recovery from fecal sludge. Colombo, Sri Lanka, International Water Management Institute (IWMI). *CGIAR Research Program on Water, Land and Ecosystems (WLE)*. 47p. (Resource Recovery and Reuse Series 2). doi: 10.5337/2014.228.

Reed, B., Scott, R., and Shaw, R., 2014. Pour-flush latrines, Guide 26. Water, Engineering and Development Centre (WEDC), Loughborough University: Leicestershire, UK. Available at https://wedc-knowledge.lboro.ac.uk/resources/booklets/G026-Pourflush-latrines-online.pdf (accessed on January 10, 2019).

Robbins, D.M., 2007. Septage management guide for local governments. *A Step-by-Step Practical Guide to Developing Effective Septage Management Programs for Cities and Municipalities*. www.rti.org/pubs/septage_management_guide_l.pdf (accessed April 2013).

Seck, A., Gold, M., Niang, S., Mbéguéré, M., Diop, C. and Strande, L., 2015. Faecal sludge drying beds: Increasing drying rates for fuel resource recovery in Sub-Saharan Africa. *Journal of Water, Sanitation and Hygiene for Development*, 5(1), pp. 72–80.

Sedlak, R., 1991. *Phosphorus and Nitrogen Removal from Municipal Wastewater: Principles and Practice,* 2nd Edition. Lewis Publisher: New York. ISBN 0-87371-683-3.

Seidu, R., and Drechsel, P., 2010. *Cost-effectiveness analysis of interventions for diarrhoe disease reduction among consumers of wastewater-irrigated lettuce in Ghana*, pp. 261–283. doi: 10.22004/ag.econ.127726

Semiyaga, S., Okure, M.A., Niwagaba, C.B., Katukiza, A.Y. and Kansiime, F., 2015. Decentralized options for faecal sludge management in urban slum areas of Sub-Saharan Africa: A review of technologies, practices and end-uses. *Resources, Conservation and Recycling, 104*, pp. 109–119.

Strande, L., Ronteltap, M., and Brdjanovic, D., 2014. *Fecal Sludge Management: Systems Approach for Implementation and Operation*. IWA Publishing: London, 404 p.

Strauss, M., Heinss, U. and Montangero, A., 2000. On-site sanitation: When the pits are full–planning for resource protection in faecal sludge management. *Schriftenreihe des Vereins fur Wasser-, Boden-und Lufthygiene, 105*, pp. 353–360.

Strauss, M., Larmie, S.A. and Heinss, U., 1997. Treatment of sludges from on-site sanitation: Low-cost options. *Water Science and Technology, 35*(6), pp. 129–136.

Strauss, M. and Montangero, A., 2002. *Faecal Sludge Management: Review of Practices, Problems and Initiatives*. Water and Sanitation in Developing Countries EAWAG/SANDEC: Dübendorf, Switzerland.

Syed-Hassan, S.S.A., Wang, Y., Hu, S., Su, S. and Xiang, J., 2017. Thermochemical processing of sewage sludge to energy and fuel: Fundamentals, challenges and considerations. *Renewable and Sustainable Energy Reviews, 80*, pp. 888–913.

Tayler, K., 2018. *Faecal Sludge and Septage Treatment: A Guide for Low and Middle Income Countries*. Practical Action Publishing: Rugby, Warwickshire.

Tilley, E., Ulrich, L., Lüthi, C., Reymond, P. and Zurbrügg, C. 2014. *Compendium of Sanitation Systems and Technologies*, 2nd Revised Edition. Swiss Federal Institute of Aquatic Science and Technology (Eawag): Dübendorf, Switzerland.

Tsai, W.T., 2012. An analysis of waste management policies on utilizing biosludge as material resources in Taiwan. *Sustainability*, *4*(8), pp. 1879–1887.

USAID (United States Agency for International Development), 2008. Operations manual on the rules and regulations governing domestic sludge and septage. USAID, Philippine Department of Health: Washington, DC, 42 p. http://open_jicareport.jica.go.jp/pdf/11948882_24.pdf.

USAID (United States Agency for International Development), 2015. Implementer's guide to lime stabilization for septage management in the Philippines. Available at http://pawd.org.ph/wp-content/uploads/2015/10/LimeStabilization-Sept02-Final-lowres.pdf (accessed on January 10, 2019).

USEPA (United States Environmental Protection Agency), 1993a. Standards for the use or disposal of sewage sludge. Title 40 (Protection of Environment) of the Code of Federal Regulations (CFR), Part 503. Available at https://www.ecfr.gov/cgi-bin/R?gp=2&SID=3ba5c96eb4bfc5bfdfa86764a30e9901&ty=HTML&h=L&n=pt40.30.503&r=PART (accessed on February 6, 2019).

USEPA (United States Environmental Protection Agency). 1993b. A guide to the federal EPA rule for land application of domestic septage to non-public contact sites. EPA 832-B-92-005, September 1993. USEPA: Washington, DC, 52p. Available at https://www.epa.gov/biosolids/guide-federal-eparule-land-application-domestic-septage-non-public-contactsites (accessed on February 6, 2019).

USEPA (United States Environmental Protection Agency), 1994. Guide to septage treatment and disposal. EPA/625/R-94/002; US.

USEPA (United States Environmental Protection Agency), 1999. Decentralized systems technology fact sheet: Septic tank-soil adsorption systems. EPA 932-F-99-075, US EPA.

Ward, B.J., Yacob, T.W. and Montoya, L.D., 2014. Evaluation of solid fuel char briquettes from human waste. *Environmental Science & Technology*, *48*(16), pp. 9852–9858.

Wald, C., 2017. The new economy of excrement. *Nature*, *549*(7671), pp. 146–148.

WBCSD FLP 2014. *Accenture – Integrated Performance Management*. https://www.wbcsd.org/Projects/Education/Leadership-program/Resources/WBCSD-FLP-2014-Accenture-Integrated-Performance-Management.

World Health Organization, 2006. WHO guidelines for the safe use of wasterwater excreta and greywater (Vol. 1). World Health Organization.

Index

Note: Bold page numbers refer to tables and *italic* page numbers refer to figures.

acidic pre-treatment 243
acid thermal treatment 265, **266**
activated sludge process (ASP) 3–4
adsorbent (CAZ) 25
adsorption 170
adsorption-based treatment 22–25, **26**
advanced oxidation processes (AOPs) 86
   applications 94, **94–95**
   comprehensive analysis 86
   Fenton-based 88–91
   forms 87, **87**
   ozon-based 92–94
aerated CWL 122
aerated lagoons 329
aerobic digestion (AD) 213
aerobic granulation technology
   aerobic starvation phase 39
   biochemical processes 45, *46*
   case study 47, *48*, **49**
   characteristics 44
   dissolved oxygen 40
   divalent metal ions 39
   full-scale applications 44
   granular sludge 38
   lab-scale applications 41–44, **42, 43**
   microbiology 46–47
   organic loading rates 40
   SBR technologies 45
   selection pressure 39
   sequencing batch reactor 38
   temperature 40
   vs. anaerobic granulation 40–41, *41*
aerobic pre-treatment 242
aerobic treatment
   activated sludge process 3–4
   fixed bed bioreactor 5
   moving bed bioreactor 4–5
Agrobacterium 46
agrochemicals 160, **160**
algae-based sequencing batch suspended biofilm
   reactor (A-SBSBR) 66
alkaline pre-treatment 243
alkali thermal treatment 265
anaerobic baffled reactor (ABR) 325
anaerobic digester (AD) 326
anaerobic digestion (AAD) 212–213
anaerobic digestion (AD) 235, 258
   municipal solid waste 283
   principles 258–259

anaerobic pre-treatment 242
anaerobic sludge blanket reactors 6–7
anaerobic treatment 6–7
anammox process 58–59
AOPs *see* advanced oxidation processes (AOPs)
apoptosis 156
Aquabacterium sp. 46
aqua privy 304
arsenic 153, 156
A-SBSBR *see* algae-based sequencing batch
        suspended biofilm reactor (A-SBSBR)
Australia 309
   municipal solid waste 287–289

baffled subsurface-flow CWLs 122
belt filter press (BFP) 217
BFT *see* bio-floc technology (BFT)
biochemical processes 45, *46*
biodegradation
   biotechnological approaches 186–187
   emerging contaminants 170
   hydraulic retention time 171–172
   immobilized enzymes 188–189, **190**, 191, *191*
   invention of novel genes 192–193
   metabolic engineering approaches 191–192
   microbial community 173
   microbiological electrochemical systems *187*,
      187–188
   nanoremediation 186
   in natural systems 174
   $_p$H 172
   photocatalysis 184–186
   physicochemical treatment 183–184
   redox condition 172
   solids retention time 171
   sonochemical methods 186
   suspended *vs.* attached growth process 173
   temperatures 172–173
biofilms 141
bio-floc technology (BFT) 64
biogas production
   anaerobic digestion 236–240
   of different organic waste 235, **235**
biological nitrogen removal process 55–57,
    56, 57
biological nutrient 42, **43**
biological oxygen demand (BOD) 87
biological pre-treatment 241–243
biological stabilization 212–213

biological treatment 222
  aerobic treatment 3–5
  anaerobic treatment 6–7, **8**
  of organics from wastewater 41–42, **42**
  traditional treatment methods 2–3
bioremediation 183, 193–194
biosorption, of dyes and heavy metals 42, **43**
biotechnological approaches 186–187
biotransformation **174**, 174–175, *175*, 184
bisphenols 158
Black Soldier Fly Larvae (BSFL) 332
BTEX 290

Cambi thermal hydrolysis process (THP)
        263, *263*
cancer 159–162, **160**, *161*
capital expenditure (CapEx) 127
carbon nanotubes (CNTs) 23, 186
cardiovascular diseases (CVDs) *154*, 154–155
CASS *see* cyclic activated sludge system
        (CASS)
Central Pollution Control Board (CPCB)
        205, 272
Central Public Health and Environmental
        Engineering Organization
        (CPHEEO) 272
centrifugal sludge dewatering 218–219
centrifugal thickening 216
centrifuges 328
CFP *see* classical Fenton process (CFP)
chemical precipitation 14, **15**, 16–17
chemical pre-treatment 243–244
chemical treatment 222
chitosan-coated magnetic nanoparticles
        (cMNPs) 25
Chlorella vulgaris 3
classical Fenton process (CFP) 88–89
Co-incineration 332
combined pre-treatment 246–247
composting 281–283, 330
constructed wetland (CWL) 120
  advantages 127, **128**
  in agriculture 127–129
  challenges 120, 129–130
  cost analysis 129
  depth of water 124
  design and operation 122–124, *123*
  enhanced 122
  environmental conditions 124
  feeding modes 125
  free water surface-flow 121
  hydraulic retention time 125
  integrated microbial fuel cells with 129
  operational factors 120
  operational parameters 124–125
  pollutant removal mechanisms 125–127
  subsurface-flow 121–122

conventional thermal pretreatment 26**0**, 259–260
CWL *see* constructed wetland (CWL)
cyclic activated sludge system (CASS) 61

Dechloromonas 46
decreased acetylcholinesterase activity 156
denitrifying phosphorus accumulating organisms
        (DPAO) 60–61
dewatering process 216, *217*
  belt filter press 217
  centrifugation 218–219
  drying beds 219
  electro-dewatering process 220
  lagoons 219–220
  rotary press 218
  screw press 218
dichlorodiphenyltrichloroethane (DDT) 141
diffuse source pollution 145–146
dimethyl sulfoxide (DMSO) 64
dissolved air flotation (DAF) thickening 215
DPAO *see* denitrifying phosphorus accumulating
        organisms (DPAO)
dried sludge 330

EBPR *see* enhanced biological phosphorus
        removal (EBPR)
effluent discharge 313, 321, **322**
effluent treatment 328–329
EFP *see* electro-Fenton process (EFP)
electrical and electronics equipment (EEE) 287
electrochemical oxidation 20–21
electrochemical technologies 104
  electrocoagulation *112*, 112–113
  electrodeposition 107, *109*, 109–110
  electrodialysis 110–112
  electro-oxidation 104–107, *105*, *106*, **108**
electrochemical treatment 17, **18**, **19**, 19–22, **22**
electrocoagulation 17, **18**, 19, **19**, *112*, 112–113
electrodeposition 107, *109*, 109–110
electro-dewatering (EDW) process 220
electrodialysis (ED) 21–22, **22**, 110–112
electro-Fenton process (EFP) 90–91
electroflotation 19–20
electro-oxidation 104–107, *105*, *106*, **108**
emerging contaminants (ECs) 167; *see also*
        micropollutants (MPs)
  biodegradation (*see* biodegradation)
  biotransformation **174**, 174–175, *175*
  removal mechanisms 169–170
  sources, treatment, and potential
        receptors *169*
emerging micropollutants (EOMs) 184
emulsion immobilization 189
endocrine-disrupting compound 157–158
enhanced biological phosphorus removal (EBPR)
        59–60
enhanced CWL 122

Environment Protection Agency (EPA) 287
enzyme pre-treatment 242
17β-Ethinylestradiol (EE2) 143
European Environmental Agency 182
Exelys™ process 263–264, *264*
Extended Producer Responsibility (EPR) 287

faecal sludge (FS)
    agricultural and non-agricultural land
        application 316, **317**
    aquaculture 320–321
    capture and containment 302, **303–306**
    characteristics 299, **300**, *301*, 323
    compost application 319–320
    effluent discharge 321, **322**
    emptying and transportation 304, 306, **307**,
        307–309
    lime stabilisations 317–318
    local conditions 324
    occupational risks on farms 318, **319**
    sampling and analysis 299–301
    soil application 319
    technical and economic feasibility 323–324
    treatment 310–314, *312*, 321–323, 324–334,
        325, 326, 330
    use 314–315, *315*
FBSBR *see* fixed bed sequencing batch reactor
        (FBSBR)
Fenton-based AOPs
    classical Fenton process 88–89
    electro-Fenton process 90–91
    heterogeneous Fenton catalysis 91
    photo-Fenton process 89–90
Fenton pre-treatment 243
fixed bed bioreactor (FBBR) 5
fixed bed sequencing batch reactor (FBSBR) 68
Flavobacterium 47
free water surface-flow CWL 121
fungal pre-treatment 242–243

GAC-SBR *see* granular-activated carbon-SBR
        (GAC-SBR)
Gammaproteobacteria 46
geobags 327
Ghana, municipal solid waste 291–292
global water shortage 120
granular-activated carbon-SBR (GAC-SBR) 65
gravity belt (GB) thickening 215–216

hazardous waste 273
heterogeneous Fenton catalysis 91
high-pressure homogenization pre-treatment 245
Hong Kong Environmental Protection
        Department (HKEPD) 292
Hong Kong, municipal solid waste 292–293
horizontal-flow CWL 121
hospital waste 274

hybrid CWL 122
hydraulic retention time (HRT) 5, 125, 171–172
hydrothermal carbonisation 333
hydroxyl radicals 86, *87*

ICEAS *see* intermediate cycle extended aeration
        system (ICEAS)
IFAS-SBR *see* integrated fixed-film-activated
        sludge sequential batch reactor
        (IFAS-SBR)
imhoff tank 327
immobilized enzymes 188–189, **190**, 191, *191*
incineration 226–227
India 309
    solid wastes 278–281
Inorganic Chemical Conditioning (ICC) 221
integrated fixed film activated sludge process
        (IFAS) 4
integrated fixed-film-activated sludge sequential
        batch reactor (IFAS-SBR) 68
integrated microbial fuel cells with CWLs
        (CWL-MFCs) 129
Integrated Product Policy (IPP) 287
integrated waste management (IWM) 286–287
intermediate cycle extended aeration system
        (ICEAS) 62
International Water Association (IWA) 38
ion exchange 12–14
ion-exchange membrane (IEM) 188
ionic liquid pre-treatment 244

Ladepa 332
lagoons 219–220
land application 225–226, 333–334
landfilling 227–228, 285
    disposal 334
    gas recovery 284
lead 153, 158–159
Leptothrix sp. 46
Life Cycle Assessment (LCA) 286
lime stabilisations 317–318
lipid peroxidation 156

macrophytes 123–124
Malaysia 308
Malaysian Water Association 307
mathematical modeling practices 44
MBSBR *see* moving bed sequencing batch
        reactor (MBSBR)
mechanical pre-treatment 244–245
membrane-based treatment 7–11, **11**
membrane-coupled sequencing batch reactor 69
mercury 153, 155
metabolic engineering approaches 191–192
metallic contaminants 209, *210*
microbial community 173

microbiological electrochemical systems (MES)
187, 187–188
Microcystis aeruginosa 3
micro-electrolysis process 66–67
micropollutants (MPs)
biofilms 141
biotransformations **142**
carcinogenic effects 159–162, **160**, *161*
cardiovascular effects *154*, 154–155
convention and regulation 163, **164**
diffuse source pollution 145–146
environmental awareness 163
gastrointestinal effects 153
in human health 142–143, **143**
industrial effluents 140
main groups 152
manmade organic contaminants 139, *140*
measures 162–163
minimata disease 157
neurological effects **155**, 155–156
occurrence 146–147, 152
pharmaceutical drugs 140
point source pollution *144*, 144–145, **145**
reproductive health 157–159, **158**
silent killer in Nigeria 159, *159*
sources 152, **152**
Toroku arsenic pollution 156–157
toxicity **141**
microwave-based alkali pretreatment 266–267
microwave irradiation 261–262
microwave pre-treatment 245
minimata disease 157
mitochondrial dysfunction 156
moving bed bioreactor (MBBR) 4–5
moving bed sequencing batch reactor
(MBSBR) 68
municipal solid waste (MSW)
anaerobic digestion 283
in Australia 287–289
composting 281–283
in Ghana 291–292
in Hong Kong 292–293
incineration 283–284
in kingdom of Saudi Arabia 284–285
life cycle 275–276
in Nigeria 285–286
sanitary landfills and landfill gas
recovery 284
in Spain 289–290

nanoremediation 186
Nereda 4
Nereda system 39
neurological disorders **155**, 155–156
Nigeria, municipal solid waste 285–286
nitrogen removal pathways 126
Nitrosomonas europaea 185

Nitrospira sp. 46
non-Hodgkin lymphoma 160

Ohtaekwangia 47
operational cost (OpEx) 127
organic contaminants 209, 211
organic pollutants 126
ozonation at high pH 94
ozonation pre-treatment 244
ozon-based AOPs 92–94

palm oil fuel ash (POFA) 24
partial nitrification anammox (PNA) 4–5
pathogenic organisms 211
pelletisers 331
peroxone process (O_3/H_2O_2) 93
persistent bioaccumulative and toxic
contaminants (PBT) 142
persistent organic pollutants (POPs) 154, *154*,
161, 161–162
pesticides 182
PFSBR *see* photo-fermentative sequencing batch
reactor (PFSBR)
pharmaceuticals and personal care products
(PPCPs) 167
phenanthrene bioremediation 192
phenoxy herbicides 161
photocatalysis 184–186
photocatalytic hybrid sequencing batch reactor
(PHSBR) 70–71
photo-Fenton process 89–90
photo-fermentative sequencing batch reactor
(PFSBR) 70
photolysis 170
photo-sequencing batch reactors (PSBRs) 69–70
PHSBR *see* photocatalytic hybrid sequencing
batch reactor (PHSBR)
phthalates 158
physical pre-treatment 244–246
physicochemical treatment 183–184
physiochemical techniques 183
physiochemical treatment
adsorption-based treatment 22–25, **26**
chemical precipitation 14, **15**, 16–17
electrochemical treatment 17, **18**, **19**, 19–22, **22**
ion exchange 12–14
membrane-based treatment 7–11, **11**
point source pollution *144*, 144–145, **145**
pollutants 1
Pollution Pay Principle (PPP) 287
polyaromatic hydrocarbons (PAHs) 142
pour flush latrine 304
predicted non-effective concentration (PNEC) 147
preliminary treatment 312
pressurized sequential batch reactor 66
pre-treatment methods 240
advantages 258

biological pre-treatment 241–243
challenges 247–249
chemical pre-treatment 243–244
classification *241*
combined pre-treatment 246–247
physical pre-treatment 244–246
PSBRs *see* photo-sequencing batch reactors
    (PSBRs)
Pseudomonas luteola 5
Pseudomonas putida 193, 194
Pseudomonas sp. 183
pulse electric field pre-treatment 245–246
pyrolysis 332–333

radiation treatment 222–223
reutilization 227
reverse osmosis (RO) 76, 79, *79*, 169
rotary drum (RD) thickening 216
rotary press 218

salt-concentrating techniques 76
sanitary landfills 284
Saudi Arabia, municipal solid waste 284–285
SBR *see* sequential batch reactor (SBR)
screw press 218
segregation 281–282
septic tank 304
sequential batch reactor (SBR) 38, *67*, 67–68
    aerobic granular sludge *67*, 67–68
    airlift loop 66
    algae-based sequencing batch suspended
        biofilm reactor 66
    anammox process 58–59
    automation control 55
    biological nitrogen removal process 55–57,
        56, *57*
    controlling and monitoring mechanics 55
    cyclic activated sludge system 61
    developments 62, **63**, 64–65
    enhanced biological phosphorus removal
        59–60
    fixed bed sequencing batch reactor 68
    integrated fixed-film-activated sludge 68
    intermediate cycle extended aeration
        system 62
    membrane-coupled sequencing batch reactor 69
    micro-electrolysis process 66–67
    moving bed sequencing batch reactor 68
    nitrogen and phosphorus 60–61
    operational parameters 54
    phases 54, *54*
    photocatalytic hybrid sequencing batch
        reactor 70–71
    photo-fermentative sequencing batch reactor 70
    photo-sequencing batch reactors 69–70
    pressurized aeration process 66
    short-cut nitrogen removal process 58

simultaneous nitrification-denitrification
    process 57–58
ultrasound-induced 69
UNITANK system 61–62
settling-thickening tank 327
sewage sludge management
    biological treatment methods 204
    characteristics 207, **208**
    characterization **234**
    conditioning 220–222, **221**
    contaminants 209
    current practices 223, **224**, 225
    current status *205*, 205–206
    dewatering process 216–220, *217*
    disposal options 225–228
    metallic contaminants 209, 210
    organic contaminants 209, 211
    pathogenic organisms 211
    pathogen removal 222–223
    stabilization 211–213, *212*
    thickening **214**, 214–216, *215*
    unit operations 206, *206*, **207**
short-cut nitrogen removal process 58
simultaneous nitrification-denitrification (SND)
    process 57–58
single-pit latrine 302
sludge drying beds 219
Society for Environmental Toxicity and
    Chemistry (SETAC) 286
solar drying 331
solar sludge oven 331
solid dewatering/drying 313
solid–liquid separation 313, 327
solids retention time (SRT) 171
solid wastes (SW) 272–273
    collection of 280
    energy equivalent 278
    hazardous waste 273
    health impacts 274–275
    hospital waste 274
    in India 278–281
    municipal 273, 275–276, **276**
    sampling and analysis 276–278
    types 273
sonochemical methods 186
Spain, municipal solid waste 289–290
Sphingopyxis 47
stabilization 211–213, *212*, 312–313
substrate media 124
subsurface-flow CWL 121–122

temperature-phased anaerobic digestion (TPAD)
    260–261
tetrabromobisphenol A (TBBPA) 140
tetra-methyl ammonium hydroxide (TMAH) 64
Tetrasphaera 47
thermal conditioning 221–222

thermal hydrolysis 262–264
thermal pre-treatment 246, 258
    anaerobic digestion 258–259
    conventional 260, 259–260
    hydrolysis 262–264
    microwave irradiation 261–262
    thermochemical methods 264–267
thermal treatment 222
thermal zero liquid discharge technology 78, *78*
thermochemical methods 264–267
thiamine deficiency 156
thickening
    centrifugal 216
    dissolved air flotation 215
    gravity 214, 215
    gravity belt 215–216
    rotary drum 216
Thiothrix 46
3-MLD SBR plant 47, **49**
total phosphate (TP) removal 126–127
toxic substances, degradation of 42, **43**
trace organic compounds (TOrCs) 86
trenching 333–334
Trichosporon 46

ultrasonic pre-treatment 246
ultrasonography 186
ultrasound-induced sequential batch reactor 69

UNITANK system 61–62
United Nations Environment Program (UNEP) 282
United States Environmental Protection Agency
    (USEPA) 283
up-flow anaerobic sludge blanket (UASB) 6

ventilated improved latrine 302
vermicomposting 282
vertical-flow CWL 121–122
volatile fatty acids (VFA) 59–60
volatilization 169

waste Stabilisation ponds 329
wastewater treatment technologies
    adopting and implementing technology 2
    biological treatment 2–7, **8**
    physiochemical treatment 7–26

zero liquid discharge (ZLD) technology
    basic process flow diagram *76*
    challenges 82
    components **77**
    factors influencing *77*
    importance 80–82, **81**
    with membrane-based techniques 79–80, *80*
    reverse osmosis 76, 79, *79*
    thermal 78, *78*
ZLD *see* zero liquid discharge (ZLD) technology

For Product Safety Concerns and Information please contact our EU
representative GPSR@taylorandfrancis.com
Taylor & Francis Verlag GmbH, Kaufingerstraße 24, 80331 München, Germany

www.ingramcontent.com/pod-product-compliance
Lightning Source LLC
Chambersburg PA
CBHW060803220326
41598CB00022B/2526